For Sue Once Again

Table of Contents

Volume One

Tome One

Preface

The work I have undertaken in this and the succeeding volumes grew from a more modest idea that came to me as I served on the committee for publishing Source Books in the History of the Sciences, namely to produce a Source Book in Ancient Egyptian Science consisting of enough extracts to illustrate some of the aspects of that science. However, when I considered the matter I realized that a few documentary extracts were insufficient to give a historian of science without any special knowledge of the Egyptian language and culture a well-rounded view of the growth and development of that science. Hence I decided to add substantial essays to introduce the documents. This resulted in a work independent of the Source Books series, a work whose first volume is published here.

It will be evident to the reader that the first section comprising Chapter One and Documents I.1-I.9 attempts to assay the importance for the development of Egyptian science--and its practitioners and institutions--of the invention and maturation of the art of writing in Egypt during the three thousand years or so after 3000 B.C. The first chapter supplies a general and connected account, and the documents present some detailed evidence to support that account. I have paid particular attention to the so-called Palermo Stone as my first document, since few if any efforts have been made to evaluate its content for the beginnings of Egyptian numerical, metrical, and calendaric techniques. Furthermore, no translator has included all the fragments together and at the same time rendered all of the many numbers that appear in the text. Previous authors have generally passed over the difficult numerical passages. My careful attention to numbers has yielded evidence of a previously undetected, rudimentary place-value system. The relevance of the rest of the documents to those points I have made in the introductory essay should be clear without further elaboration here.

The second section has the same general format as the first. Chapter Two attempts to show in a coherent way the fundamental religious context of Egyptian cosmogonic and cosmological ideas. The various schemes of the world and its creation are detailed and organized according to the temple centers in which they developed. Again, ample documentary material has been given as supplementary support for the general account. I have purposely left the more technical astronomical considerations for the next volume, in which I shall examine in detail calendars, astronomy, and mathematics. A third volume will treat Egyptian medicine and biology and will close with a detailed presentation of Egyptian techniques for representing nature.

A few remarks are in order concerning the English translations of the documents. For most of the documents I have had the help of translations into modern languages made by competent scholars and I have often followed them closely. However, in many cases I have rendered the texts in my own way, mainly to bring some consistency to translation. A case in point is my effort to translate 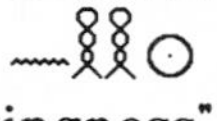 and 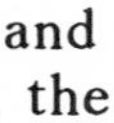 almost always as "eternity" and "everlastingness", the first being the eternal past and the second the eternal future. Also,

I have at times had available to me a more complete Egyptian text than the earlier translators had. This is evident in the case of some of the translations of tomb inscriptions by Breasted. Revised texts by Sethe have allowed me to give a more complete interpretation. As I have already mentioned, my rendering of Document I.1, the so-called Palermo Stone, is more complete than any of the efforts in other modern languages, since it is based on all of the fragments and attempts to render all of the numbers. On the other hand, it will be clear to the reader familiar with the documents given in the second section, like the *Pyramid Texts* and the *Coffin Texts*, that these documents of mine are merely short excerpts from long originals, presented by me to illustrate sundry cosmogonical and cosmological concepts which I believe to be important. Certainly I make no pretense of giving definitive versions of even those passages that I have presented. More definitive versions would necessitate a careful rendering of the whole document to see if recent philological treatments of similar passages throughout the document throw further light on the passages in question. I am hoping that a student of the history of science coming to Egyptian culture for the first time will derive benefit from having these translations immediately available to him, incomplete though they might be.

A final word may be said about the title of this first volume: *Knowledge and Order*. It translates a pair of crucial Egyptian words: *rekh* (⟨⟩) and *maat* (⟨⟩). It will be noticed as we examine the documents of the first section that the scribal craft embraced an ideal of the knowledgeable man, who by his writing abilities was able to measure, count, and record. It is this ideal that is epitomized by *rekh*. Similarly, if we peruse the second section with its documents, it will be evident that the concept of cosmic rightness or order, one of the meanings of *maat*, lies at the heart of the ancient Egyptian efforts to describe the cosmos and its birth, together with the role of the gods and the king therein. By using these separate terms "knowledge" and "order" I do not mean to imply a more modern scientific method whereby systematic order in nature was intentionally and primarily sought by the careful gathering of knowledge. I merely mean that the concepts of "knowledge" and "order" arose as important aspects of the Egyptian intellectual achievement, and without their development Egyptian science, rudimentary as it was, would have taken some other form.

This volume and its successors are the result of many trips to Egypt and much reading, and I hope they will be of some use to budding historians of science, and perhaps even to those who have fully blossomed. I know that the study of Egyptian science has given me great pleasure, for I have turned aside from my many years of detailed Archimedean studies to an earlier cultural area of perennial fascination. Fortunately my colleague Otto Neugebauer not only gave me his entire collection of Egyptian reprints but has shared with me his learning at all stages of this work. This will perhaps be more clearly evident in the technical volumes that follow. I am as well indebted to Robert Bianchi for reading my manuscript with a careful eye and for the many pleasant and (to me) profitable hours we had together in Egypt and to Erik Hornung for his keen analytical and textual studies that have proved so useful to me in composing Section Two and for his cheerful letters. My research assistant Mark Darby read this work with his usual care and it has benefited from the close attention he has given it. Similarly my wife Sue, an editor by profession, has exercised her craft towards its improvement. Further, I owe much to her for photographing countless objects in the museums of Egypt, Europe, and America, and some of her many photographs have been included in the pages that follow. I must also thank my friend Dr. Alison Frantz for the skill and artistry with which she has copied and improved many older photographs that before her attention were scarcely readable.

The reader will notice that, for easy access, I have grouped all of the figures, i.e. the line drawings and photographs, at the end of the separately printed second tome, which also includes documents for the second section, a chronology, a bibliography and bibliographic abbreviations, and indexes. Permissions for and acknowledgments of the use of the illustrations are included in the legends accompanying them. Permission by the University of California Press to use the translations by Miriam Lichtheim of the Great Hymns to Osiris and Aten, the Bentresh Stela, and the Song from the Tomb of King Intef (i.e. The Song of the Harper) is gratefully acknowledged.

I initially composed this volume (on sundry different computers) with Nota Bene, a superior word-processing program, designing, by means of the font program Lettrix, some 500 hieroglyphs and a phonetic font needed to represent the glyphs. These fonts were designed for a dot matrix printer and are illustrated in my article "Computer-generated Hieroglyphs," *Proceedings of the American Philosophical Society*, Vol. 131 (June, 1987), pp. 197-223. Later I acquired a laser printer and converted both my manuscript and its glyphs for use with that printer, employing the font program Fontrix and its complementary printing program Printrix (mentioned and briefly described in the above-noted article). I also found it necessary to compose a font which includes the accented letters that appear so frequently in the quotations, notes, and bibliography of my work. This was necessary because the Fontrix fonts do not include the higher ASCII characters representing the accented letters. Though no ambiguity is present, I regret the manner in which some of my accented capital letters dangle below the normal base line. On the whole, Printrix did a yeomanly job with the often complex and highly formatted text and only failed on occasion to impose proper spacing and justification, in some cases leading to loose lines. Also notice that the dots inserted to connect quoted passages after an omission are sometimes preceded and sometimes followed by spaces added to aid justification. But, as usual, three dots indicate that no period marking the end of a sentence is present in the omission, while four signify the presence of one or more periods. I believe that these few infelicities of Printrix's laser printing are a small price to pay for the enormous saving in publication costs.

The computer knowledge of my son Michael was always at hand, and I made frequent use of it, particulary in solving the problems of converting Nota Bene text to Printrix text. Incidentally, in making that conversion I directly converted footnotes to endnotes. But, needless to say, an endnote must continue to be thought of as a continuation of the text at the point where the note number has been inserted, and accordingly any reference to "before" or "after" in a note relates to that point in the text where the note number occurs.

My secretary Ann Tobias has also participated in every step of preparing the initial manuscript and she it was who perfected many of the original glyphs. Needless to say, she has my warmest thanks, as does the staff of the Institute for Advanced Study and its library for supplying me with much of the equipment and books I needed. Indeed, the Institute has provided me with the ideal academic home for a quarter of a century. Finally, I must once again thank the American Philosophical Society for bringing another complex work of mine to light.

Marshall Clagett
Professor Emeritus
The Institute for Advanced Study

Section One

Knowledge

Section One

Chapter One

The Fruits of Scribal Activity
in Ancient Egypt

The fundamental role played by scribes in the development of the various professions in ancient Egypt was well recognized in a prayer addressed by a scribe to the god Thoth, the patron of his profession:[1]

> Come to me, O Thoth, august ibis, O god beloved of Khmun (i.e. Hermopolis), secretary of the Ennead (i.e. the nine gods),...come to me, to advise me; give me skill in thy craft which is better than all other crafts, for men have found that he who is skilled therein becomes a nobleman (or dignitary). I have seen many for whom thou didst act and [now] they are members of the Thirty (i.e. the traditional Grand Jury of Egypt), strong and wealthy because of what thou hast done.

That the craft of scribe excels all other crafts was a favorite theme of later authors, as is well illustrated by Documents I.7 and I.8 below. Not only was this a belief articulated by the scribes but it is borne out by an examination of the extraordinary careers in political and administrative professions as well as in the learned professions revealed in the biographies found in the tombs of nobles and officials throughout Egypt in all periods. The majority of these important officials held some scribal position in their careers. For example, consult the biography of Metjen in Document I.2 below, perhaps the earliest of the tomb biographies, his career spanning the end of the third and the beginning of the fourth dynasty (see the Appendix in Section III for a chronology of ancient Egypt). We learn from this biography that Metjen was the son of an honorary or pensioned scribe Anubisemankh and inherited from him the headship of the scribes of a provision bureau as that bureau's administrator, and along with that held many other important titles and posts. The high regard of officials for their scribal activities and training was also reflected in the fact that a great many of the prominent officials during the whole of Egyptian history had statues of themselves sculpted in the scribal position, with knees crossed and papyri on their laps (see Fig. I.1). Also they were occasionally represented as scribes on the reliefs in their tombs, as for example in the tomb of Hesyre, the well-known scribe, dentist, and physician of the third dynasty (see Fig. I.2).

To understand why the scribe was an important person in ancient Egyptian society, we have to realize how much the highly organized state and its culture depended on the invention and development of writing. Already in the days that preceded and introduced the dynastic period at the end of the fourth millennium B.C. a rudimentary written language is evident on palettes, mace-heads, tablets, sealings, stelas, and vases. The mace-head of the predynastic king Scorpion, who wears the tall white crown of Upper Egypt (see Fig. I.3), has before the head of the king a rosette (probably symbolizing the title of the king). It stands above the hieroglyph of a scorpion (apparently the king's name). Though we do not know the phonetic rendering of the name, that is, the syllables for which the hieroglyph stands, it is almost certain that the glyph did stand for his name rather than another kingly title as Baumgartel suggested,[2] for we notice that, on a vase found in Hieraconpolis (along with Scorpion's mace-head), in several places above the hieroglyph of the scorpion is the conventional Horus falcon, itself the hieroglyph indicative of one title of the king throughout dynastic history (see Fig. I.4). It is unlikely that the vase was to be identified by two titles rather than by a title and a name, as was usually the case. Such use of hieroglyphs to render proper names reflects the key step taken in the invention of writing: the use of pictorial signs not just to represent the things pictured but rather to represent names (and later abstract ideas) which have a similar sound. This is the rebus principle familiar to players of charades since time immemorial. A whole series of signs was devised to represent one, two or more consonants (see Figs. I.5-I.7). The signs for single consonants are sometimes called an alphabet, for in theory one could represent the consonants of all Egyptian words by using these signs alone. But note that vowels were not written down. Despite this development of phonetic representation of consonants, the pictorial element of hieroglyphics remained strong through the whole of dynastic history. Not only did a single sign often represent two or more consonants but in countless cases a single sign was used to express the whole word regardless of its length. Furthermore many such pictorial signs were used as determinatives following the already represented consonantal elements of the word. In this position the determinative sign was unpronounced and acted to distinguish the class to which the word written in consonants belonged. Such determinatives were needed because, without the writing of vowels, the consonants alone were not sufficient to distinguish words that shared the same consonants but had differing vowels and above all different meanings or to distinguish words that had the same vowels as well as consonants but still had different meanings. So, with the persistence of a large number of pictorial signs over and beyond those we can call alphabetic, the scribe was forced to learn a larger repertoire of signs--in all some 800 signs (which in the Ptolemaic and Roman periods multiplied tenfold), as well as the cursive forms of these signs (the so-called hieratic script) used to write on papyrus and bits of pottery and stone (and from the seventh century B.C. a still more cursive script named demotic).

More certain than the signs on the Scorpion mace-head as evidence of writing are the hieroglyphs and pictorial signs appearing on the famous palette celebrating the victory of King Narmer over the Delta people and thus the union of Upper and Lower Egypt at the beginning of or just before the first dynasty (Fig. I.8). Sir Alan Gardiner summarizes the significance of the recto of the palette for revealing an early form of Egyptian writing:[3]

> This is one of the oldest specimens of Egyptian writing known. The name of the king, written with the $n^c r$-fish and the mr-chisel, occupies the rectangle between the Hathor-heads. The other small hieroglyphs give the names or titles

of the persons over whose heads they are written; the captured chieftain may have been named Washi (harpoon w^c, pool *š*). The group at top on the right was probably intended as explanation of the picture in the centre; at this early date the gist of complete sentences could apparently be conveyed only by symbolical groups of which the elements suggested separate words. The conjectural meaning is: The falcon-god Horus (i.e. the king) leads captive the inhabitants of the papyrus-land (*Tᵌ-mḥw* 'the Delta').

The verso of this palette shows one figure with accompanying hieroglyphs that is of considerable interest for our account of scribal activity. In front of the king is a figure wearing a long wig who appears to have something that looks like scribal paraphernalia thrown over his shoulder. The figure is accompanied by the hieroglyphs standing phonetically for *ṯt*. There has been considerable discussion of the meaning of *ṯt*. Four different translations have been proposed. The first is "scribe"; this was supported by the appearance of the above-mentioned paraphernalia and by a very much later use of the term *ṯt* as the "staff" of the scribal-religious-medical organization commonly known as "The House of Life" (see below). The second translation assumes that *ṯt* is archaic writing for *wṯt* and thus renders it as "the begotten one" or "son", and the third takes it for an old form for *ṯᵌty* (i.e. "vizier"). All three of these translations are cogently rejected by

H. Kees, who believes that _tt_ is archaic writing for _'tt_
(i.e. tutor).[4] This same figure appears on the mace-head
of Narmer (see Fig. I.9) along with the king's sandal
bearer, who was also on the Narmer palette. He stands
behind the king, who wears the crown of Lower Egypt
and is seated on a throne in a pavilion at the top of
some stairs. The whole scene may well represent the
ceremony of Narmer's coronation as King of Lower
Egypt after his victory over the North. Notice that in
the top register the standards which are being carried
(and presumably represent his allies and thus the
worldly "followers of Horus") are the same as those on
the verso of Narmer's palette: two of Horus, one of the
wolf-god Wepwawet, and one with the royal placenta
(or so this is usually identified). Finally notice that in
the lowest register are recorded 400,000 oxen,
1,422,000 goats, and 120,000 prisoners, no doubt the
fruits of Narmer's victory. This is clear evidence that
the Egyptians had already invented the signs for larger
powers of 10, namely 1,000, 10,000, 100,000, and
1,000,000 (see Fig. I.10). Incidentally the lesser powers
(i.e. 1, 10, and 100) appear with signs for 1,000 and
10,000 on the bases of two statues of the
second-dynasty king Khasekhem, found at Hieraconpolis
in inscriptions which tell us that 47,209 or 48,205
enemies from the North have been slain (see Fig. I.11).

There is also evidence of the new writing on
vases and leather bag sealings. On one of the latter we
find two wavy lines (an old form for the sign meaning
"water") preceded above by the sign for _m_ (an owl),
which acts as a phonetic complement. We can thus
read the whole word as _mw_ which means "water" (see
Fig. I.12). The written language was also used early to
express the names (royal and noble) found on countless

stelas in or near the first- and-second-dynasty tombs at
Abydos and Saqqara. The most perfect one artistically
is certainly the celebrated stela of the first-dynasty
Horus Djet (or better, Wadji) from Abydos, now in the
Louvre (see Fig. I.13). There, in a serekh (the
rectangular palace-facade design; see Document I.1, n.
24), surmounted by a falcon (=Horus), we see a serpent,
which is the hieroglyph for _d_ and apparently is an
abbreviated writing for Djet or for Wadji. Another
interesting one is the stela of the second-dynasty king
Horus Reneb (or Nebre) from Saqqara and now in the
Metropolitan Museum (see Fig. I.14a). In the serekh we
see the sun sign ☉ *(r^c)* signifying the sun-god Re above
the basket sign ▽ *(nb)* signifying "lord", giving the
name the meaning "Re is the lord". See also the stela of
Meryetneith, an early first-dynasty queen, in Fig. I.14b.
It has the hoe sign for *mr* (standing for *mryt* "beloved")
coupled with the sign of crossed arrows that is to be
read as "Neith", and thus the whole may be translated as
"Beloved of Neith".

Also of interest for the early stages of writing is
the large number of labels or tablets that record events
and thereby years of the reigns of the early dynastic
kings. The lack of formal syntax in the "statements" on
these tablets makes their exact translation difficult, and
the archaic form of the hieroglyphs in which these
statements are couched increases the difficulty of
understanding them precisely. Nevertheless, let us look
briefly at five such tablets. The first (see Fig. I.15) is a
tablet from Naqada of the reign of Horus Aha, either
the first or second king of the first dynasty. In the
first register we have the name Horus Aha repeated
twice (i.e., the falcon clutches the mace and shield

which together mean "The Fighter"), then to their right a boat transporting the king in a shrine-like cabin (with a vulture or a falcon flying above), then the king's Horus name in a serekh, and finally a shrine (enclosed by three lines) which contains the signs for the Lady Cobra (of the North) and the Lady Vulture (of the Southland), signs which introduce the second, so-called *nebti* or "Two-Ladies" name of a king (for the king's titulary, see below Document I.1, n. 24). The name that follows seems to be *Men*, that is *Meni* or *Menes*, the traditional founding king of the united land. At first it was thought that this was to be read as the *nebti* name of Horus Aha written to the left of it and that this juxtaposition of the two names settled the question of which king (Narmer or Aha) was Menes. However, it seems equally plausible that since the *nebti* name *Men* is enclosed in lines that appear to form a shrine, the whole should rather be read as "the Shrine of *Nebti Men*" and we could accordingly translate the first register in the following manner: "Horus Aha proceeds by boat to the Shrine of *Nebti Men*", that is, to the shrine of his predecessor Menes,[5] who could then be identified with Narmer. Skipping the other two registers, which are more difficult to render with any precision, we pass to a second tablet, i.e. a wooden label, also from the reign of Aha (see Fig. I.16). It is replete with significant symbols. In the top register we see the Horus name of Aha; then the fox-skins sign for *ms* ("birth" or "making"); next a sign of a pelt on a rod later read as *'Imy-wt* ("He who is in the bandaging room"), which probably here designates Anubis (as it did from at least the third dynasty) and was produced in model form as a cult object of the god (see Document I.1, n. 19, below); farther to the right are two (?) boats, and below them the depiction of a sanctuary with the crossed-arrows symbol of the goddess Neith erected in the middle. Hence the whole register may be translated: "[Celebration of] the Birth (or Making) of the pelt (cult object of Anubis?) by the Horus Aha and his visit by two (?) boats to the Sanctuary of Neith [at Sais?] ". The second register appears to record further pilgrimages, translatable, from right to left: "And to the Sanctuary of the Heron of Buto (cf. Document I.1, n. 101), as well as to the Running of the [Apis?] Bull [at Memphis?] (cf. *ibid.* n. 46). The king makes an offering at Nekhen (i.e. Hieraconpolis)." This register then seems to be telling of pilgrimages to both the North and the South of Egypt.

Once more I skip the two lower registers and go on to the third tablet (Fig. I.17), an ebony tablet of the Horus Den (perhaps to be read Wedimu). On the right we see the crooked, stripped, and notched palm-branch sign ⌠ for *rnpt* ("year") and to its left the king seated in a shrine-like pavilion at the top of the stairs before a course indicated by three semicircles (course markers) at each end, a course which the king is shown to be running, all of this representing part of the Sed Festival of renewal of the king traditionally (but not invariably) held in the thirtieth year of rule (see Document I.1, n. 41). Then beyond a vertical bar we see the Horus name: Den or Wedimu, and to the left of that the name of an official: "Controller of the King of Lower Egypt, Hemaka". Thus this tablet has so far told us that it was made for Hemaka in the year of the Sed Festival of King Den. Again I pass over the rest of the tablet except to note that in the third register we see the hieroglyphs for *nsw-bit* (literally meaning "belonging to the sedge and the bee" but usually translated in the king's titulary as "King of Upper and Lower Egypt"; see Document I.1, n. 24). This introduces the so-called *nesu-bit* or *insibya* name of the Horus Den, that is Semti or Zemti, as the two sandy-hill-country signs are usually read. This tablet is of particular interest, for with its year sign distinguishing the king's activities of this year from the rest of the tablet it obviously directly reflects the practice by the royal scribes of keeping some record of the activities that gave to each year of the king's reign its name. Some such annals were probably composed for each reign, and together they provided the sources for Document I.1 below, the stone *Annals* composed in the fifth dynasty and usually called the Palermo Stone from its principal fragment. This latter document I have given in full as an example of one of the earliest systematic documents that resulted from scribal activity and as one which showed the learned scribe in his early creativity. For in it he collected several kinds of data, not the least interesting of which, for our recounting of scientific activity, being the Nile heights for each year of the various reigns (at least from the time of Horus Djer onward). As we shall see, it also included the establishment and celebration of various festivals of the gods and the royal attendance at them, and it has further interest because it reflects old calendaric information and tells us something about the numerical systems used in Egypt at the earliest period of dynastic history.

The fourth tablet is an ivory label of Horus Den's reign (Fig. I.18a) used to identify some sandals. It is dated to the "[Year of] the First Smiting of the East" (this can be read at the right edge of the tablet). It appears to refer to a year of Den's reign recorded in the *Annals* (see Document I.1, note 40). On the left we see the king smiting his enemies in a figure that is repeated throughout Egyptian history from the time of the Palette of Narmer to that of the temple pylons of the Greco-Roman period. Farther left is the name of the owner of the sandals marked by the tablet and it may be read something like *Akain*.

Finally let us look briefly at an ivory label from the reign of Semerkhet, the penultimate king of the first dynasty (Fig. I.18b). At the far right we again see the stripped palm branch indicating the word "year" and with it the activity giving the year its name: "Year of the Following of Horus [in which took place at] the 'Mansion of the Great Ones' [the festivals of] the Great White One and Sokar" (for similar events recorded in the reigns of Djer and Den, see Document I.1, nn. 11, 18, and 33). Left of the vertical line we see the priest-like hieroglyph representing both the *insibya* and *nebti* names of Semerkhet, a name that some have read as Semenptah and others as Iryneter.[6] Though the rest of the tablet is by no means clear, it is evident from the hieroglyphs that it belonged to "the Royal Carpenter (or Maker of Axes) the Chief Count Henuka".

An equally if not more important source for early writing is the large number of seals and sealings of the first dynasties found in museums everywhere. I shall limit myself here to mentioning four (see Fig. I.19a). The top seal is of a priest of Neith named Shesh; the second one carries the name of Neithhotep, the wife of the Horus Aha; the third is a sealing from the time of "Seth Peribsen", of the second dynasty, and the last is one from the final king of that dynasty: "Horus-Seth Khasekhemuy in whom both gods are at rest", as can be read in the serekh. The last two sealings reflect the apparent contest (some would say civil war) between the followers of the God Horus and those of Seth.[7]

Another result of scribal activity that became a characteristic feature of Egyptian culture and funeral activity was the development of tomb biographies, in which are recorded the multiple activities of nobles and high officials, most of whom started as scribes. Needless to say, we shall not include here biographies that are largely confined to political activities but only a few of the many that reflect some kind of cultural and quasi-scientific activity. How did these tomb biographies arise? They have their roots in the stelas or marking stones of the protodynastic period that began, we have already seen, with the simple recording of a name (again see Figs. I.13 and I.14). But the biographies which interest us did not develop from the royal stelas but arose rather from private ones. No doubt the earliest of these stelas contained only the names of the courtiers as did the royal stelas the names of the kings. But very early the title or titles of the deceased began to be added along with the name. Of the 150 or so private stelas at Abydos (see Fig. I.20 for 43 of them) which date in great part from the first dynasty, we see that many of them also included titles.[8] In the context of this chapter I mention only two of them, the first being that represented in Fig. I.20, No. 43, where we read that the deceased was "Controller of Lower Egypt" and some kind of scribe (note that this is indicated by the sign for a scribe's outfit in the second register, consisting of a palette, a bag for the powdered pigments, and a reed holder).[9] On the whole these stelas are smaller than the royal ones. They are quite irregular in shape, though many of them have a curved top like the royal stelas. Presumably the earlier ones were erected separately in the ground around or near the kings' tombs and served largely as identification of the deceased's tomb and perhaps as a reminder of the name of the deceased, which had to be repeated in offering services. A good many of the stelas have the figure of the deceased in a sitting position, which no doubt also served as a hieroglyphic determinative for the name of the deceased. Some of the stelas had standing rather than seated figures. Indeed the most complex stela of these early ones (the second stela to be singled out) includes such a standing figure. This is the one of Sabef in the Cairo Museum (Fig. I.20, No. 48; Fig. I.21), probably dating from the reign of Ḳaᶜa, the last king of the first dynasty. Sabef stands at the bottom of the stela, holding a *kherep* scepter (signifying "to have control" or "to be at the head") in his right hand and a walking stick in his left hand. Behind him is his name, Sabef, in hieroglyphics, and above him are two registers of titles. They are by no means understood precisely. The first title may mean: "Controller of Festivals [at] the mansion called 'The protector is behind [us]'". Among the other titles are those that probably can be read as: "Companion in the Palace", "Controller of the Secrets of Decrees", and "Priest of Anubis".

In the North (Saqqara, Helwan, and Giza) quite a different development of stelas took place.[10] There the stelas were ordinarily rectangular, were almost all embedded in walls or ceilings, and had as their purpose the magical representation of the offering scene. Hence they all have the deceased seated before an offering table. In front of and around the table are representations of the offerings and in addition a list of cloth or linen materials that were to be provided for the deceased in the other world. In the upper register or registers were the titles. There is some variation in the location of the three major elements of the tomb stelas of the north. I give here a few examples of these stelas. The first is one from the end of the first or the beginning of the second dynasty and was found at Helwan (see Fig. I.22). The second is from Saqqara and dates from the reign of Ḳaᶜa at the end of the dynasty. The deceased was the nobleman Merka (see Fig. I.23). The third offering stela is that of a princess of the second dynasty (Fig. I.24 left) and it may be compared with another stela from the same dynasty (Fig. I.24 right). The fourth is the stela of Tetenankh (see Fig. I.25), from the third dynasty, and the fifth that of Wepemnofret from the fourth dynasty (Fig. I.26). It was not until the third dynasty that the traditional elements of the offering scene were clearly separated into titles, food-libation-purification offerings, and linen offerings. I merely note here that in Tetenankh's stela the name and titles are written in the horizontal and vertical arms of the gnomon that embraces the square including the deceased seated before an offering table and surrounded by the various items of offering (all properly identified by their hieroglyphic signs). Beyond the vertical arm of the gnomon is a column, which at the top includes the traditional list of linen materials to be offered to the deceased and below that more offerings. The formulistic request for 1000 of each offering item is evident under the offerings, where we see the lotus signs that mean "1000" (Gardiner, *Egyptian Grammar*, 1957, p. 480, sign M 12). The name and titles are of interest to us in the context of this chapter: (vertical column) "Controller of scribes, Scribe of petitions in the House of the Nome, Tetenankh" and in the horizontal column "Scribe of petitions, Familiar of the King, Tetenankh". By the third and fourth dynasties these offering scenes and particularly the title lists had expanded greatly into the various rooms of the tombs. We see them expand (1) into the offering niches of the third-dynasty tombs like the panels in Hesyre's tomb (see Fig. I.2), which were found in six niches of the western wall of the tomb, and like the reliefs in the offering niche of the tomb of Khabawsokar (Fig. I.27), and (2) into the false doors which dominate tombs for the remainder of ancient Egyptian history (Fig. I.28). These were "doors" above the burial chambers that allowed the deceased to come forth and return to his tomb ("false doors" because the doors in fact were mere recesses in the walls in which they were constructed). Around these doors titles and family information began to be included, while the offering scene itself became fixed above the door. We cannot study these developments in detail, but we can look briefly at the two panels from Hesyre's tomb given in Fig. I.2. The left one reports his titles, among which we particularly note "Head of Royal Scribes" and "Magnate of the Ten of Upper Egypt".[11] The right panel has a list of titles above the offering scene. The first title is the one of most interest to us: "Chief of Dentists and Physicians" (wr ibḥ swnw). The list also includes the already mentioned titles of "Head of Royal Scribes" and "Magnate of the Ten of Upper Egypt".

It will be obvious that in describing the development of the title lists, I have throughout this chapter (both in regard to the cases illustrated here and in the documents I have included) emphasized those officials who were scribes. Before listing some of the various kinds of scribes, I ought to mention an interesting list of scribal names which dates from the time of King Djoser in the third dynasty (see Fig. I.29). Each column includes two scribes, first giving the title "Scribe" and then the name. We can report the names as follows, starting from the left and reading the names at the tops of the columns first: (top) ---, ---, Tjenet-ty, Senedjemib, Kanefer, ---(Wepwawetsaes ?), Amkhent; (bottom) Mesa, Samery, Inkhemu, Ikau, Wedjwer, Serefka, Shepseska. What the purpose of this list was, I do not know, but it is interesting that such a group should be singled out in some way, perhaps not surprising in a reign that saw the building of the great stone funerary complex of King Djoser at Saqqara.

It is not enough to say that a great many occupants of the tombs were at one time or another scribes, for the title lists report a truly extraordinary variety of scribes:[12]

> Scribe of the Royal Documents, Overseer of Scribes of the Pyramid of Khufu, Overseer of Scribes of the Crews, Scribe of a Phyle, Judge and Overseer of Scribes, Tutor of the King's Sons, Inspector of Scribes of the Granary, Scribe of Divine Book(s), Judge and Boundary Official, Judge and Overseer of Scribes of the Great Court, Director of Scribes and Members of the Great Jury, Scribe of the Royal Documents in the Presence, Inspector of Bookkeepers, Overseer of the Scribes of the Lands, Overseer of the Scribes of the Royal Sealed Documents, Scribe of the Royal Linen, Scribe of the

Expedition, Director of Scribes of the Great Broad Hall, Overseer of Scribes of the King's Repast, Scribe of the Treasury of the Estate 'Mansion of Menkaure', Bookkeeper of the Royal Documents, Overseer of the Scribes of the Land in the Two Houses of Upper and Lower Egypt, Director of the Scribes of Petitions, Director of Scribes of the House of the Master of the Largesse, Scribe of the King's Expedition in Wenet, Serer, Tep and Ida, Scribe of the Archives, Judge and Overseer of Scribes of the Herds, Scribe of the Troops of the Lord of the Two Lands, Scribe of the Altar of the Temple of Ptah, Scribe of the Lord of the Two Lands, Scribe of the Treasury of the Pyramid of Merykare, Scribe of the Two Granaries and of the Pyramid-town, Overseer of Royal Scribes of the Army of Upper and Lower Egypt, Scribe of the Treasury of the Estate of Amun, Scribe and Counter of the Grain in the Granary of Divine Offerings of Amun, Scribe of Divine Writings in the estate of Amun, Counter of Cattle of the God's Wife of Amun, Scribe of Recruits of the Ramesseum in the Estate of Amun, Scribe of the Divine Offerings of the Gods of Thebes, Scribe of the Royal Documents of the Granary, Scribe of

the Altar of the Royal Apartments,
Scribe of Divine Offerings of the
Temple of Ptah, Scribe of the
Documents of the Chief Steward,
Scribe of the Counting of Cattle of
the House of the Overseer of the
Seal, Great Scribe of the Counting of
Amun, Keeper of the Writing
Material of the King, Scribe of the
Silver and Gold of Ptah, Royal Scribe
of Accounts of Everything in the
Temple of Imhotep, Son of Ptah,
Scribe of the Two Granaries of the
Temple of the Aten in Memphis,
Scribe of the Doors of the Temple of
Neith, Mistress of Sais, Scribe of the
Great Prison, Overseer of Writing in
the Palace, Overseer of Writing in
the House of Life, Overseer of
Scribes of the Fields, Temple Scribe
and Chief Accountant of the Divine
Adoratrice Nitocris, and many, many
others.[13]

In addition to their scribal activities, a great
many of the occupants of the tombs had titles which
indicated some attention to "quasi-learned" professions,
particularly to those we shall encounter in the
remaining chapters of this work. The most common of
such professions was that of physician *(swnw)*. The
title of physician may go back to the first dynasty.[14]
At any rate, we recall that by the third dynasty the
above-mentioned Hesyre built a tomb at Saqqara with
celebrated wooden reliefs on which he is entitled: "Chief
of Dentists and Physicians" (see Fig. I.2). Among the

98 physicians listed by Jonckheere 42 were from the Old Kingdom, 16 from the Middle Kingdom, 29 from the New Kingdom, and 11 from the Late Period, a list corrected and elaborated by Ghalioungui to include 52 names of persons from the Old Kingdom, 20 from the Middle Kingdom, 40 from the New Kingdom, and 15 from the Late Period.[15] Jonckheere, Ghalioungui, and von Känel report the following titles: Physician, w^cb-Priest of Sekhmet (a kind of physician; see below, n. 24), Overseer of w^cb-Priests of Sekhmet, Conjurers and Protective Magicians of Selket (physicians specializing in stings and bites; *ibid.*), Chief Physician and/or Chief of Physicians, Overseer of Physicians, Chief of Physicians of Lower Egypt, Chief of Physicians of Upper and Lower Egypt (or of Lower and Upper Egypt), Chief of the Physicians of the Lord of the Two Lands (i.e. of the Pharaoh), Chief of Physicians of the King, Physician or Chief of the Physicians of the Great House (i.e. the Royal Palace), Physician of the House (i.e. Temple) of Amun, Inspector of Physicians (and Inspector of Physicians of the Great House or of the King), Dean of Physicians and Dean of the Physicians of the Great House, Chief of the Physicians of the Mansion of Life ($Hwt \ ^cnh$ instead of $Pr \ ^cnh$ translated as "House of Life"), Physician of the Company (or Troop), Overseer of the Two Teams of Physicians of the Great House (*lit.* Overseer of the Two Sides of the Barge of Physicians of the Great House),[16] Physician of the Queen's House and Chief of the Physicians of the Queen's House, Directress of Female Physicians (a title given to one Peseshet in the Old Kingdom, the only known female physician),[17] Physician of the Eyes (i.e. oculist), Physician of the Eyes in the Great House and Chief of the Physicians of the Eyes in

the Great House, Controller of the Physicians of the Eyes in the Great House, Physician of the Stomach and the Eyes, Physician of the Stomach, Physician of the Stomach in the Great House, Dentist, Chief of Dentists and Chief of Dentists of the Great House, Interpreter of the Secret Art (liberally translated by Junker as "Interpreter of the Internal Organs of the Body"),[18] Interpreter of Liquids (urine?; *lit.* "water") inside the *ntnt.t* (bladder?), Guardian (or Shepherd) of the Anus (i.e. Proctologist). I should observe that often physicians had more than one of these titles, and in one case the physician Irenakhty II (end of the Old Kingdom or First Intermediary Period) had the following titles on his stela: Physician of the Great House, Inspector of the Physicians of the Great House, Physician of the Eyes in the Great House, Physician of the Stomach in the Great House, Dean of Physicians of the Great House, Shepherd of the Anus, and Interpreter of Water inside the *ntnt.t*.[19] A short biography of one successful physician of the fifth dynasty is given as Document I.3 wherein is recounted the reward for his career: two false doors for his tomb provided by the king.

No doubt allied with the physician were those simply called "savants" (*rhw* or *rhw-ht*), and sometimes they were surely the same person. Such would appear to be the case of the savants appealed to in the late-period story of the curing of the princess Bentresh, the daughter of the prince of Bakhtan:[20]

> It happened in year 23, second month of summer, day 22, while his majesty was in Thebes—the—Victorious, the mistress of cities, performing the rites for his father Amen-Re, lord of Thrones—of—the—

Two-Lands, at his beautiful feast of Southern Ipet, his favorite place since the beginning, that one came to say to his majesty: "A messenger of the prince of Bakhtan has come with many gifts for the queen" (i.e. *Nefrure*, his daughter). He was brought before his majesty with his gifts and said, saluting his majesty: "Hail to you, Sun of the Nine Bows! Truly, we live through you!" And kissing the ground before his majesty, he spoke again before his majesty, saying: "I have come to you, O King, my lord, on account of Bentresh, the younger sister of Queen *Nefrure*. A malady has seized her body. May your majesty send a learned man *(rḫw-ḫt)* to see her!"

His majesty said: "Bring me the personnel of the House of Life and the council of the residence." They were ushered in to him immediately. His majesty said: "You have been summoned in order to hear the matter; bring me one wise of heart with fingers skilled in writing from among you." Then the royal scribe Thothemheb came before his majesty, and his majesty ordered him to proceed to Bakhtan with the messenger.

The learned man reached Bakhtan. He found Bentresh to be

possessed by a spirit: he found him to be an enemy whom one could fight.[21] Then the prince of Bakhtan sent again to his majesty, saying "O King, my lord, may your majesty command to send a god [to fight against this spirit. The message reached] his majesty in year 26, first month of summer, during the feast of Amun while his majesty was in Thebes. His majesty reported to Khons-in-Thebes-Neferhotep, saying: "My good lord, I report to you about the daughter of the prince of Bakhtan." Then Khons-in-Thebes-Neferhotep proceeded to Khons the Provider, the great god who expels disease demons. His majesty spoke to Khons-in-Thebes-Neferhotep: "My good lord, if you turn your face to Khons-the-Provider, the great god who expels disease demons, he shall be dispatched to Bakhtan." Strong approval twice. His majesty said: "Give your magical protection to him, and I shall dispatch his majesty to Bakhtan to save the daughter of the prince of Bakhtan." Very strong approval by Khons-in-Thebes-Neferhotep. He made magical protection for Khons-the-Provider-in-Thebes four times. His majesty commanded to let Khons-the-Provider-in-Thebes proceed to the great bark with five boats and a chariot, and many horses from east to west.

This god arrived in Bakhtan at the end of one year and five months. The prince of Bakhtan came with his soldiers and officials before Khons-the-Provider. He placed himself on his belly, saying: "You have come to us to be gracious to us, as commanded by the King of Upper and Lower Egypt, *Usrmare-sotpenre*!" Then the god proceeded to the place where Bentresh was. He made magical protection for the daughter of the prince of Bakhtan, and she became well instantly.

As is apparent from this quotation and from the evidence in Chapter Two below and in Volume Three, Chapter Four, magic played an important part in the fabric of religious, cosmological, and medical thought. This is illustrated in a literary context by the magical tales recounted in the Westcar Papyrus (see Document I.6) where it was the lector-priest (*ḥr.ḥb* or *ḥry.ḥbt*; lit. "he who carries the ritual roll") who was cast as a magician, no doubt because he was the priest charged with the production and knowledge of the sacred books of ritual and theology which had as their major objective the protection of the king from evil, and the preservation of the right order of the world *(m¹ᶜt)*. Furthermore, leaving the imaginary world of story, we note that it was the lector-priest who was summoned, along with the physician, when in the fifth dynasty the king's vizier, chief judge, and architect Washptah fell ill and died during a visit by the king to a building being erected by the architect (see Document I.4). We find that some physicians themselves also bore the title of Lector-priest.[22]   Other professionals involved in the use of magic had the titles of Magician (ḥkⁱ or ḥkⁱy), Overseer of Magicians (imy-r ḥkⁱw), Prophet of the God Heka (or the Goddess Hekat), Conjurer (or authoritative one) of the Goddess Selket (ḥrp Srḳt), and Protective Magician (sⁱw) or Protective Magician of Selket (sⁱw Srḳt).[23]   We know of physicians (swnw) who also bore these various titles of magicians, the most important being the last mentioned Conjurers and Magicians of Selket, both of whom pale in medical importance before the wᶜb-Priest of Sekhmet, who was widely mentioned in medical contexts and strictly speaking is a physician.[24]   Thus in the Edwin Smith Surgical Papyrus (Document V.1, Case 1, Gloss A) we read: "Now if the priest of Sekhmet or any physician puts his hands (or) his fingers [upon the head, or upon the back of the] head, upon the two hands, upon the pulse, upon the two feet, [he] measures (ḥⁱy) the heart...." Furthermore in the Ebers Papyrus (Document V.2, Sect. XCIX) the same passage concerning the pulse as the measurer of the heart is repeated, except that to the priest of Sekhmet and physician in the Smith Papyrus has been added the magician designated as sⁱw (-Srḳt?) in the Ebers Papyrus.   Among the other professionals of ancient Egypt who interest us is the hour-watcher or astronomer (imy-wnwt or simply wnwty),[25] whose activities will concern us in Volume Two, Chapter Three.   And finally we should mention the reckoner or calculator (ḥsb), or scribe of counting or reckoning (sš ḥsb or sš nt ḥsb, or nisw), responsible not only for accounting and the recording of all kinds of things but also no doubt for the mathematical papyri we shall examine in Volume Two, Chapter Four.

References to institutions in which these scribes and derivative professionals were active go back at least to the Old Kingdom. On the stela of Wepemnofret, King Khufu's son (see Fig. I.26), we see him described as a "Priest of Seshat presiding over the House of Book(s) *(ḥm nṯr Sš't ḫntt pr mḏ't)*. This same institution is mentioned in connection with a scribe represented in the tomb of Kanenesut in the early fifth dynasty,[26] and is also found in the tomb of Tetu, where we find the name and title of his son given as "Nefereshemka, Inspector of the Scribes of the House of Book(s)".[27] Presumably this is the organization called the "House of Sacred Book(s)" mentioned in the tomb of Chief Lector-priest Medunefer, where he is entitled as "Scribe of the House of Sacred Book(s) *(sš pr mḏ't nṯr)*. It may be that it is also the same as the "Place of Records" *(st-ꜥ)* which King Djedkare (Dynasty V) visited (see Document I.5). I should not be surprised if the "chest of writings" that King Neferirkare had brought to him when Washptah fell ill and died was kept in just such a house of books (see Document I.4). I also suspect that it was this place that in the late Old Kingdom came to be called a House of Life and appeared in a number of locations in the course of Egyptian history. At any rate, we should examine this institution at some length.[28]

The "House of Life" is no doubt to be distinguished from the Old Kingdom "Mansion of Life" (Gardiner, *JEA* 24, p. 83). Still the expression "House of Life" is found in some sources of the Old Kingdom, though whether it was applied to one or more organizations at this early time is not known. The god

ANCIENT EGYPTIAN SCIENCE

Khnum is called "Lord of the House of Life" already in the fifth dynasty, and we find references in two decrees of Pepi II (*ibid.*, p. 160) to "the apparatus of the House of Life," presumably including "papyrus, reed-pens, ink, medical instruments, perhaps even the whole pharmacopoeia". Among Middle Kingdom references to the House of Life we note a stela (from Abydos) which calls the prince Mentuhotep (time of Sesostris I) "Master of the Secrets of the House of Life," and we see (Gardiner, p. 160; Ward, p. 45) the deceased Iha speaking of himself as "one who sees to the propitiation of the gods, [who is] overseer of writings in the House of Life, to whom all private matters are revealed". On a Middle-Kingdom scarab appears the title "Instructor of the House of Life" (Ward, p. 149), though references to teaching in the House of Life are scarce and it is doubtful that any regular courses were taught in the houses of life. Also in the Middle Kingdom we find the earliest use of the title "Scribe of the House of Life" for one Keku (Gardiner, p. 160; Ward, p. 160, No. 1380). Proceeding to the eighteenth dynasty, we can mention a brick that came from an actual House of Life in Amarna, a discovery of Pendlebury's reported by Gardiner (p. 160):

> The 'House of Life' consists of two buildings, Q.42.19 and Q.42.20, which lie 400 metres to the south of the great temple and 100 metres east of the small temple and royal estate that are themselves to the east of the Palace. The 'House of Life' abuts upon the so-called Records Office....'The place of correspondence of Pharaoh',...as befits the similarity

of these activities....The only object
of any importance found [here] was a
fragmentary funerary papyrus...now
in the Ashmolean Museum.

Particularly interesting as illustrative of the
variety of activity in the House of Life is a reference
in Tomb 111 at Thebes (Dynasty XIX) belonging to
Amenwahsu, who is called "scribe of the sacred book(s)
in the house of Amun, w^cb-priest of Sekhmet in the
house of Amun...scribe who wrote the annals of the
gods and the goddesses in the House of Life" (Gardiner,
p. 161; von Känel, pp. 45-46). In the same tomb we find
the statement that "this inscription has been written in
this tomb by the scribe of the House of Life
Amenwahsu in his own hand" (von Känel, p. 46) and
indeed elsewhere Amenwahsu is entitled as "engraver
($s\check{s}$ t^iy $m\underline{d}^it$) in the House of Life". Before leaving
Amenwahsu, we should note that a royal scribe with
the name Khaemope, whom Gardiner identified as
Amenwahsu's son of the same name (Gardiner, p. 161), a
claim disputed by von Känel (p. 48), engaged in an
activity similar to that of Amenwahsu when he (i.e.
Khaemope) "wrote the annals of all the gods in the
House of Life". In view of these statements, it is
tempting to imagine that the annals that appear as
Document I.1 were composed in a House of Life since,
of course, the King himself was the "Good God" and a
great many of the events by which the years are
designated in the annals are festivals celebrating the
"births" of the various gods (see below, Document I.1, n.
6).

The great Ramesses II did research on Thebes as
the original land of creation in the writings of the
House of Life (see Chap. 2 below, n. 3). In the reign of

Ramesses IV a stela reports the king as investigating "[the annals?] of Thoth who is in the House of Life; I have not left unseen any of them at all, in order to search out both great and small among the gods and goddesses, and I have found...the entire Ennead,[29] and all of thy [Osiris's] forms are more mysterious than theirs" (Gardiner, p. 162). Ramesses' love of such recondite learning produced in the House of Life is also extolled in inscriptions in the Wadi Hammamat. The king is said to be "excellent of understanding like Thoth, and he hath penetrated into the annals like the maker thereof, having examined the writings of the House of Life" (ibid.). Among other documents of this period we find that Amenope, the author of the Onomasticon given below as Document I.9 (a great list of names arranged under the various categories in which the author or general Egyptian tradition saw the world), has the title "scribe of the sacred book(s) in the House of Life." His Onomasticon gives a hint of the variety of the activity of the members of the House of Life. A pursuit similar to that of Amenope seems to be implied in the title given to the specialized god Horus in Snwt, "Who is known to have magical powers...[and] be a master of words [and] of exalted rank in the House of Life, a creator in the library" (Gardiner, p. 164). These titles ascribed to the god reflect, I suppose, what the scribes of the House of Life were doing.

The magical activity of the House of Life is also clear from a magical papyrus wherein we read: "The first spell of all water-enchantments--now the head ones [i.e. the lector-priests] have said with regard to it, Open the heart to no strangers concerning it--a true secret of the House of Life" (ibid.). Furthermore, one scene in a painted coffin of a scribe Ankhefenamun links Isis (well

known for her magical powers) with the House of Life
(*ibid.*, p. 165).

In the Saite period (Dynasty XXVI) Peftuauneith,
the Chief Physician of King Apries, had recorded on his
statue (Louvre A 92) an inscription that mentions his
rebuilding of the temple of Osiris at Abydos, apparently
during the reign of Apries' successor Amasis. Following
this account the inscription further tells us:

> I restored the House of Life after
> (its) ruin. I renewed the sustenance
> of Osiris, and put all of his (or its?)
> ordinances in their proper place
> (Gardiner, p. 165).

Similarly under the king Darius I the Chief Physician
Udjahorresne was sent from Persia back to Egypt

> in order to restore the department(s)
> [or establishment] of the House(s) of
> Life...[30] after they (it?) had
> decayed....I did as His Majesty had
> commanded me; I furnished them
> with all their staffs consisting of
> persons of rank, not a poor man's son
> among them. I placed them in the
> charge of every learned man [in
> order to teach them?] all their crafts.
> His Majesty had commanded [me] to
> give them every good thing that
> they might carry out all their crafts.
> I supplied them with everything
> useful to them, with all their
> equipment that was on record, as
> they had been before. His Majesty
> did this because he knew the worth
> of this guild in making well those

that are sick, in making endure
forever the names of all the gods,
their temples, their offerings, and the
conduct of their festivals.[31]

From this statement we can perhaps deduce that
not only were magic and medicine important subjects at
the House of Life, but perhaps also (1) astronomy in
connection with the determination of festivals, (2) the
preparation and preservation of offering rituals, and (3)
the development and recording of the detailed aspects
of the gods and their temples.

It was the ideal House of Life at Abydos which
was described in a mysterious work of the Late Period
or the Ptolemaic Period contained in Papyrus Brit. Mus.
10051 (Salt 825): [32]

The House of Life. It shall be
in Abydos and composed of four
[side] bodies and an interior body
[made] of covered reeds. *The four
[bodies called] Houses and [the
central body called] the Living One.*
The Living One is Osiris, while the
four Houses are Isis, Nephthys, Geb,
and Nut (! Horus and Thoth?). Isis is
on one side and Nephthys is on the
other. Horus is on the third side and
Thoth on the last one. These are the
four corners [as shown in Fig. I.30].
Geb is the ground floor of it (the
House of Life) and Nut its ceiling.
The hidden one who rests within is
the Great God. The four outer
bodies are of stone which is from the
Two Houses (?). The ground floor is

of sand. The exterior is pierced in all by four doors (two times): one to the south, another to the north, a third to the west, and a fourth to the east.

It shall be very, very, secret, not known and not seen. It is only the solar disk which looks upon its mystery.

The people who enter into it are the staff of Re: and they are the scribes of the House of Life. The people who are inside are (1) the "Shaven-headed Priest" *(fkty)*, who is Shu; (2) the "Slaughterer" *(ḫnty)*, who is Horus, who slays those who rebel against his father Osiris; (3) the "Scribe of the Sacred Book(s)," who is Thoth, who recites [the ritual of] glorification every day, without being seen or heard. [Properly] silent of mouth, their mouths and bodies covered, they are far removed from any repulsive violence.

No Asiatic [i.e. foreigner] will be allowed to enter there; he shall not see it and you will be far removed from it.

The books which are inside [the House of Life] are the "Emanations of Re," wherewith to keep alive the god and to overthrow his enemies. The staff of the House of Life, who are in it, are the

"Followers of Re" who protect his son every day....They (the gods or their priestly representatives at the House of Life) are personally occupied in protecting the god, in defending the king in his palace, and in overturning those who rebel against him.

The editor of this papyrus submits this passage and indeed the whole text to a profound and subtle analysis. The detailed rituals described therein show that the major objective of the House of Life is the defense of the gods, the king, and the right order by which society and the world run smoothly, that is *maat*.[33] He paraphrases the ideal-physical description of the House of Life at Abydos as follows (see Fig. I.30):[34]

The House of Life ought to be found at Abydos. It ought to be built of four bodies plus an interior body made of covered reeds (like a pavilion)....The four exterior bodies are of stone and form an enclosure. The ground floor of the House of Life is made of sand and the exterior is pierced by four doors....It ought to be impenetrable to profane gaze. It cannot be known and cannot be seen. Only the solar disk ought to gaze on its secret.

Surrounded by these walls or bodies (each assigned to a god), the House of Life is apparently open to the sky except for the reed-covered pavilion in the center occupied by a statuette of "Osiris, First of the

Westerners", i.e. the dead. The scribes of the House of Life, that is the "Staff of Re" or "Followers of Re", produce and guard the books, called "Emanations (i.e. *Bas*) of Re", which contain the details of the rites necessary to its purpose, namely to revivify the king and Osiris, and the magical charms and spells made use of in those rites. According to Derchain's belief, the House of Life also contained:[35]

> treatises of medicine [presumably like those of Volume Three, Chapter 5 below], of astronomy [cf. Vol. Two, Chap. 3], of mathematics [cf. Chap. 4], of mythological narratives [cf. Chap. 2] (like the annals of gods and goddesses), catalogues of divinities, manuals of the decoration of temples, rituals for embalming, exemplar manuscripts of the Book of the Dead from which copies could be made, without counting the more diverse manuals among which are those of dream interpretation, and various commentaries, all of the works constituting the information for which a ritualist might have need. Doubtless one ought to find there other works as well, e.g. geographical notes describing the nomes of Egypt....Finally, as the care of preparing royal tombs fell upon the personnel of the House of Life, it ought also to have contained manuscripts of cosmogonic compositions which we know to be

sculpted on the walls of the rock-cut
tombs in the Valley of the Kings [cf.
Chap. 2]....Finally it is also there [at
the House of Life] that the
descriptions of Egypt which we call
Onomastica were compiled [cf. Doc.
I.9]. In a word, one ought to find
there the complete totality of all the
philosophic and scientific knowledge
of the Egyptians.

I should add that the evidence assuring us that
the various works listed by Derchain were in fact
prepared and/or contained in the House of Life is scarce
but not wholly missing (e.g. see particularly the
quotation given above from the biographical statement
concerning Udjahorresne). Even before the study made
by Derchain, the wide variety of subjects considered at
the House of Life was described by Volten.[36] Since his
main objective was the study of Dream
Interpretation,[37] he stressed the connection of this
subject with the House of Life. Volten particularly
emphasized the importance of magic and medicine[38] at
the House of Life, but he also thought that astronomy
and mathematics were important in connection with
preparation and determination of the great religious
festivals. After listing the various subjects pursued at
the House of Life, Volten concludes (and this Gardiner
had specifically denied) that the House of Life was a
kind of university. Though this may be doubtful, that
it functioned as a scriptorium for the production of
many kinds of works like those that we shall study is
scarcely to be gainsaid, just as we cannot deny its great
importance in the determination of ritual.

But probably the House of Life was not the only

institution to serve as a scriptorium. We have already mentioned the Record Hall at Amarna that was located in the vicinity of the House of Life. Furthermore we note references to a "hall of writing" and an "office of writing" in a letter from the scribe Hori (of about the time of the reign of Ramesses II) to the scribe Amenemope, where he describes himself as:[39]

> The scribe of noble parts, patient in discussion, at whose utterances men rejoice when they are heard, skilled in hieroglyphics; there is nothing he does not know. He is a champion in valour and in the art of Seshyt [i.e. Seshat, the goddess of writing]; servant of the lord of Khmun [i.e. Thoth, lord of Hermopolis] in the hall *(is)* of writing; assistant-teacher (or, *in a variant reading*, teacher of subordinates) in the office *(ḥꜣ)* of writing.[40] First of his companions, foremost of his fellows, prince of his contemporaries, without his peer. His merit is proved in every stripling [i.e., he is a good teacher]....Swift to inscribe empty rolls....Unraveling the obscurities of the annals like him who made them.

In connection with the production and preserving of books, we should mention finally the temple libraries. It appears that normally each temple had a House of Books *(pr mḏ' t)*. The only ones now extant are in the temples at Philae and Edfu.[41] These consisted of a single niche (Philae) or a small room (Edfu). In both cases there was very little space and hence it is a safe

conclusion that these "libraries" contained only books
that were needed for the festivals at the temple. They
thus contrast with the libraries at the House of Life,
which were more like research libraries. But in terms
of its purpose, the library at Edfu can be clearly tied in
with a House of Life (one presumably at Edfu). Its
reliefs show that it had some of the same objectives as
those of a House of Life like that already described at
Abydos: the annihilation of the king's enemies and the
protection of the life of the king. For example, a rite is
described which is entitled "Overthrowing the enemies
of his father Osiris each day".[42] It is executed on
behalf of "Osiris, the First of the Westerners, the Great
God, Lord of Abydos, who made resplendent the House
of Life by means of his ka".[43] Established in the
library is the god Horus *imy Snwt* elsewhere mentioned
as "the lord of books...the master of discourse, the Old
One in the House of Life, who is installed in the House
of Books...the redoubtable lord of the House of
Books".[44] Furthermore Seshat, the goddess of writing,
appears in the friezes on the four walls of the library at
Edfu and also lower on the east wall of that library
(see Fig. I.31b), and she on occasion has the titles of
"Lady of the House of Books" and "Lady of the House
of Life".[45] In conclusion note that a catalogue of more
than thirty books of the library is given on the south
wall.[46]

To this point I have briefly sketched the
invention of writing and scribal activities. Using the
fruits of that invention for recording and describing,
certain people became specialized in the nascent
scientific activities mentioned here but reserved for
detailed treatment later. Now we may proceed to the
first set of documents.

Notes to Chapter One

1. A. H. Gardiner, *Late-Egyptian Miscellanies* (Brussels, 1937), p. 60; R. A. Caminos, *Late-Egyptian Miscellanies* (London, 1954), p. 232; and P. Montet, *Everyday Life in Egypt in the Days of Ramesses the Great* (Philadelphia, 1981), p. 275.

2. E. J. Baumgartel, *The Cultures of Prehistoric Egypt*, Vol. 2 (London/New York/Toronto, 1960), p. 103, and "Scorpion and Rosette and the Fragment of the Large Hierakonpolis Mace Head," *Zeitschrift für ägyptische Sprache und Altertumskunde*, Vol. 93 (1966), pp. 9-13. For the possible identification of the rosette as a designation of the king, see W. Helck, *Untersuchungen zu den Beamtentiteln des ägyptischen Alten Reiches* (Glückstadt/Hamburg/New York, 1954), p. 24.

3. A. H. Gardiner, *Egyptian Grammar,* 3rd ed. (London, 1957), p. 7.

4. H. Kees, "Archaisches 𓅓𓄿𓏏𓏏 [tt=ˀtt] 'Erzieher'?" *Zeitschrift für ägyptische Sprache und Altertumskunde,* Vol. 82 (1957), pp. 58-62.

5. A. H. Gardiner, *Egypt of the Pharaohs* (Oxford, 1961), pp. 405-07. For a review of the literature on the meaning of the ivory tablet from Naqada, see J. Vandier, *Manuel d'archéologie égyptienne*, Vol. 1 (Paris, 1952), pp. 827-31.

6. W. B. Emery, *Archaic Egypt* (Harmondsworth, England, 1961), p. 84, adopts Semenptah as his *nebti* and *nesu-bit* names. Grdseloff reads the name as Iryneter (see Vandier, *Manuel*, Vol. 1, p. 856).

7. Evidence of writing in these seals and sealings, and indeed in all of the tablets, vases, and the

like which we have briefly examined to this point, serve only to illustrate the early period of scribal activity. The reader who wants fuller evidence may consult the invaluable collection of inscriptions from the early period presented and discussed by P. Kaplony, *Die Inschriften der ägyptischen Frühzeit*, Vols. 1-3 (Wiesbaden, 1963), *Supplementband* (Wiesbaden, 1964), together with his *Kleine Beiträge zu den ägyptischen Frühzeit* (Wiesbaden, 1966). See also Fig. I.19b for some inscriptions in early cylinders.

8. Vandier, *Manuel*, Vol. 1, pp. 731-32. We are of course primarily dependent on W. M. F. Petrie, *The Royal Tombs of the First Dynasty*, Part I (London, 1900), plates XXXI-XXXVI, and *The Royal Tombs of the Earliest Dynasties*, Part II (London, 1901), plates XXVI-XXXA.

9. See signs Y 3 and Y 4 in Gardiner, *Egyptian Grammar*, p. 534. It appears to be sign Y 4 that occurs on this stela, with the reed holder now missing because of the deteriorated state of the stela.

10. Vandier, *Manuel*, Vol. 1, pp. 733-74, gives a very valuable, detailed description of the development of these stelas.

11. For remarks on the Magnates of the Ten of Upper Egypt, presumably an administrative and judicial council, see J. H. Breasted, *A History of Egypt*, 2nd ed. (New York, 1912), pp. 79-80. I have followed the conventional rendering of this title (cf. Helck, *Untersuchungen zu den Beamtentiteln des Alten Reiches*, pp. 18-19). But see H. Goedicke, "Die Laufbahn des Mtn," *Mitteilungen des Deutschen Archäologischen Instituts, Abteilung Kairo*, Vol. 21 (1966), pp. 57-62, who reviews this title (and the literature on it) in detail and comes to the translation "Grosser des Katasters von

Oberägypten" (Great One of the Cadastre of Upper Egypt). Perhaps it was already a title of a rank rather than of a specific administrative office. See K. B. Gödecken, *Eine Betrachtung der Inschriften des Meten im Rahmen der sozialen und rechtlichen Stellung von Privatleuten im ägyptischen Alten Reich* (Wiesbaden, 1976), p. 80, who believes that the shift from office-title to rank-title had taken place by the early fourth dynasty.

12. In listing the various scribal titles, I have primarily used three standard works: B. Porter and R. L. B. Moss, *Topographical Bibliography of Ancient Egyptian Hieroglyphic Texts, Reliefs, and Paintings*, Vols. 1-3, 2nd ed. (Oxford, 1960-78); M. A. Murray, *Index of Names and Titles of the Old Kingdom* (London, 1908); and W. A. Ward, *Index of Egyptian Administrative and Religious Titles of the Middle Kingdom* (Beirut, 1982). H. Goedicke, "Diplomatic Studies in the Old Kingdom," *Journal of the American Research Center in Egypt*, Vol. 3 (1964), pp. 31-41 (whole article), has (p. 31) some wise statements concerning the rise of the office of king's scribe: "In the III Dynasty the position of scribe for the king develops into an established office, and the appearance of a 'leader of the king's scribes' suggests the formation of a royal chancery. The early holders of this position, to judge from their burials, were persons of highest social status. From this early title subsequently develops a more specific *zš* ^{c}*-nswt*, 'scribe of the King's documents,' a title of which various compounds are attested." The authority for these remarks is Helck, *Untersuchungen zu den Beamtentiteln des Alten Reiches*, pp. 71 ff., and, in addition, the above noted work of Kaplony, *Die Inschriften der ägyptischen*

Frühzeit, Vol. 1, p. 583; for a list of titles involving scribes, see Vol. 2, p. 1215 (under *zš*, given alone and in compound).

13.   See the long list of additional titles in Ward, *Index*, pp. 156-68, which I make no attempt to reproduce here; see also Porter and Moss, *Topographical Bibliography*, Vol. 3, 2nd ed., pp. 928, 934, and the long list of titles given in the index (under *sš*) of F. von Känel, *Les prêtres-ouâb de Sekhmet et les conjurateurs de Serket* (Paris, 1984), pp. 324-25.

14.   Petrie, *Royal Tombs*, Part I, p. 38.

15.   F. Jonckheere, *Les médecins de l'Égypte pharaonique. Essai de prosopographie* (Bruxelles, 1958) and P. Ghalioungui, *The Physicians of Pharaonic Egypt* (Cairo, 1983).   See L. Habachi and P. Ghalioungui, "Notes on Nine Physicians of Pharaonic Egypt," *Bulletin de l'Institut d'Égypt*, Vol. 51 (1969), pp. 15-24.   See also Porter and Moss, *Topographical Bibliography*, Vol. 3, 2nd ed., p. 932, and von Känel, *Les prêtres-ouâb de Sekhmet*, p. 324 (*swnw*).   Incidentally, I have followed Gardiner, Faulkner, and Ghalioungui in transcribing the hieroglyphs used for "physician" as *swnw* rather than as *sinw* used by Jonckheere throughout, though a case can be made for *zinw* in the Old Kingdom (see Gardiner, *Egyptian Grammar*, p. 512, sign T 11).

16.   Jonckheere, *Les médecins*, pp. 116-17.   Note that von Känel, *Les prêtres-ouâb de Sekhmet*, p. 227, substitutes for Jonckheere's title "Physician of the Troop" the reading "Physician and Magician of Selket (*sˀw Srḳt)*."

17.   Peseshet's stela is reproduced in Jonckheere, *Les médecins*, Fig. 9.   He incorrectly translates it "Directrice des médecins" as if the reading were *imyt-r swnw* when in fact it is *imyt-r swnwt* and hence it is

obvious that her title should be translated "Directress of Female Physicians". Cf. Ghalioungui, *The Physicians of Pharaonic Egypt*, p. 18.

18. Jonckheere, *Les médecins*, p. 67, n. 1.

19. *Ibid.*, p. 25, and Fig. 2 for the stela, cf. von Känel, *Les prêtres-ouâb de Sekhmet*, pp. 167-68.

20. M. Lichtheim, *Ancient Egyptian Literature*, Vol. 3 (Berkeley/Los Angeles/London, 1980), pp. 91-92. For printing purposes I have removed, both in the text and the next footnote, the hyphens in the gods' names as given in Miss Lichtheim's translation, but I have restored them by hand.

21. This conclusion reflects ordinary medical procedure, whereby the physician after diagnosing the case comes to the conclusion "an ailment to be treated" or "an ailment not to be treated". In the case of Princess Bentresh, the learned man saw that as a case of possession her ailment could be treated by a specialist (who it turned out was the cult image of the god called Khons–the–Provider–in–Thebes). Miss Lichtheim's note on the god is of interest: "The Theban god Khons was worshipped under several distinct manifestations, with Khons–in–Thebes–Neferhotep occupying the leading position, while the outstanding trait of Khons...[the-Provider-in–Thebes] was that of healer."

22. Jonckheere, *Les médecins*, pp. 126-27.

23. A. H. Gardiner, "Some Personifications. I. ḤĪKE, The God of Magic," *Proceedings of the Society of Biblical Archaeology*, Vol. 37 (1915), pp. 253-62; "Professional Magicians in Ancient Egypt," *ibid.*, Vol. 39 (1917), pp. 31-44, 138-40, and the work of von Känel cited in note 13.

24. Gardiner, "Some Personifications," pp. 261-62;

Jonckheere, *Les médecins*, pp. 127-28. For a recent and detailed study of w^cb-priests of Sekhmet and the Charmers of Serket as medical personnel, see von Känel, *Les prêtres-ouâb de Sekhmet.*

25. Ward, *Index*, pp. 9, 58, 86; O. Neugebauer and R. A. Parker, *Egyptian Astronomical Texts*, Vol. 3 (Providence, Rhode Island, and London), 1969, p. 213; A. H. Gardiner, *Ancient Egyptian Onomastica*, Vol. 1 (Oxford, 1947), pp. 61*-62*.

26. G. A. Reisner, *A History of the Giza Necropolis*, Vol. 1 (Cambridge, Mass., and London, 1942), p. 341.

27. *Ibid.*, p. 502. For the succeeding reference to Medunefer, see p. 491.

28. Many of the initial references to the House of Life have been taken from A. H. Gardiner, "The House of Life," *Journal of Egyptian Archaeology*, Vol. 24 (1938), pp. 157-79 (abbreviated in the next paragraphs simply as "Gardiner"). The first reference (abbreviated as "Gardiner, *JEA* 24") is to another article in the same volume of the journal: "The Mansion of Life and the Master of the King's Largess," *ibid.*, pp. 83-91. I have also in the passages occasionally cited Ward's *Index* merely as "Ward" and von Känel's *Les prêtres-ouâb de Sekhmet* as "von Känel."

29. The name Ennead is one that embraced the nine gods at Heliopolis: Atum, Shu, Tefnut, Geb, Nut, Osiris, Seth, Isis, and Nephthys. For the variations in composition and number of the Ennead, see E. Hornung, *Conceptions of God in Ancient Egypt: The One and the Many* (Ithaca, 1982), pp. 221-23, where the pertinent literature is cited. See also below, Chap. 2, n. 29.

30. I have generally followed Lichtheim, *Ancient Egyptian Literature*, Vol. 3, pp. 39-40, but occasionally

have preferred the reading of Gardiner, pp. 157-58. The translation here of "departments" instead of "establishment" is a case in point. Gardiner goes on to say that the missing part may have completed the phrase as follows: "the department(s) of the House(s) of Life dealing with medicine". Miss Lichtheim objects: "But medicine was only one of several crafts practiced by the members of the House of Life."

31. It is this last sentence that gives us concrete evidence of the wide variety of subjects pursued at the House of Life, at least at this late date.

32. P. Derchain, *Le Papyrus Salt 825 (B.M. 10051), rituel pour la conservation de la vie en Égypte, Académie Royale de Belgique, Classe des Lettres, Mémoires*, Vol. 58 (Brussels, 1965), pp. 139-40; Part III, pp. 7*-10*. For the possible (but doubtful) location of Abydos' House of Life in the Second Osiris Hall of the Temple of Sethos I, see R. David, *A Guide to Religious Ritual at Abydos* (Warminster, Wilts., England, 1981), pp. 147-48.

33. Derchain, *op. cit.*, Part I, *pass.*

34. *Ibid.*, p. 49.

35. *Ibid.*, pp. 56-57.

36. A. Volten, *Demotische Traumdeutung (Pap. Carlsberg XIII und XIV verso), Analecta aegyptiaca*, Vol. 3 (Copenhagen, 1942), pp. 17-44.

37. *Ibid.*, p. 40.

38. *Ibid.*, p. 30. Note that L. Habachi and P. Ghalioungui discovered Priests of Sekhmet and physicians apparently connected with a House of Life at Bubastis: "The 'House of Life' of Bubastis," *Chronique d'Égypte*, Vol. 46 (1971), pp. 59-71. See also P. Ghalioungui, *The House of Life: Per Ankh. Magic and Medical Science in Ancient Egypt* (Amsterdam, 1973),

pp. 66-67.

39.   A. H. Gardiner, *Egyptian Hieratic Texts: Series I. Literary Texts of the New Kingdom. Part I. The Papyrus Anastasi and the Papyrus Koller* (Leipzig, 1911), p. 6*. I have omitted Gardiner's notes except for the note I have numbered as 40.

40.   *Ibid.*, n. 6: "*O.P.* [Petrie Ostracon] continues differently thus: 'Iskilled in (?)] his profession; knowing the secrets of heaven and earth....; there is none who repels(?) ....in writing. First of his fellows in the midst of his neighbors; chief of his contemporaries, they are not equal [to him]; teacher of subordinates in the office of writing; his merit is proved in every stripling. Lamp....' (end)."

41.   Derchain, *op. cit.* in n. 32, p. 58. Both libraries are discussed at some length by Derchain, pp. 58-61.

42.   E. Chassinat, *Le temple d'Edfou*, Vol. 3 (Cairo, 1928), p. 345, E'o. 1 d., lin. 1.

43.   *Ibid.*, p. 346.

44.   Derchain, *op. cit.*, p. 60.

45.   *Ibid.* See Chassinat, *Le temple d'Edfou*, Vol. 3, p. 350, E'e. 2 g. I, lin. 2, and Gardiner, p. 164 (23). See also Document I.1, n. 44.

46.   Chassinat, *Le temple d'Edfou*, Vol. 3, pp. 347(E's. 2 d., lin. 2)-348(E's. 2 d., lin. 6) and 351(E's. 2 g., lin. 3-6). See H. Brugsch, "Bau und Maasse des Tempels von Edfu," *Zeitschrift für ägyptische Sprache und Alterthumskunde*, Vol. 9 (1871), pp. 43-45 (full article pp. 32-45), and M. Weber, *Beiträge zur Kenntnis des Schrift- und Buchwesens der alten Ägypter* (Cologne, 1969), pp. 131-34. Most of the titles in the catalogue are unknown elsewhere, many are in the form of abbreviated titles, and many are difficult to translate

precisely. The first set of works (E's. 2 d., lin. 2-6) includes the following: "Many Chests Containing Books Together with a considerable [Number of] Rolls of Leather: (1) The Overthrowing of Seth; (2) Repelling of Crocodiles; (3) Protection of the Hours; (4) Guarding of the Sacred Bark; (5) Opening of the *Nam.t*-Bark; (6) Book of Making the King Rise (i.e. his coronation?); (7) Guides to Ritual; (8) Great Glorifications of Him Who is on the Deathbed; (9) The Protection of a City; (10) The Protection of a House; (11) The Protection of the White Crown; (12) The Protection of Thrones; (13) The Protection of the Year; (14) The Protection of Graves (or Beds); (15) The Appeasement of Sekhmet; (16) The Guardians (*not listed by Weber*); (17) The Totality of Books (or the Book of the Totality?) Concerning the Hunting of a Lion, the Repelling of Crocodiles, the Guarding of ____(?), and the Repelling of Snakes; (18) The Craft of all Priests of the Temple-Kitchen (or Knowledge of All Secrets of the Laboratory [for the production of unguents and the like]); (19) The Knowledge of Your Offerings to the Gods in All Their Images and to Your Families on Your (Birth?)days and Every Inventory of Your Secret Godly Forms and the Forms of Your Ennead." The second set of titles (E's. 2 g., lin. 3-5) is: "(20) The Book of What is in the Temple; (21) The Book of *Rth* (Dread?); (22) Book of All Writings Concerning Arms (or Battles or Tools?); (23) Book of the Management (or Plan) of a Temple; (24) The Book of Guards (or Officials) Who Belong to the Temples; (25) Rules for Painting the Wall and the Observation (or Protection) of Body Forms; (26) Book of the Protection of the King in his Palace; (27) Spells for the Repelling of the Evil Eye; (28) Knowledge of the Periodic Returns of the Two Celestial Spirits: the

Sun and the Moon; (29) The Governing of the Periodic Returns of the Stars; (30) The Numbering of All the Sepulchers and the Knowledge of Those Who are in Them; (31) All Consultation-books of the Summoning of Your Majesty to Your Temple in Your Festivals." It is obvious that items (17), (22), and (31) seem to refer to more than one book. Weber in the above-cited work (pp. 113-31) gives other lists of titles, and S. Sauneron, *Les prêtres de l'ancienne Égypte* (Paris, 1957), p. 136, translates some of the titles in the Edfu catalogues somewhat differently. Sauneron also mentions the remains of a similar catalogue at Tod (southeast of Luxor), and he tells us (p. 137) that the works themselves from one of these temple libraries have been discovered in the small village of Tebtunis (in the Fayyum).

Document I.1: Introduction

The Early Egyptian *Annals* on Stone, Generally called the Palermo Stone

Our first document is, as I have stated in my general remarks above, the earliest historiographical document extant from ancient Egypt's long history. It is one of the most important by-products of the Egyptian development of a written language. It consisted of a rectangular block of black diorite, on both sides of which were inscribed the principal events of the reigns of the kings of the first four dynasties and a part of the fifth. It was composed in the fifth dynasty, perhaps from now lost annals of early kings, or if not from formal annals at least from many tablets and labels giving the year names in terms of events.[1] From this one (or less likely more than one)[2] stone seven fragments have been found (Figs. I.32-I.43): one now in Palermo, five in Cairo, and one in London. Valiant efforts have been made to reconstruct the size and even the content (at least in terms of the number of years in each reign) of the whole stone on the basis of these fragments and various extant lists of kings (see the literature listed below and Fig. I.44). But it is fair to say that none of these constructions is entirely convincing, for they are all based in some significant degree on unproved and implausible assumptions.[3] Still, the most useful are the reconstructions of Kaiser, Helck, and Giustolisi (see Figs. I.42 and I.43 for the latter

two).   The size of the stone has been variously
estimated: the length ranging from six to ten feet and
the height agreed upon as somewhat more than two
feet.[4]

Details of the document have been discussed in
the notes that accompany it, but we can briefly
summarize here its most important aspects.

(1) The stone represents the first partially extant,
extensive set of annals in Egyptian history, which
covered in all likelihood about 600 years of pharaonic
rule (from ca. 3000 B.C. to ca. 2400 B.C.). As I have
said above, its author of the fifth dynasty probably had
access to annals prepared in the earlier dynasties.

(2) The annals of the first three dynasties consist
almost exclusively of events that give the years their
names.   Thus for those dynasties the document is
largely a list of year names rather than true annals of
the most important royal events in the reign of each
king.   In these early years two events were generally
sufficient to identify the year.   The first event was
often a repeated event such as the "Following of Horus",
which was either a festival celebrating the conquest of
Egypt by Horus and his allies or a biennial tour of
Egypt by the king and his officials, perhaps for tax
purposes (see Document I.1 below, note 11).   Since this
event was repeated every other year, a second event
was needed to identify the year more completely.   The
second event was frequently (though not exclusively)
the celebration of a religious festival, often entitled the
"Birth of Such-and-Such God" (see Document I.1, n. 6).
Occasionally an event that happened only once (or that
the original annalist thought would happen only once)
was chosen by itself as the name of the year.   For
example, the third year of Horus Djer's reign in the

first dynasty was named exclusively as "The Year of
[the Festival of] the Birth of the Two Children of the
King of Lower Egypt" (see *ibid.*, n. 16). An alternative
to taking a single unusual event as the name of the
year was the practice of taking a more common event
and using an ordinal number to precede it. So, Year 9
of Djer's reign was named "The Year of the First
Occurrence of the Feast of Djet (Uto *or* Wadjet)" (*ibid.*,
n. 22). As I have said, sometimes the unusual single
event by which a year was named had no religious
significance but was a secular event. Thus the Year X
+ 2 of the Horus Den in the first dynasty was named
"The Year of the Smiting of the Asiatics" (*ibid.*, n. 34).
Similarly the Year Y + 2 of the same king was called
"The Year of the Smiting of the Troglodytes [i.e.
Easterners]" (*ibid.*, n. 40). Needless to say, even when
the event was secular it concerned an action of the
king, for these are royal annals. We should note that
the first year of a king's reign is characterized by the
celebrations accompanying his assumption of the throne,
namely, ceremonies of "the Union of the Two Lands and
of the Circuit of the Wall" (*ibid.*, end of n. 13).

(3) While the principal purpose of the *Annals*
reflected in the first three dynasties was the recording
of events which gave the names of the years, an
important supplementary datum was included for each
year, the Nile height at full flood, whose measurement
or at least recording apparently started by the time of
the Horus Djer (*ibid.*, n. 14). This is important as
illustrating Egyptian concern from a very early date
with regular measurements, and it illustrates the early
use of the linear units of cubit, span, palm, and finger
and their fractions. In fact, the flooding of the Nile
was of such importance that Year Y + 4 of the reign of

ANCIENT EGYPTIAN SCIENCE

Horus Den apparently received its name from an unusually high flooding: "Year of the Filling [i.e. Flooding] of all the Lakes [or Nomes] of the Rekhyt-Folk in the West and the East of Lower Egypt [i.e. in the Delta]" (*ibid.*, n. 42).

Another important "measuring" event that is reflected in the early years of the *Annals* is that connected with laying out the ground plan or "stretching the cord" as the first step in the construction of temples or mansions. Year Y + 7 of Den's reign was called "The Year of Stretching the Cord at the Great Door of the Castle [called] 'Seats of the Gods' by the Priest of Seshat" (*ibid.*, n. 44). Similarly the Year X + 2 of Horus Ninetjer in the second dynasty was named "The Appearance of [the king as] the King of Upper Egypt [at] the Stretching of the Cord for the Mansion [called] 'Hor-ren'" (*ibid.*, n. 57). In the next dynasty, Year 4 of King Djoser's reign, the king appears at the "Stretching of the Cord for the Mansion [called] 'The Refreshment of the Gods'".

(4) By the second dynasty the biennial survey called the "Following of Horus" attracts to itself the supplementary characterization of the "counting" of the wealth of the land. The significant point is that along with the specification of the "counting" an ordinal number began to be used. For example, the Year X + 3 (=8?) of Ninetjer is called "The Year of the Following of Horus [in which took place] the Fourth Occurrence of the Counting" (*ibid.*, n. 58). Two years later it was the "fifth" occurrence of the counting, two years after that the "sixth", and successively every two years, the "seventh", "eighth", and "ninth" occurrences of the counting. The general "counting" without specification of what is being counted which originated in the second

dynasty seems to have spawned specific countings in the third. For example, in Year X + 1 (=3?) of Horus Zanakht's reign we read of the "First Occurrence of the Counting of Gold" (*ibid.*, n. 66). This was followed in the next year by the "Second Occurrence of the Counting", which probably was confined to the counting of the cattle and the cultivated land. By the Year Y + 3 (=14?) of this reign the counting of the gold and that of the fields were merged and the year was called "The Year of the Following of Horus [in which took place] the Seventh Occurrence of the Counting of the Gold and the Fields" (*ibid.*, n. 70). Compare also Year Y + 5 (=16?). Though these biennial countings were not specifically mentioned in the next reign, that of the great King Djoser who was responsible for building the step pyramid at Saqqara, no doubt the "countings" were subsumed by the biennial "Followings of Horus". Mention of the biennial occurrences of the countings was restored in the records of the fourth dynasty, in the first reign, that of Sneferu (see *ibid.*, n. 86). The references to the countings in Sneferu's reign appear for the most part to have been general references, but note that in the year X + 1 (=4?) the annalist speaks of the "Second Occurrence of the Counting of Silver and Lapis lazuli" (*ibid.*, n. 81).

There is one peculiarity concerning the system of "countings" in the reign of Sneferu. In all previous references to the countings, at least to the nonspecific countings, the countings are always recorded for alternate years, that is, the system was clearly a biennial one. But notice that in the reign of Sneferu the eighth counting follows in the next year after the seventh counting (*ibid.*, n. 90). This seems to have been an aberration only, for the biennial system was still in place later in the fourth and also in the fifth and sixth dynasties. One further point is worth noting in connection with Sneferu's reign. In Year W + 1 the designation of the counting includes "small cattle", and presumably also oxen (see below C-4 r, line 2, Nr. 1).

The next most important step in freeing the system from identifying regnal years by names and instead substituting a completely numerical system can be noted in the *Annals* of the fifth dynasty when the Year X + 1 (=3?) of Weserkaf's reign is designated as "The Year *after* the First Counting of the Oxen" (*ibid.*, n. 97).[5] By adding *after*, which I have italicized, the chronographers have achieved a system which ties every regnal year (except the first) to an ordinal numerical count, so that from the second year the years of a reign would be successively numbered: Year of first occurrence of the counting, Year after the first occurrence, Year of second occurrence, Year after the second occurrence, and so on. I have said "except the first" because it is generally believed that the Year of the first occurrence of the counting was the second year of a reign. Hence we would have these equations: Year of the assumption of the throne = Year 1, Year of the first occurrence of the counting = Year 2, Year after the first occurrence = Year 3, and so on. It is not known exactly when the practice of numbering the even and odd years in the manner just described became a system in which every year was simply given its own number.[6] This was achieved either by shifting to an annual counting or by simply dropping the use of the system of cattle counting as the basis of numbering the years. What is known is that the designation of an odd-numbered year as the "The Year *after* the *n*th occurrence of the counting" was dropped (it appears that the word *khet* or *m-khet*, meaning "after", disappeared in Pepi II's reign at the end of the sixth dynasty). Furthermore it seems clear that at some time after Dynasty VI the expression "The Year of the *n*th occurrence of the counting" in its abbreviated form "Year of the *n*th occurrence" came simply to mean "Regnal year *n*". [7]

(5) Though the system of numerical regnal years took many centuries to develop, it is evident from these *Annals* that the divisions of "months" and "days" as the basic units of the year were in effect from at least the first dynasty. From the listing of the number of months and days of the last calendar year which Aha served before his death, followed by the number of months and days of the first calendar year of Djer's reign, it is obvious that some kind of calendar year was in use (*ibid.*, n. 13). We cannot know from the data the length of this year. It could well be that it was the civil year of 12 months of 30 days plus five epagomenal days that was demonstratively in effect in the Old Kingdom and was evident in these *Annals* in the fourth dynasty at the end of Menkaure's reign and the beginning of that of Shepseskaf (*ibid.*). I shall return to calendaric questions later in Volume Two, Chapter Three, at which time I shall discuss the problem of the time of the installation of the civil calendar.

(6) Beginning with the fourth dynasty, the *Annals* now begin to resemble chronicles of which the *raison d'être* is to recount a multiplicity of memorable events in each year of a reign. This change is not unconnected with the rise of a quasi-numerical system of designating the regnal years, so that the text itself is freed from the purpose of simply giving an event by which the year can be named and can be devoted to

several important events of the year. For example, the Year Z + 2 of Sneferu in the *Annals* records the following events: "The Building with *meru*-wood of a 100-cubit Dewatowe ship and 60 sixteen-[oared?] barges, The Smiting of Nubia and taking of prisoners numbering 4,000 men and 3,000 women, together with 200,000 large and small cattle, the Building of the Wall of the Southland and the Northland [called] 'The Houses of Sneferu', The Transporting of 40 ships filled with *ash*-wood (so-called 'cedar')" (*ibid.*, nn. 87 and 88). Not only are the variety and extent of the information notable but also the fact that all of these events are secular activities. The same is true for the following year of Sneferu's reign, though the fragments of other years of his reign contain a mixture of secular and religious events. This was also true even of the fifth dynasty, in which religious donations predominated, for in the Year Z + 1 (=15?) of the reign of Sahure we find reported the importation of copper from the Malachite Country and myrrh and metals from Punt.

(7) The flurry of secular events reported in the previous paragraph should not blind us to the dominance of the reporting of festivals and other religious activities throughout the *Annals*, as the reader will see if he peruses the text at random. Note the following references to gods or their heralds (the citations in parentheses are either to the document's notes or to its marginal numbers).

First dynasty: Anubis (nn. 6, 9, 21, 25; C-1 r, lin. 3, Nr. 4), Sokar (nn. 18, 28, 37, 43), Yamet (n. 19; or probably the cult-object later called *Imy-wt*, "He who is in the bandaging room", i.e. Anubis--n. 21), Min (n. 20), Wadjet (n. 22; Pr, lin. 3, Nr. 5), Sed (n. 27; Pr, lin. 3, Nr. 11), Ha (n. 29), Neith (? n. 30), The Great White One

(? n. 33), Saw or Ptah (n. 39), Seshat (nn. 44, 47), Harsaphes (Pr, lin. 3, Nr. 9), the Apis Bull (n. 46), Mafdet (n. 47); *second dynasty*: the Apis Bull (n. 59; Pr, line 4, Nr. 10), Sokar (Pr, lin. 4, Nr. 6; Nr. 12), Horus of Heaven (Pr, lin. 4, Nr. 8), Wadjet (n. 61), Nekhbet (n. 61); *third dynasty*: Min (Pr, lin. 5, Nr. 10); *fourth dynasty*: the Apis Bull (? C-4 r, lin. 2, Nr. 2), Sed (n. 92), the Horus Bull (C-2, lin. 2, Nr. 1), Bastet (n. 94), Seshed (Pv, lin. 1, Nr. 2), the Two Wepwawets (*ibid.*); *fifth dynasty*: the Souls of Heliopolis (C-1 v, lin. 2, Nr. 2; Pv, lin. 2, Nr. 2; Pv, lin. 4, Nr. 2; Pv, lin. 5, Nr. 2), Re (C-1 v, lin. 2, Nr. 1; Pv, lin. 2, Nr. 2; Nr. 3; lin. 3, Nr. 1; lin. 4, Nr. 1; Nr. 2; lin. 5, Nr. 1; Nr. 2), Hathor (same as for Re, except for the third citation and the last two), the Gods of the Sun-Temple called "Sep-Re" (Pv, lin. 2, Nr. 2), Gods of the Heron of Djebakherut (?*ibid.*), Sepa (*ibid.*), Nekhbet (*ibid.*, lin. 2, Nr. 2; lin. 3, Nr. 1), Wadjet (*ibid.*, lin. 2, Nr. 2; lin. 3, Nr. 1; lin. 5, Nr. 2), Gods of the Shrine of the South (Pv, lin. 2, Nr. 2), Min (*ibid.*, Nr. 3), Mesenkhet (*ibid.*, lin. 3, Nr. 1), Semkhet (*ibid.*), Khenti-yawatef (*ibid.*), the White Bull (*ibid.*), Divine Ennead (*ibid.*, Nr. 2; lin. 4, Nr. 2), Seshat (*ibid.*, lin. 4, Nr. 1), the Gods of Kheraha (*ibid.*, Nr. 2), Re-senef (*ibid.*), and Ptah (*ibid.*, lin. 5, Nr. 2).

There are also numerous festivals and ceremonies which are general or ambiguous, or where the gods to whom they are dedicated are not known. For example, mention is made of a Festival of Desher (nn. 15, 31), of the "Censing the Decapitated Folk" (n. 17), of the "Birth of the Two Children of the King of Lower Egypt" (dedicated to the gods Shu and Tefnut[?]; see nn. 16, 85).

We should also single out the reference to the building out of stone of a temple called "The Goddess abides" (n. 69), for if the reign in which this took place is correctly identified as Nebka's, then we have here the earliest reference to a stone building in ancient Egypt (though stone was used for flooring, stairs, gates, and portcullis blocking in the first dynasty). I should note that there are references to pyramids (n. 96; C-1 v, lin. 2, Nr. 1; Nr. 2) and to Sun-Temples (Pv, lin. 2, Nr. 2; lin. 3, Nr. 1; lin. 5, Nrs. 1-2). There is also reference to the deeply religious Sed-festival, the festival at which the king's rule was renewed (n. 41).

(8) While I forbear citing detailed references to other temples and shrines in the *Annals*, I must stress the historical importance of the *Annals* for its plethora of information concerning the granting of regular offerings and of arable land to a host of temples and shrines in the fifth dynasty, and above all I single out the gifts to Re (see the references given to Re in the preceding paragraph), reflecting as they do Re's primacy in that dynasty.

(9) The gifts of land mentioned in the *Annals* also yield precious information to the historian of Egyptian numeration and measurement. The measures of land in these grants involve *ḥꜣ* (a 10-aroura measure), *stꜣt* (unit-aroura measure), *rmn* (1/2 aroura), *ḥsb* (1/4 aroura), *sꜣ* (1/8 aroura), *mḥ* (a cubit-area, i.e. 1/100 aroura). If we start with a cubit-area (i.e. an area 1 cubit wide and 100 cubits long), then 1/8 aroura = 12 1/2 cubit-areas, 1/4 aroura = 25 cubit-areas, 1/2 aroura = 50 cubit-areas, 1 aroura = 100 cubit-areas, and 10 arouras = 1000 cubit-areas (which reveals why the lotus sign [Gardiner sign M 12] read as *ḥꜣ*, meaning 1000, was used for the 10-aroura measure; see Fig. I.50). In expressing the various areas of the land grants, the annalist has sometimes made use of a primitive place-value system, a

fact which I have not seen mentioned before (see Document I.1, nn. 99, 107). In this system we find numbers where the 10-aroura measures are numbered first in units without an expected preceding 10-aroura sign, then followed by a unit-aroura sign which is itself followed by the counting of those unit-arouras in units. We also find numbers where 100-aroura measures are counted in units without any preceding 100-unit sign, then followed by the 10-aroura sign which is itself followed by the counting of 10-aroura measures in units. But the annalist was not always consistent in the use of the place-value technique (*ibid.*, nn. 99, 103) and in fact the use of this technique in later times was rare.

Texts and Studies of the Fragments

I. *The Palermo Fragment.* Abbreviated as Pr and Pv for the recto and verso.
    Text:
        H. Schäfer, "Ein Bruchstück altägyptischer Annalen," *Abhandlungen der Königlichen Preussischen Akademie der Wissenschaften, 1902, Phil.-hist. Abh.* (Berlin, 1902), pp. 3-41, with plates of Pr and Pv (see Figs. I.32, I.34) and a German translation.
        K. Sethe, *Urkunden des Alten Reichs*, Vol. 1, 2nd ed. (Leipzig, 1933) (= *Urkunden* I), pp. 235-49, a revised text of the parts containing Dynasties IV and V; this includes additions from fragments C-1 through C-4.
    Studies and Translations:
        A. Pellegrini, "Nota sopra un'iscrizione egizia del Museo di Palermo," *Archivio storico*

siciliano, n. s., Anno xx (1895), pp. 297-316, with reproductions of Pr and Pv (see Figs I.33, I.35).

E. Naville, "La Pierre de Palerme," *Recueil de travaux relatifs à la philologie et à l'archéologie égyptiennes et assyriennes*, 25me année (1903), pp. 64-81, with a French translation of Pr and a discussion of Pv.

K. Sethe, *Untersuchungen zur Geschichte und Altertumskunde Ägyptens*, Vol. 3: *Beiträge zur ältesten Geschichte Ägyptens* (Leipzig, 1905).

L. Borchardt, *Die Annalen und die zeitliche Festlegung des Alten Reiches der ägyptischen Geschichte* (Berlin, 1917).

E. Meyer, *Ägyptische Chronologie* (Berlin, 1904).

J. H. Breasted, *Ancient Records of Egypt*, Vol. 1 (Chicago, 1906), pp. 51-72, a translation based on Schäfer's text and thus of the Palermo fragment alone.

II. *The Cairo Fragments*. Abbreviated as C-1, C-2, C-3, C-4, and C-5.

H. Gauthier, "Quatre nouveaux fragments de la Pierre de Palerme," *Le Musée Égyptien*, Vol. 3 (1915), pp. 29-53, with photographs, partial transcriptions, and some translations of fragments C-1 to C-4 (see Figs. I.36-I.39).

G. Daressy, "La Pierre de Palerme et la chronologie de l'Ancien Empire," *Bulletin de l'Institut Français d'Archéologie Orientale*, Vol. 12 (1916), pp. 161-214, with further transcriptions, corrections, and suggestions regarding fragments C-1 to C-4.

F. W. Read, "Nouvelles remarques sur la

Pierre de Palerme," *ibid.*, pp. 215-22.

J. H. Breasted, "The Predynastic Union of Egypt," *ibid.*, Vol. 30 (1931), pp. 709-24; he established that line 1 in C-1 r contains the names of predynastic kings of a united Upper and Lower Egypt.

See also K. Sethe, *Urkunden*, Vol. 1, as above, who makes use of C-1 to C-4 for his text of the parts of the *Annals* concerning the fourth and the fifth dynasties.

J. L. de Cenival, "Un nouveau fragment de la Pierre de Palerme," *Bulletin de la Société Française d'Égyptologie*, Vol. 44 (1965), pp. 13-17, reporting C-5 (see Fig. I.40).

III. *The London Fragment.* Abbreviated as L (see Figs. I.41a-b).

W. M. F. Petrie, "New Portions of the Annals," *Ancient Egypt*, 1916, pp. 114-20, including a reproduction of Lr.

C. N. Reeves, "A Fragment of Fifth Dynasty Annals at University College London," *Göttinger Miszellen. Beiträge zur ägyptologischen Diskussion*, Heft 32 (1979), pp. 47-50, with photographs and a transcription of Lv (see Fig. I.41c).

H. M. Stewart, *Egyptian Stelae, Reliefs and Paintings from the Petrie Collection. Part Two: Archaic Period to Second Intermediate Period* (Warminster, England, 1979), p. 6.

IV. *Other studies of all or several fragments.*

A. Weigall, *A History of the Pharaohs*, Vol. 1 (New York, 1925), includes a dogmatic interpretation of the *Annals* which is now usually rejected.

G. Godron, "Quel est le lieu de provenance de la 'Pierre de Palerme'?" *Chronique d'Égypte*, Vol. 27 (1952), pp. 17-22; his tentative conclusion: Memphis.

W. Helck, *Untersuchungen zu Manetho und den ägyptischen Königslisten* (Berlin, 1956), especially pp. 9-10, 12, 77-81.

W. Kaiser, "Einige Bemerkungen zur ägyptischen Frühzeit. II. Zur Frage einer über Menes hinausreichenden ägyptischen Geschichtsüberlieferung," *ZÄS*, Vol. 86 (1961), pp. 39-61.

V. Giustolisi, "La 'Pietra di Palermo' e la cronologia dell'Antico Regno," *Sicilia archeologica*, Anno I, Nr. 4, Dec. 1968, pp. 5-14; Nr. 5, March, 1969, pp. 38-55; Anno II, Nr. 6, June, 1969, pp. 21-38, being a good overall summary of all the fragments, with an Italian translation of the *Annals* (except for the difficult sections of the fifth dynasty including the donations to shrines and temples) (see my Fig. I.43).

W. Helck, "Bemerkungen zum Annalenstein," *MDAIK*, Vol. 30 (1974), pp. 31-35 (see my Fig. I.42).

P. F. O'Mara, *The Palermo Stone and the Archaic Kings of Egypt* (La Canada, Calif., 1979) and *The Chronology of the Palermo and Turin Canons* (La Canada, Calif., 1980), studies which I believe to be entirely unsatisfactory since the basic point of departure is that the stone commences with the reign of Horus Den, which is easily refutable.

W. Barta, "Die Chronologie der 1. bis 5. Dynastie nach den Angaben des rekonstruierten Annalenstein," *ZÄS*, Vol. 108 (1981), pp. 11-23.

A. Roccati, *La Littérature historique sous l'ancien empire égyptien* (Paris, 1982), "Annales des IVe et Ve dynasties," pp. 36-58, with a number of translations that differ from mine. Note that Roccati interprets the month and day designations as specific dates rather than as the number of months and days in the first or last regnal year, as I have done following Borchardt.

D. B. Redford, *Pharaonic King-Lists, Annals and Day-Books: A Contribution to the Study of the Egyptian Sense of History* (Mississauga, Ontario, Canada, 1986). Has no translation or detailed analysis of the Palermo Stone, but sets it in an illuminating study and analysis of the ancient Egyptian approach to history. See particularly pp. 86-90, 134-35. Attributes the *Annals* to "a heightened consciousness of the past and a desire to edit it...evidenced from the reign of Neferirkare".

The English Translation

In my translation I have kept my eye on the previous efforts of Breasted, Schäfer, Gauthier, Daressy, Cenival, Giustolisi, and others. It is, I believe, the most complete translation of all the fragments yet made in any language; it attempts to translate all of the numerical details of the donations mentioned in the course of the fifth dynasty. I have added brackets both for doubtful but probable readings and for additional phrases that make the account less syncopated than the original text without (I trust) altering the intent of the original annalists.

The various fragments are noted in the left margins of the translation by using the abbreviations mentioned in the description of sources given above (P, C-1, C-2, C-3, C-4, C-5, and L, supplemented by r or v for recto or verso where appropriate). The additional citations of Nr. 1, Nr. 2, etc. represent the numbers of the boxes for the successive years of the reigns, starting from the right. This will be clear to the reader if he examines Figs. I.32-I.43 where the fragments are depicted.

The detailed notes following the document indicate some of the difficulties involved in translating accurately the early forms of glyphs found in the *Annals*. They also give general information about early dynastic culture that might be helpful to the non-specialist.

Notes to the Introduction of Document I.1

1. W. M. F. Petrie, *The Making of Egypt* (London and New York, 1939), p. 98, suggested a successive composition of the *Annals*: "The primary text was issued under Sneferu at the close of the IIIrd (!beginning of the IVth) dynasty; and it continued to receive additions until nearly halfway through the Vth dynasty." Earlier in his *History of Egypt*, Vol. 1 (XIth ed. rev., London, 1924), p. 9, Petrie had declared that the *Annals* were composed in the reign of Userkaf (Weserkaf). "After that Sahura and Neferkare (!Neferirkare) scratched their annals slightly upon the stone." Cf. Redford, *Pharaonic King-lists*, pp. 134-35.

2. Helck, "Bemerkungen," p. 33 (see literature below), has presented a convincing argument that all of the fragments are from the same stone, answering all of the arguments presented by Kaiser, who believed that Fragment C-4 alone was from a different stone. Earlier Gauthier, "Quatre nouveaux fragments," p. 30, believed that Fragments P, C-1, C-2, and C-3 were from the same stone. But Borchardt, *Die Annalen*, pp. 22-23, argued that P and C-1 were from different stones. In opposition to this, Breasted, after presenting new measurements of P and C-1 and answering Borchardt's arguments one by one, concludes ("The Predynastic Union of Egypt," p. 719): "A fair consideration of the measurements, however, both horizontal and vertical, makes it very probable that the two fragments [i.e. C-1 and P]...formed parts of the same monument. In any case the identical divisions and disposition of the surface of the stone in laying out the first five rows

make it perfectly certain that the records on the two fragments are parts of the same document, and that they may be employed to supplement each other in our endeavors to reconstruct the complete record once inscribed on the front of the slab." Kaiser, "Einige Bemerkungen," p. 44, n. 2, essentially supporting Gauthier, concludes: "Allein Fragment K 4 [= C-4] stammt sowohl nach der Stärke der Platte wie der Art der Feldereinteilung mit Sicherheit von einem anderen Annalenstein und gibt so den wertvollen Hinweis, dass mehrere, und zwar unterschiedlich gestaltete Annalenplatten existiert haben." Helck opposes the arguments supporting this conclusion in the article mentioned above.

3. The devastating review which T. E. Peet made of Borchardt's *Die Annalen* in the *Journal of Egyptian Archaeology*, Vol. 6 (1920), pp. 149-54, can be extended to the assumptions of others who have attempted to reconstruct the stone with precision from the fragments. I must admit that, after attempting somewhat similar constructions of my own, I have complete sympathy with the remarks of W. F. Edgerton, "Critical Note," *The American Journal of Semitic Languages and Literature*, Vol. 53, (1937), pp. 187-97, and especially p. 197, n. 23: "In my opinion the charge of inaccuracy in the underlying measurements [which Borchardt later made against his own earlier reconstruction], while certainly true, was of minor importance. What makes the reconstruction completely worthless is the fact that it rested on assumptions which are at best unprovable, not to say improbable. This was pointed out in detail by Peet....On its face Peet's criticism eliminates not merely the reconstruction of the Old Kingdom Annals which Borchardt published

in 1917 but every similar reconstruction which may be published in the future, no matter how accurate the underlying measurements may be."

4. Breasted, *Ancient Records of Egypt*, Vol. 1, p. 52: "some seven feet long and over two feet high." Daressy, "La Pierre de Palerme," p. 180: "La longueur du tableau...serait ainsi de 2m. 65cent." (=8.69 feet). The complicated calculations of Borchardt which have been criticized so severely by Peet in the review mentioned in the previous note need not be reported in detail here. But we can say that his calculation of the length of the Palermo Stone (which he distinguished from the stone from which Fragment C-1 came) was a little over eight feet (ca. 8.05 feet), while the Cairo stone he computed to be somewhat longer (ca. 8.50 feet) (see Borchardt, *Die Annalen*, pp. 13-30 and plates 1 and 2).

5. A. H. Gardiner, "Regnal Years and Civil Calendar in Pharaonic Egypt," *Journal of Egyptian Archaeology*, Vol. 31 (1945), pp. 14-15, quotes a somewhat earlier use of the expression of the year after a counting: "*the 17th day of the second month of Winter in the year after the first occasion*, i.e. year 3" taken from the tomb of Meresankh III, the wife of King Chephren (misidentified by Gardiner as the niece of Chephren's wife).

6. K. Sethe, *Untersuchungen*, Vol. 3, p. 88, gives several examples from the Middle Kingdom where there is no mention of the "counting" but only the regnal year. See also below, Document I.5, n. 3, where a year is simply designated by a number without any reference to an "occurrence" (of a counting).

7. *Ibid.*, pp. 69-100, for the classic history of regnal dating in ancient Egypt. See the subsequent contributions of E. Edel, "Zur Lesung von 𓈖𓏤

'Regierungsjahr'," *Journal of Near Eastern Studies*, Vol. 8
(1949), pp. 35-39, J. von Beckerath, "Die Lesung von 𓇳
'Regierungsjahr': Ein neuer Vorschlag," in *ZÄS*, Vol. 95
(1969), pp. 88-91, and W. Barta, "Das Jahr in
Datumsangaben und seine Bezeichnungen," *Festschrift
Elmar Edel: 12. März 1979* (Bamberg, 1979), pp. 35-41,
tracing the readings for the expression "regnal year".
"Regnal year" was used in both the eighth and eleventh
dynasties (see Barta, p. 39).

Document I.1

Early Egyptian Stone Annals:
A Translation of the Fragments of
Palermo, Cairo, and London

[The Gods as Rulers][1]
(*no part on fragments*)

[Kings of Upper Egypt][2]
(*no part on fragments*)

Pr, line 1
Kings of Lower Egypt:[3] ___pu, Seka, Khayu, Teyew, Tjesh, Ni-heb, Wadjenedj, Mekh, __a,_____
C-1 r, line 1
[Some Predynastic] Kings of Upper and Lower Egypt.[4]

[First Dynasty]

[Horus Narmer]
(*no part on fragments; may be predynastic*)

[Horus Aha?][5]

C-5, line 1, Nr. 1
[Year X + 1] [The Year of....in which took place the Festival of] the Birth of Anubis.[6]

[Nile height:] *not recorded.*[7]

Nr. 2

[Year X + 2] [The Year of....in which took place ....] Bull.[8]

[Nile height:] *not recorded.*

Nr. 3

[Year X + 3] [The Year of....in which took place the Festival of] the Birth of....[9]

[Nile height:] *not recorded.*

[After four years or so, the Palermo Fragment continues the same reign as follows.][10]

Pr, line 2, Nr. 1

[Year X + 7 (?) + 1] The Year of the Following of Horus [in which took place the Festival of] the Birth of Anubis.[11]

[Nile height:] *not recorded.*

Nr. 2

[Year X + 7 (?) + 2] [The last civil year of the reign of the King, of which he reigned the first] six months and seven days.

[Nile height:] *not recorded.*

[Horus Djer?][12]

Nr. 3

[Year 1] [The first civil year of the reign of the king in which he reigned the last] four months and thirteen days [after his enthronement and in which took place the ceremonies of] the Union of the Two Lands and of the Circuit of the Wall.[13]

[Nile height:] 6 cubits.[14]

Nr. 4

[Year 2] The Year of the Following of Horus [in which took place] the Festival of Desher.[15]

[Nile height:].....

Nr. 5

[Year 3] The Year of [the Festival of] the Birth of the Two Children of the King of Lower Egypt.[16]

[Nile height:] 4 cubits, 1 palm.

Nr. 6

[Year 4] The Year of the Following of Horus [in which took place the Festival of] the Censing of the Decapitated Folk.[17]

[Nile height:] 5 cubits, 5 palms, 1 finger.

Nr. 7

[Year 5] The Year of [the First Occurrence of] the Festival of Sokar [celebrated] behind [or at] the Castle [called] "Companion of the Gods".[18]

[Nile height:] 5 cubits, 5 palms, 1 finger.

Nr. 8

[Year 6] The Year of the Following of Horus [in which took place the Festival of] the Birth of the Goddess Yamet (or the cult-symbol for Anubis).[19]

[Nile height:] 5 cubits, 1 palm.

Nr. 9

[Year 7] The Year of the Appearance of the [king as] King of Upper Egypt [at the Festival of] the Birth of Min.[20]

[Nile height:] 5 cubits.

Nr. 10

[Year 8] The Year of the Following of Horus [in which took place the First Occurrence of the Festival of] the Birth of Anubis.[21]

[Nile height:] 6 cubits, 1 palm.

Nr. 11

[Year 9] The Year of the First Occurrence of the Festival of Djet (Uto or Wadjet).[22]

[Nile height:] 4 cubits, 1 span.
Nr. 12
[Year 10]....(*destroyed*).[23]

C-1r, above line 2

Horus Djer, King and [Horus] of Gold Itet, whose mother was Khened (priestess of?) Hap.[24]

C-1r, line 2, Nr. 1

[Year X + 1 (=20?)] The Year of the Following of Horus [in which took place the Second Occurrence of the Festival of] the Birth of Anubis.[25]

[Nile height:]....

Nr. 2

[Year X + 2 (=21?)] The Year of the Festival of Traversing the Two Lakes (?).[26]

[Nile height:]....

Nr. 3

[Year X + 3 (=22?)] The Year of the Following of Horus [in which took place the Festival of] the Birth of the God Sed (?).[27]

[Nile height]....

Nr. 4

[Year X + 4 (=23?)] The Year of [the Second Occurrence of] the Festival of Sokar [celebrated] behind (*or* at) the Castle [called] "Companion of the Gods".[28]

[Nile height:]....

Nr. 5

[Year X + 5 (=24?)] The Year of the Following of Horus [in which took place the Festival of]...

[Nile height:]....

Nr. 6

[Year X + 6 (=25?)] The Year of the Appearance of [the king as] the King of Upper Egypt [at the

Festival of] the Birth of [the desert God] Ha.[29]
    [Nile height:]....
Nr. 7
[Year X + 7 (=26?)]   The Year of the Following
of Horus [in which took place the Festival of] the
Birth of Neith(?).[30]
    [Nile height:]....
Nr. 8
[Year X + 8 (=27?)]   The Year of the Festival of
Desher [held] behind (*or* at) the castle [called]
"Companion of the Gods".[31]
    [Nile height:]....
Nr. 9
[Year X + 9 (=28?)]   The Year of the Following
of Horus [in which took place the Festival of] the
Birth of....[32]
    [Nile height:]....

[Horus Wadji]
(*no part on fragments*)

C-5, Pr, above line 2
            Horus Den (Wedimu),
        [King Zemti whose] mother
        [was Me]ryet[-Neith].

C-5, line 2 (=Pr line 3), Nr. 1
    [Year X + 1 (=18?)]   The Year of ....The Great
White One (?)[33]
        [Nile height:]....
    Nr. 2
[Year X + 2 (=19?)]   The Year of the Smiting of
the Asiatics.[34]
        [Nile height:] 5 cubits

-71-

Nr. 3

[Year X + 3 (=20?)]   The Year of the [Festival of the] Birth of the pelt [of Anubis (?) later called Imy-wt] at the *Senuty*-Sanctuary (i.e. the Double Shrines).[35]

[Nile height:]....cubits.

Nr. 4

[Year X + 4 (=21?)]   The Year of the Smiting of the Wolf (?) People.[36]

[Nile height:] 6 cubits, 1 palm, 2 fingers.

Nr. 5

[Year X + 5 (=22?)]   The Year of the Festival of Sokar [celebrated] behind (*or* at) the Castle [called] "Companion of the Gods".[37]

[Nile height:]....cubits....

[*After five years or so, the Palermo Fragment continues the same reign.*][38]

Pr, line 3, Nr. 1

[Year Y + 1 (=28?)]   The Year of the [King's?] Sojourn (*or* Station) in the Temple of Saw (*or* Ptah?) in the City of Heka....[39]

[Nile height:] 3 cubits, 1 palm, 2 fingers

Nr. 2

[Year Y + 2 (=29?)]   The Year of the Smiting of the Troglodytes (*Intyw*).[40]

[Nile height:] 4 cubits, 1 span.

Nr. 3

[Year Y + 3 (=30?)]   The Year of the Appearance of [the king] as King of Upper Egypt and [his] Appearance as King of Lower Egyp[t] [at] the [Royal] Sed Festival.[41]

[Nile height:] 8 cubits, 3 fingers.

Nr. 4

[Year Y + 4 (=31?)]   The Year of the Filling [i.e.

Flooding] of all the Lakes [or Nomes] of the Rekhyt-Folk in the West and the East of Lower Egypt [i.e. in the Delta].[42]

[Nile height:] 3 cubits, 1 span.

Nr. 5

[Year Y + 5 (=32?)] The Year of the Second Occurrence of the Festival of Djet (Uto or Wadjet).

[Nile height:] 5 cubits, 2 palms.

Nr. 6

[Year Y + 6 (=33?)] The Year of the Festival of Sokar [celebrated] behind (or at) the Castle [called] "Seats of the Gods".[43]

[Nile height:] 5 cubits, 1 palm, 2 fingers.

Nr. 7

[Year Y + 7 (=34?)] The Year of Stretching the Cord at the Great Door of the Castle [called] "Seats of the Gods" by the Priest of Seshat.[44]

[Nile height:] 4 cubits, 2 palms.

Nr. 8

[Year Y + 8 (=35?)] The Year of the Opening of the Lake of the Castle [called] "Seats of the Gods" and of Shooting the Hippopotamus.[45]

[Nile height:] 2 cubits.

Nr. 9

[Year Y + 9 (=36?)] The Year of the [King's?] Sojourn (or Station) on the Lake of the Temple of Harsaphes in Heracleopolis.

[Nile height:] 5 cubits.

Nr. 10

[Year Y + 10 (=37?)] The Year of the Journey to Sahseteni and of the Smiting of Wer-ka.

[Nile height:] 4 cubits, 1 span.

Nr. 11

[Year Y + 11 (=38?)] The Year of [the Festival of] the Birth of the God Sed.

[Nile height:] 6 cubits, 1 palm, 2 fingers.

Nr. 12

[Year Y + 12 (=39?)] The Year of the Appearance of [the king as] the King of Lower Egypt [at] the First Occurrence of the Running of the Apis Bull.[46]

[Nile height:] 2 cubits, 1 span.

Nr. 13

[Year Y + 13 (=40?)] The Year of the [Festival of the] Birth of Seshat and [that of the Birth of] Mafdet.[47]

[Nile height:] 3 cubits, 5 palms, 2 fingers.

Nr. 14

[Year Y + 14 (=41?)] The Year [of the Appearance of the king as] the King of Upper Egypt [at the Festival of] the Birth of....

[Nile height:]....

[Horus Andiyeb or Adjib (Enezib),
King of Upper and Lower Egypt,
Merpibia?][48]

C-1 r, line 3, Nr. 1
[Year X + 1]....[49]
[Nile height:]....

C-1 r above line 3

Horus Semerkhet (?), King of
Upper and Lower Egypt and
Favorite of the Two Ladies
Semenptah(?), whose mother is
Baterits(n?)[50]

C-1 r, line 3, Nr. 2

[Year 1] The Year of the Appearance of [the king as] the King of Upper Egypt and [his] Appearance [as] the King of Lower Egypt [at the Ceremony] the Union of the Two Lands.[51]

[Nile height:] 4 cubits, 4 palms.

Nr. 3

[Year 2] The Year of the Following of Horus [in which took place]....

[Nile height:] 4 cubits, ...4 (?) palms.

Nr. 4

[Year 3] The Year of the Appearance of [the king as] the King of Upper Egypt [at the Festival of] the Birth of Anubis.

[Nile height:] 4 cubits,....palms.

Nr. 5

[Year 4] The Year of the Following of Horus [in which took place]....

[Nile height:] 4 cubits,....palms.

Nr. 6

[Year 5] The Year of....[and of the Festival of] the Birth of....

[Nile height:]....

Nr. 7

[Year 6] The Year of the Following of Horus [in which took place the Festival of] the Birth of ....

[Nile height:] 4 cubits, 4 palms.

Nr. 8

[Year 7] The Year of the Appearance of [the king as] the King of Upper Egypt and [his] Appearance [as] the King of Lower Egypt [at]....

[Nile height:] 4 cubits,....palms.

Nr. 9

[Year 8] The Year of the Following of Horus [in which took place]....
　　　[Nile height:] 4 cubits,....palms.
Nr. 10
[Year 9] The Year of the Appearance of [the king as] King of Upper Egypt and [his] Appearance [as] King of Lower Egypt [at the Festival of] the Birth of....[52]
　　　[Nile height:]....
　　　　　　　[Horus Ḳaᶜa?][53]
Nr. 11
[Year 1] The Year of....[54]
　　　[Nile height:]....

[The Second Dynasty]

[Horus Hetepsekhemuy]
(*no part on fragments*)

[Horus Nebre]
(*no part on fragments*)

Pr, above line 4
　　　　　Horus Ninetjer,
　　　Son of Nub-[Nufer]....[55]
line 4, Nr. 1

[Year X + 1 (=6?)] The Year of the Following of Horus [in which took place the Third Occurrence of the Counting].[56]
　　　[Nile height:]....
Nr. 2
[Year X + 2 (=7?)] The Appearance of [the king as] the King of Upper Egypt [at] the Stretching of

the Cord for the Mansion [called] "Hor-ren".[57]

    [Nile height:] 3 cubits, 4 palms, 2 fingers.

Nr. 3

[Year X + 3 (=8?)] The Year of the Following of Horus [in which took place] the Fourth Occurrence of the Counting.[58]

    [Nile height:] 4 cubits, 2 fingers.

Nr. 4

[Year X + 4 (=9?)] The Year of the Appearance of [the king as] the King of Upper Egypt and [his] Appearance [as] King of Lower Egypt [at] the Running of the Apis Bull.[59]

    [Nile height:] 4 cubits, 1 palm, 2 fingers.

Nr. 5

[Year X + 5 (=10?)] The Year of the Following of Horus [in which took place] the Fifth Occurrence of the Counting.

    [Nile height:] 4 cubits, 4 palms.

Nr. 6

[Year X + 6 (=11?)] The Year of the Appearance of [the king as] the King of Upper Egypt [at] the Second Occurrence of the Festival of Sokar.

    [Nile height:] 3 cubits, 4 palms, 2 fingers.

Nr. 7

[Year X + 7 (=12?)] The Year of the Following of Horus [in which took place] the Sixth Occurrence of the Counting.

    [Nile height:] 4 cubits, 3 fingers.

Nr. 8

[Year X + 8 (=13?)] The Year of the First Occurrence of the Festival of the Worship of Horus of Heaven [in which also took place] the Destruction of the Mansion of Shem-Re and of the Mansion of the North.[60]

ANCIENT EGYPTIAN SCIENCE

[Nile height:] 4 cubits, 3 fingers.
Nr. 9
[Year X + 9 (=14?)] The Year of the Following of
Horus [in which took place] the Seventh
Occurrence of the Counting.
[Nile height:] 1 cubit.
Nr. 10
[Year X + 10 (=15?)] The Year of the Appearance
of   [the king as] the King of Lower Egypt [at]
the Second Occurrence of the Running of the
Apis Bull.
[Nile height:] 3 cubits, 4 palms, 3 fingers.
Nr. 11
[Year X + 11 (=16?)] The Year of the Following of
Horus [in which took place] the Eighth
Occurrence of the Counting.
[Nile height:] 3 cubits, 5 palms, 2 fingers.
Nr. 12
[Year X + 12 (=17?)] The Year of the Appearance
of [the king as] the King of Lower Egypt [at] the
Third Occurrence of the Festival of Sokar.
[Nile height:] 2 cubits, 2 fingers.
Nr. 13
[Year X + 13 (=18?)] The Year of the Following
of Horus [in which took place] the Ninth
Occurrence of the Counting.
[Nile height:] 2 cubits, 2 fingers.
Nr. 14
[Year X + 14 (=19?)] The Year of the Appearance
of [the king as] the King of Lower Egypt [at] the
...(?) Festivals of Djet (Uto *or* Wadjet) and
Nekhbet.[61]
[Nile height:] 3 cubits.
Nr. 15

[Year X + 15 (=20?)] The Year of the Following of Horus [in which took place] the Tenth Occurrence of the Counting.
    [Nile height:]....
Nr. 16
[Year X + 16 (=21?)]....
    [Nile height:]....

[Horus Sekhemib and Seth Peribsen][62]
(*partly on C-1, but not readable*)[63]

[The last kings of the IInd Dynasty: Sened, Neferka, Neferkaseker, Khasekhem, and Khasekhemuy][64] (*no part on fragments*)

[Third Dynasty]

[Horus Zanakht, King Nebka?][65]

Lr, line 1, Nr. 1
    [Year X + 1 (=3?)] The Year of......[in which took place] the First Occurrence of the Counting of the Gold.[66]
    [Nile height:]....
Nr. 2
[Year X + 2 (=4?)] The Year of the Following of Horus [in which took place] the Second Occurrence of the Counting.
    [Nile height:] 3 cubits, 6 palms, 2 fingers.
Nr. 3
[Year X + 3 (=5?)] The Year of....the Rekhyt-Folk.[67]
    [Nile height:] 3 cubits, 1 palm.
Nr. 4

[Year X + 4 (=6?)] The Year of [the Following of Horus in which took place the] Third [Occurrence of the Counting].[68]

[Nile height:]....[2+1] cubits....

Pr, line 5, Nr. 1

[Year Y + 1 (=12?)] The Year of the Following of Horus [in which took place] the Sixth Occurrence of the Counting.

[Nile height:] 2 cubits, 4 palms, 1 1/2 fingers.

Nr. 2

[Year Y + 2 (=13?)] The Year of the Appearance of [the king as] King of Upper Egypt and [his] Appearance as King of Lower Egypt [at] the Building out of Stone of [the Temple called] "The Goddess abides".[69]

[Nile height:] 2 cubits, 3 palms, 1 finger.

Nr. 3

[Year Y + 3 (=14?)] The Year of the Following of Horus [in which took place] the Seventh Occurrence of the Counting of the Gold and the Fields.[70]

[Nile height:] 3 2/3 cubits.

Nr. 4

[Year Y + 4 (=15?)] The Year of the Making of a Copper [statue called] "Exalted is King Khasekhemuy".[71]

[Nile height:] 2 cubits, 6 palms, 2 1/2 fingers.

Nr. 5

[Year Y + 5 (=16?)] The Year of the Following of Horus [in which took place] the Eighth Occurrence of the Counting of Gold and Fields.

[Nile height:] 4 cubits, 2 palms, 2 2/3

fingers.
Nr. 6
[Year Y + 6 (=17?)] The Year of the Fourth Occur‑
rence of Bringing the Wall (?) of Dewadjefa and
of Shipbuilding.[72]

> [Nile height:] 4 cubits, 2 palms.

Nr. 7
[Year Y + 7 (=18?)] The [last] year [of King Nebka
(?) in which he reigned only] two months and 23
days [of the whole civil year].[73]

[Horus Netjeri-khet, King Djoser][74]

Nr. 8
[Year 1] The Appearance of [the king as] the
King of Upper Egypt and [his] Appearance [as]
King of Lower Egypt [at the Ceremonies of] the
Union of the Two Lands and of the Circuit of the
Wall.

> [Nile height:] 4 cubits, 2 palms, 2 2/3

fingers.
Nr. 9
[Year 2] The Appearance of [the king as] the
King of Upper Egypt and [his] Appearance [as]
King of Lower Egypt [at] the Double Shrines.[75]

> [Nile height:] 4 cubits, 1 2/3 palms.

Nr. 10
[Year 3] The Year of the Following of Horus [in
which took place the Festival of] the Birth of
Min.

> [Nile height:] 2 cubits, 3 palms, 2 3/4

fingers.
Nr. 11
[Year 4] The Year of the Appearance of [the

king as] the King of Upper Egypt and [his]
Appearance as King of Lower Egypt [at] the
Stretching of the Cord for the Mansion [called]
"The Refreshment of the Gods".

[Nile height:] 3 cubits, 3 palms, 2 fingers.
Nr. 12
[Year 5] The Year of the Following of Horus [in
which took place]....[76]

[Nile height:]....

[Horus Sekhemkhet, King Djoser-tety]

(*apparently all six years are on Fragment C-1 r,
line 4, but are quite unreadable*)[77]

[Horus Khaba(?), King --ka]

(*apparently the first two or so years of the
six-year reign of this king, whose complete king's
name is not known and whose Horus name is
merely guessed at, are also on Fragment C-1 r,
line 4, but are unreadable*)[78]

[King Huny]
(*no part of this reign is on any fragment*)[79]

[Fourth Dynasty]

[Horus Nebmaat, King Sneferu,
the son of Merlisankh.[80]

C-4 r, line 1, Nr. 1
[Year X + 1 (=4?)]   The Year of....Sneferu....[in
which took place] the Second Occurrence of the

Counting of Silver and Lapis lazuli.[81]

    [Nile height:] 3 cubits

Nr. 2

[Year X + 2 (=5?)] [The Year of the following events: the Appearance of the king as the King of Upper Egypt at] the Shrine Per-wer [and his Appearance as King of Lower Egypt at] the Shrine Per-nu,[82] and of fashioning in copper of a statue of the Horus Nebmaat (i.e. King Sneferu),[83] and of....

    [Nile height:] 3 cubits, 5 palms.

L, line 2, Nr. 1

    [Year Y + 1 (=10?)] The Year of....[84]

    [Nile height:]....

Pr, line 6, Nr. 1

    [Year Z + 1 (=12?)] [The Year of....and of the Ceremony of the Birth of] the Two Children of the King of Lower Egypt.[85] [The Sixth Occurrence of the Counting.][86]

    [Nile height:]....

Nr. 2

[Year Z + 2 (=13?)] The Year of [the following events:] The Building with *meru*-wood of a 100-cubit Dewatowe ship[87] and 60 sixteen-[oared?] barges. The Smiting of Nubia and the taking of prisoners numbering 4,000 men and 3,000 women, together with 200,000 large and small cattle. Building of the Wall of the Southland and the Northland [called] "The Houses of Sneferu". The Transporting of 40 ships filled with *ash*-wood (so-called "cedar").[88]

    [Nile height:] 2 cubits, 2 fingers (*or perhaps* 1 1/2).

Nr. 3

[Year Z + 3 (=14?)] [The Year of the following events:] The Erection of 35 Houses, and also enclosures (?) for 122 oxen. The Building of one 100-cubit Dewatowe ship of "cedar" and two 100-cubit ships of *meru*-wood. The Seventh Occurrence of the Counting.

[Nile height:] 5 cubits, 1 palm, 1 finger.

Nr. 4

[Year Z + 4 (=15?)] [The Year of the following events:] The Erection of "Exalted is the White Crown of Sneferu" upon the Southern Gate and of "Exalted is the Red Crown of Sneferu" upon the Northern Gate.[89] The making of "cedar" gates for the Royal Palace. The Eighth Occurrence of the Counting.[90]

[Nile height:] 2 cubits, 2 palms, 2 3/4 fingers.

Nr. 5

[Year Z + 5 (=16?)]....

[Nile height:]....

C-4 r, line 2, Nr. 1

[Year W + 1] The Year of....[and of the....Counting of Oxen] and small Cattle.

[Nile height:]....cubits, 2 palms.

Nr. 2

[Year W + 2] The Year of [the following events:] The Appearance of [the king as] the King of Upper Egypt [at] the Fourth Occurrence of the Running of the Oxen (*or* Apis Bull?). The Fashioning of A Gold Statue of the Horus Nebmaat for honoring the Gods. Bringing from Libya 1,100 prisoners and 13,100 large and small cattle. Going forth to the Building of the Fortress, *T-t* (?)....

[Nile height:]....

Pr, line 7

[Horus Medjed,
King of Upper and Lower Egypt
Khufu (=Cheops) whose] mother
[was Hetepheres].[91]

C-2, line 1
[Year W + 1] The Year [of the following events:]
....14 statues....Khufu....100 cubits....
[Nile height:] 3 cubits, 6 palms, 3 1/2 fingers.

C-1 r, line 7
[Year X + 1] The Year of Fashioning a Statue of the Horus....
[Year X + 2] The Year of....Sneferu....
[Year X + 3] The Year of....[there is none] beside him....
[Nile heights for these years unobtainable]

C-4, line 3
[Year Y + 1] The Year of [the following events of] the King of Upper and Lower Egypt Khufu....He made [it] as his monument for....made of Lapis lazuli....and as his monument for....
[Nile height:]....

C-2, line 2, Nr. 1
[Year Z + 1] The Year of [the following events:] The King of Upper and Lower Egypt [Khufu]....Building....The Establishing of Sed[92]....Birth of the Horus Bull....Opening the mouth of all (?) the Gods[93]....
[Nile height:]....

Nr. 2

[Year Z + 2]   The Year of [the following events:]
The Appearance of [the king as] King of Lower
Egypt [at] ....The Following....Names....
    [Nile height:]....

[King Djedefre]

C-3, line 2, Nr. 1
    [Year X + 1]   The Year of [the following events:]
    [Building] 20+ (?) barks....Engraving of the Double
    Shrine....[A block] 14 cubits, 2 fingers, in length.
    Granite from [the quarries of] Hatnub.
        [Nile height:]....
    Nr. 2
    [Year X + 2 (or is it a part of X + 1?)]   The Year
    of [the following events of] the King of Upper
    and Lower Egypt Djedefre. He made [it] as his
    monument for his (?) mother(?) Bastet....in....(?
    Bubastis?).94
        [Nile height:]....

[King Menkaure (Mycerinus)]

C-1 v
    [Year X + 1]....
        [Nile height:]....3/4.95
Pv, line 1, Nr. 1
    [Year Y + 1]   The [last civil] year [of which
    Menkaure reigned] [4] months and 24 days.

[King Shepseskaf]

    Nr. 2
    [Year 1] [After the king's enthronement there

remained of the civil year] [3] + 4 months and 11
days [and the following events took place:] The
Appearance of [the king as] the King of Upper
Egypt and [his] Appearance [as] the King of
Lower Egypt [at the Ceremonies of] the Union of
the Two Lands and the Circuit of the Wall. The
Festival of Seshed. The [Festival of the] Birth of
the two Wepwawets. The Year of the King's
Following of the Gods who United the Two
Lands. The Establishing of the Estate of the
Garden Lake and the Selection of the Place for
the Pyramid [called] "Shepseskaf is
Purified".[96]....20 [portions] every day for the
Double Shrines of the South and the North....
1,624.... 600 (300 + 300?) ....

[Nile height:] 4 cubits, 3 palms, 2 1/2
fingers.

[Fifth Dynasty]

[King Weserkaf]

L-1 v,line 2, Nr. 1
[Year X + 1 (=3?)] The Year [of the following
events of the King of Upper and Lower Egypt
Weserkaf:] .... Coming from the first journey (or
way) .... gifts which they brought to the Pyramid
of Weserkaf [called] "Purest of Places" as well as
70   women of foreign countries. The Year after
the First Counting of the Oxen.[97]

[Nile height:]....
Nr. 2
[Year X + 2 (=4?)] The Year of [the following
events of] the King of Upper and Lower Egypt

Weserkaf. He made [it] as his monument for (I) the Souls of Heliopolis:[98] the establishing for them of the following divine invocation-offerings: 242 (?) offerings of bread and beer, 43 oxen, 4 (?) oryx, 132 fowl, 12 pintail ducks, on the First Occurrence of the Festival of *It* and on every annual return of that festival, with abundance of offerings forever; for (II) Re: [a gift of] arable land in the following quantity of arouras (Egyptian: *st'wt*): 44.10 + 8 + 1/4 + 1/8 [*total in modern terms*: 448 3/8];[99] for (III) Hathor: arable land in the following quantity of arouras: 23.10 + 5 + 1/2 [*total in modern terms*: 235 1/2]; building [of a shrine] for Hathor in the estate of the Pyramid of Weserkaf [called] "Purest of Places" and building [of a shrine] for Hathor in the estate [called] "The Lake of Weserkaf is Purified".

[Nile height:] 3 cubits, 2 palms, 2 1/2 fingers.

Pv, line 2, Nr. 1

[Year Y + 1 (=5?)] The Year of.... the Third Occurrence of the Inventory of the House of Horus-Seth.[100] The Year after the Second Occurrence of the Counting.

[Nile height:]....

Nr. 2

[Year Y + 2 (=6?)] The Year [of the following events of] the King of Upper and Lower Egypt Weserkaf. He made [it] as a monument for (I) the Souls of Heliopolis: 20 invocation-offerings of bread and beer at every festival of the Sixth Day (?), 20 invocation-offerings of bread and beer at every Festival of *It* and arable land in the following quantity of arouras: 35.10 + 3 + 1/2 +

1/4 + 1/8 [*total in modern terms*: 353 7/8] in the Domain of Weserkaf; for (II) the Gods of the Sun-Temple [called] "Sep-Re": arable land in the quantity of 24 arouras in the Domain of Weserkaf, and 2 oxen and 2 pintail ducks everyday; for (III) Re: arable land in the quantity of 44 arouras in the nomes of the Northland; for (IV) Hathor: arable land in the quantity of 44 arouras in the nomes of the Northland; for (V) the Gods of the House of the Heron of Djebakherut(?):[101] arable land in the quantity of 54 arouras [for] the erection of a chapel of his [i.e. the Heron's] temple at Buto in the Nome of Xois; for (VI) Sepa:[102] arable land in the quantity of 2 arouras [for] the building of his shrine; for (VII) Nekhbet in the Shrine of the South: 10 invocation-offerings of bread and beer every day; for (VIII) Wadjet in the Shrine of Per-nu [in Buto]: 10 invocation-offerings every day; for (IX) the Gods of the Shrine of the South: 48 invocation-offerings of bread and beer every day. The Year of the Third Occurrence of the Counting of the Oxen.

[Nile height:] 4 cubits, 2 1/2 fingers.

Nr. 3

[Year Y + 3 (=7?)] The Year of [the following events of] the King of Upper and Lower Egypt Weserkaf. He made [it] as his monument for Re: arable land in the quantity of 1,804 (?) + 1/2 + 1/4 arouras and 13 [area] cubits (?) [*in modern terms with cubits converted to arouras*: 1,804 22/25 arouras][103] in the Northland; for Min: arable land in the quantity of....arouras....in the Northland....(*and the rest of the year is lost*).

[King Sahure]

C-1 v, line 3, Nr. 1

[Year X + 1 (=1?)]   [The Year of the following events of] the King of Upper and Lower Egypt Sahure. He made [it] as a monument for....The Festival of the Opening of the Mouth in the House of Gold (ḥwt-nb) (i.e. the sculptor's or goldsmith's workshop) of statues of....and that of the Opening of the Mouth in the House of Gold of six statues of Sahure. The First Occurrence of Going to the South and Inventorying the House of Horus-Seth.[104]   The Year of the First Occurrence of the Circuiting.

[Nile height:]....

Nr. 2

[Year X + 2 (=2?)]....(*nothing decipherable from this year*)

[Nile height:]....

Pv, line 3, Nr. 1

[Year Y + 1 (=5?)]   The Year of [the following events of] the King of Upper and Lower Egypt Sahure. He made [it] as his monument for (I)....in Heliopolis....[the Shrine of ?] the Divine Bark [upon its four supports?][105]....Son of Re....[the Shrine of?] the Divine Bark [upon its four supports?], 200 wᶜb-priests of [the Shrine of?] the Divine Bark [upon its four supports?]; for (II) Nekhbet, Mistress of Per-wer: 800 daily portions of invocation-offerings of bread and beer; for (III) Wadjet, Mistress of Per-nezu (!Per-nu): 4,800 daily portions of invocation-offerings of bread and beer; for (IV) Re in the Double Shrines:

138 daily portions of invocation-offerings of
bread and beer; for (V) Re in the Shrine of the
South: 40 daily portions of invocation-offerings
of bread and beer; for (VI) Re in Tep-hut: 74
daily portions of invocation-offerings of bread
and beer; for (VII) Hathor in the Sun-Temple
[called] "Sekhet-Re": 4 daily portions of
invocation-offerings of bread and beer; for (VIII)
Re in the Sun-Temple [called] "Sekhet-Re": arable
land in the quantity of 24 arouras in the Nome of
Xois; for (IX) Mesenkhet:[106] arable land in the
quantity of 2 arouras in the Nome of Busiris; for
(X) Semkhet: arable land in the quantity of 2
arouras in the Nome of Busiris: for (XI)
Khenti-yawatef: arable land in the Nome of
Memphis in the following quantity of arouras:
2·100 + 2·10 + 8 + 1/4 + 1/8 [total in modern terms:
228 3/8];[107] for (XII) Hathor in the Ro-she of
Sahure: arable land in the Nome of the East in the
following quantity of arouras and cubits: 2·100 +
2·10 + 6 + 1/4 arouras and 4 cubits [the total in
modern terms, with cubits converted to arouras:
226 29/100 arouras]; for (XIII) Hathor in the
pyramid [called] "The Soul of Sahure Appears":
arable land of 2·100 + 2·10 + 1/4 arouras in the
Libyan Nome (No. 7); for (XIV) the White Bull:
arable land in the Near Eastern Nome (Nr. 14) in
the following quantity of arouras and cubits:
23·100 + 2·10 + 1/4 arouras and 10 cubits [total in
modern terms, with cubits converted to arouras:
2,320 7/20]. The Third Ocurrence of the Inventorying of the House of Horus-Seth. The Year
after the Second Occurrence of the Counting.

[Nile height:] 2 cubits, 2 1/4 fingers.

Nr. 2

[Year Y + 2 (=6?)] The Year of [the following events of] the King of Upper and Lower Egypt Sahure. He made [it] as his monument to the Divine Ennead in the House of God....the House of...., and for the Double Shrines and the Tep-Hut: arable land in the Nome of the West (Nr. 3) in the quantity of .... arouras.... (*the rest of the year and its Nile height are lost*).

Pv, line 4, Nr. 1

[Year Z + 1 (=15?)] The Year of [the following events of] the King of Upper and Lower Egypt Sahure. He made [it] as his monument.... (*about 7 columns under this superscription are missing*): for Re, Hathor, and the House of Seshat, in the North and the South, arable land in the quantity of....arouras for Re, [2·10?] + 8 arouras for Hathor, and 2·10 + 4 arouras for Seshat; .....everybody. There was brought from the Malachite Country 6,000 [units] of copper, from Pewenet (i.e. Punt) 80,000 measures of myrrh, 6,000 [units] of $\underline{d}^c m$-gold (i.e. electrum),[108] 2,900 [units] of sn-gold,[109] and 23,020 [measures] of..... The Year after the Occurrence of the Seventh Counting.[110] [The reign of King Sahure terminated after] nine months and 28 days [of this civil year].[111]

[Nile height:]....

[King Neferirkare]

Pv, line 4, superscription

Horus Weserkhau, King of Upper and Lower Egypt, Favorite of the Two Goddesses

Khaemsekhemu (?), [Horus ?] of Gold....
Nr. 2
[Year 1] [The first civil year of which the king
reigned the last] two months and seven days, [the
year in which took place the ceremonies of] the
Birth of the Gods,[112] the Union of the Two Lands,
and the Circuit of the Wall, [and in which] the
King of Upper and Lower Egypt Neferirkare made
[it] as his monument for (1) the Divine Ennead in
the Divine Double Shrines in the city [called]
"Neferirkare Beloved of the Ennead": arable land
in the quantity of 3·100 + 30 +....arouras in the
Nome of Memphis, under [the charge of] the
House of Neferirkare beloved of his son (?); and
for (2) the Souls of Heliopolis: in the Eastern
Nome in the city [called] "Neferirkare Beloved of
the Souls of Heliopolis" arable land in the
quantity of 110·100 + 10 arouras;[113] for (3) the
Souls of Heliopolis and (4) the Gods of
Kheraha[114]: in the Near Eastern Nome (Nr. 14)
arable land in the following quantity of arouras:
352·100 + 20 + 1 + 1/2 + 1/4 [total in modern
terms: 35,221 3/4 arouras], the one parcel under
the charge of the Two Great Seers of his
temple[115] and the other under the charge of the
Officials and Priests of his temple and both
parcels as divine offerings being exempt from
taxes like the domain of the God (i.e. the King);
for (5) Re-senef and (6) Hathor: an altar for each
in the establishment [called] "Hutty" [provided]
every day with 210 portions of
invocation-offerings of bread and beer for the
one and 204 for the other, with appropriate
magazines and serfs established for each altar; for

(7) the Festival of the Opening of the Mouth of a statue of Ihy [i.e. Son of Hathor] made of electrum and the procession [with it to the Temple of] Hathor [Mistress of] the Sycamore, in Meret-Sneferu; for (8) Re [in the shrine of] Tep-hut, making for him these....the House of the Thirty[116] and likewise ....the Way.... (*the rest of the year is lost*).

Lv

[Year 2(?)] [The year of the following events of the King of Upper and Lower Egypt Neferirkare. He made it as his monument for]... 2 (*or* 10).... 5 (*or* 13).... 31 (*or* 41).... arable land in the quantity of y·10 + z arouras....[117]

   [Nile height:] 3 cubits....

Pv, line 5, Nr. 1

[Year Y + 1 (=10?)] [The Year of the following events of the King of Upper and Lower Egypt Neferirkare. He made it as his monument for....] the Festival of the Opening of the Mouth in the House of Gold [of the statue of].... and the Festival of the Opening of the Mouth in the House of Gold [of the statue of].... the levying of taxes (?).... Neferirkare.....; for Re in the Sun-Temple [called] "Favorite Seat of Re".... making for him a Circuit of the Wall.... and making for him the Estate (?) ....; for King Huny:[118] arable land in the quantity of.... arouras. The Year of the Fifth Occurrence of the Counting.

   [Nile height:]....

Nr. 2

[Year Y + 2 (11?)] The Year of [the following events:] The Appearance of [the king as] the King of Upper Egypt and [his] Appearance as King of Lower Egypt [at].... The Erection of the Wall of the Sun Bark at the Southern Corner [of the Sun-Temple called "Favorite Seat of Re"]. The King of Upper and Lower Egypt Neferirkare made [them] as his monument for (1) Re in the Sun-Temple [called] "Favorite Seat of Re": [models of] the Evening Sun Bark and the Morning Sun Bark, each of copper and each eight cubits long; for (2) the Souls of Heliopolis:.... made of electrum....; for (3) the Shrine of Ptah-South-of-his-Wall: (?) arouras of land.....; for (4) Wadjet in the Southern City: ....made of electrum ....

[Nile height:].....

The rest of the text is lost.

{For the notes to this document, see page 97.}

Notes to Document I.1

1. Though there is no evidence for or against it, it seems possible that the original stone contained in its first line the names of the gods who were sometimes considered to be forerunners of the human kings of Egypt, as is evident in the list of kings called the Turin Canon (composed in the reign of Ramesses II, ca. 1290-1224 B. C.) and in Manetho's *History of Egypt* (3rd cent. B. C.). In the Turin Canon the dynastic kings (starting with Meni) are preceded by gods and "spirits *(ʾḥw)*, [i.e.] the Followers of Horus *(šmsw Ḥr)* (see A. H. Gardiner, *The Royal Canon of Turin* [Oxford, 1959], plate I, Col. II, lines 1-9). The "Akhu, Followers of Horus", no doubt equivalent to Manetho's "Spirits of the Dead and Demigods" (see W. G. Wadell, *Manetho with an English Translation* [London and Cambridge, Mass., 1940, repr. 1980], p. 5), may well have included divinized predynastic kings like those named here in the Palermo Stone (see L. V. Žabkar, *A Study of the Ba Concept in Ancient Egyptian Texts* [Chicago, 1968], p. 35).

2. It is a likely supposition that, since the stone contained the names of predynastic kings of Lower Egypt as well as those of predynastic kings of a united Upper and Lower Egypt (as we shall see), the stone also contained names of kings of Upper Egypt. Since the stone was composed in the fifth dynasty, and thus in the tradition of the dynastic unifiers from the south, it would make sense that names of Upper Egypt would precede those of Lower Egypt.

3. I have used this rubric without brackets because each of the names of kings here listed in the

first line of the Palermo Stone has as its hieroglyphic determinative a figure of a king wearing the red crown of Lower Egypt (see Fig. I.32, line 1). Schäfer, "Ein Bruchstück," p. 14, having only the Palermo fragment with the kings of Lower Egypt, suggested that if the stone was produced in a temple of Lower Egypt perhaps it did not include names of kings of Upper Egypt. But in view of the fact noted in the preceding footnote that the stone contained kings of a predynastic kingdom of Upper and Lower Egypt, it seems much more likely that the stone also included the names of kings of Upper Egypt before those of Lower Egypt, as I propose.

4. When the Cairo fragment (C-1) was first studied both Gauthier ("Quatre nouveaux fragments," p. 31) and Daressy ("La Pierre de Palerme," p. 162) identified the predynastic kings in the first line (Fig. I.36b) as kings of Upper Egypt, i.e., the determinatives were figures wearing the white crown of Upper Egypt [though Daressy believed that the third figure perhaps wore the crown of Lower Egypt; but he thought this to be either an error or the result of wearing]. Breasted later ("The Predynastic Union," pp. 710-12) made an exhaustive examination of the Cairo fragment and showed that the kings on the top row were wearing not merely the crown of Upper Egypt but rather the double crown of Upper and Lower Egypt; or, to put it more precisely, there are ten royal figures in the top band "of which the last three heads (at the left) have been broken off. The remaining seven (Nos. 1-7 in Fig. I.36c) all show sufficient traces of the crowns to make it quite certain that each figure is wearing the double crown". This was an interesting discovery, for it seemed to show that there was a tradition of a predynastic union of Upper and Lower Egypt, as

Newberry and Sethe had thought (*ibid.*, pp. 720-21). Incidentally, Breasted also suggested that the top band included a long list of kings of Upper Egypt preceding those of Lower Egypt.

5. Most students have concluded that the first king given in the second line of the stone is Aha. If this is so, then Narmer (whose palette we mentioned above) would perhaps have been given as the last of the predynastic kings in the preceding line. However, since we have no reliable figures for the lengths of Narmer's and Aha's reigns (the numbers given in the vestiges of Manetho's account being generally unreliable), it could well be that both reigns appeared on the second line of the Stone to the right of the Palermo Fragment. The Palermo fragment contains on its right side the last two years of what appears to be Aha's reign, and further to the right the fifth fragment of Cairo seems to contain evidence of three earlier years of the reign of Aha.

6. The word *mst* is here translated "Birth". It was also used in the sense of "making" or "fashioning" statues (see K. Sethe, *Dramatische Texte zu altägyptischen Mysterienspielen* [Leipzig, 1928], p. 49). Hence it is possible that at the various ceremonies celebrating the "birth" of some god or other, it was in fact the fashioning of his statue that was being celebrated. See Gardiner, "Regnal Years," p. 13, n. 2: "I see no reason why the 'Birth of Anubis' and the 'Birth of Min' should not be interpreted as the creation of statues of those deities; the fact that these occurrences are mentioned twice each [in Djer's reign] is hardly an objection, since more than one statue, or different types of statues, may have been made. The only case in the Palermo Stone which is at all likely to have had a different sense is

'Birth of the Gods' in Vs, 4, 3, apparently a reference to the epagomenal days......If my theory is correct we have here fresh evidence of the vast importance which was attached, in the early dynasties, to such artistic creations; this was the age in which the traditional attitudes and attributes received their stereotyped forms. However, it must be remembered that from the Egyptian point of view such events will have been regarded less as artistic achievements than as acts of piety." Later Gardiner in his *Egypt of the Pharaohs* (Oxford, 1961), p. 414, while even yet interpreting "the birth of gods" as the fashioning of cult-images, suggests that the word "birth" used in the naming of these ceremonies is "the consequence of the belief that the statues became really alive after the ceremony of 'Opening the Mouth' had been performed over them". The ceremony of Opening the Mouth performed on cult-images was the same as that performed on mummified corpses to revivify them. It is described by E.A.W. Budge, *The Mummy*, Collier Books ed. (New York, 1972), p. 172: "When everything has been brought in this chamber, and the tables of offerings have been arranged, a [Sem] priest, wearing a panther skin, and accompanied by another who burns incense in a bronze censer, approaches the mummy, and performs the ceremony of 'opening the mouth'....*un-re* [*wn-r* or *wpt-r*], while a [lector] priest in white robes reads from a roll of papyrus or leather. The act of embalming [not to say, the fact of dying] has taken away from the dead man all control over his limbs and the various portions of his body, and before these can be of any use to him in the nether-world, a mouth must be given to him, and it must be opened so that...[he] may be able to speak [and eat]. The twenty-first and twenty-second chapters of the 'Book of the Dead' refer to the giving a mouth to the deceased....In the vignette to the twenty-third chapter a priest is seen performing the operation of opening of the mouth ... *arit apt re* [*irt wpt-r*], with the [small adze-shaped] instrument...and the deceased says in the text, 'Ptah has opened my mouth with that instrument of steel (*!*) with which he opened the mouth of the gods.'" Cf. A. J. Spencer, *Death in Ancient Egypt* (Harmondsworth, England, 1982), pp. 52-54, and especially A. H. Gardiner and N. de G. Davies, *The Tomb of Amenemhet* (London, 1915), pp. 57-61. A. Erman, *A Handbook of Egyptian Religion* (London, 1907), p. 134, emphasizes that after the ceremony the deceased is able to partake of the food offerings presented to him. For a recent discussion of the ceremony and its literature, see J. G. Griffiths, *The Origins of Osiris and His Cult* (Leiden, 1980), pp. 69-74. Particularly pertinent is his remark on pp. 71-72: "In its earliest form it is a ceremony concerned with the statue, and it aims at the preparation and animation of the statue in the *ht-nb* ('House of Gold') before its transport into the shrine." See also S. Morenz, *Egyptian Religion* (Ithaca, 1973), p. 155: "But we must add that from early times onward Egyptians were not satisfied with just fashioning an image, i.e. with the creation of a work of art. On the contrary, a ritual was performed on the statues while they were still in the sculptor's workshop (the 'gold house'), as a result of which the work of human hands was thought to come alive. This ceremony of 'opening the mouth' had the purpose of making all the organs serviceable and so vitalizing the image....The surviving sources which mention the ritual performed on statues go on to mention acts and texts relevant to embalmment, sacrificial ritual and certain temple rituals." Incidentally the variety of statues and cult-images produced in the sculptor's workshop (the House of Gold) is interestingly presented on a stela of the eighteenth or nineteenth dynasty (Leyden V 1; see A. H. Gardiner, *Ancient Egyptian Onomastica*, Vol. 1 [Oxford, 1947], pp. 52-53): "He (the King) appointed me to take charge of operations when I was but a weanling, he found me estimable in his heart, and I was introduced into the House of Gold in order to fashion the forms and images of all the gods, and none of them was hidden from me. I was a master of secrets seeing Re (R^c) in his changing appearance and Atum in his true shape. Then there was Osiris, lord of Abydos, in front of the Lords of the Sacred Land, and there was Thoth, lord of Khmun .....I saw Shepsy in his mysterious secrecy, and Unwet in her changing appearances. There was Min cleaving to his beauty, and Horus dwelling in Hasroet ..... [followed by many more gods] .....It was I who caused them to rest in their eternal shrines, carrying them in the conduct of the king's festival with which I was charged (?)." The most complete treatment of the ceremony of opening the mouth is that of E. Otto, *Das ägyptische Mundöffnungsritual*, 2 vols. (Wiesbaden, 1960).

Still other explanations for "birth" in the names of individual festivals have been advanced. For example, the "Birth of Min" has been interpreted in terms of the changing status of Min in the course of a festival like that recorded on the north wall of the second court of the Temple of Ramesses III at Medinet Habu, where Min is first associated with Osiris and is thus interpreted as a remote, perhaps dead deity like Osiris and then later in the ceremony after cutting a stalk of emmer he becomes a living god identified with

Horus, the transition from the one to the other being interpreted as a "birth" (see W. J. Murnane, *United with Eternity* [Chicago and Cairo, 1980], p. 37).

Needless to say, the simplest explanation of "birth" is that a day was established as an anniversary of the god's birth, for as we shall see in Chapter Two, gods were capable of birth and death.

For specific mention of the fashioning of statues in the Palermo Stone, see Pr, line 5, Nr. 4, which describes a statue of King Khasekhemuy, and Pv, line 4, Nr. 2, a statue of the god Ihy. V. Giustolisi, "La 'Pietra di Palermo'," 3a puntata, pp. 28-29, believes that this year and the second after it are years of the Following of Horus, which he persists in calling the *Festa di Horo* (see note 11 below), because, in the later years recorded on the Palermo fragment, this Horus event is listed for one of its two years. This is reasonable, but his suggestion that the second year after this also saw the Festival of the Birth of Anubis as well as the *Festa di Horo* is unlikely because then there would be nothing to distinguish the two years X + 1 and X + 3, and the second event in each Horus year was clearly selected to distinguish the years.

7. It will be seen that from the next reign onward, the maximum height of the Nile in flood is recorded below each year. It will be evident from the years remaining here from C-5 line 1 and Pr line 2 of Aha's reign that the fifth-dynasty author of the *Annals* did not have access to the Nile heights for this reign. Perhaps no record of them was kept in the reign of Aha and before.

8. It is not clear as to what ceremony is being referred to here. The last sign is that of a bull. J. L. de Cenival, "Un nouveau fragment," p. 15, thinks it is

probably a reference to an Apis Bull. Compare the references to the Apis Bull in Pr, line 3, Nr. 12, and Pr, line 4, Nos. 4 and 10. See Fig. I.16 (second register) for the representation of a running bull (possibly Apis), and perhaps this is in celebration of the same event as here appears on the Cairo fragment.

9. Not enough of this year is visible for any more specific translation than that given here. The only thing that seems clear to me is the bottom part of the sign for *mst*. Giustolisi suggests the festival was that of the Birth of Anubis, which I believe is wrong; see note 6.

10. I suggest here that about four years separate fragment C-5 from the Palermo fragment. I have done this on the basis of the distance between the fragments in line 3 and the ratio between the sizes of the year segments in lines 2 and 3.

11. "Following of Horus" is a literal rendering of *šmsy-Ḥrw*. There has been considerable discussion of this phrase and its analogous forms. For Schäfer and Breasted it meant some ceremony of the worship of Horus. But it is now often interpreted as a biennial tour of the realm undertaken by the king and his officials, perhaps with the objective of ascertaining the wealth of the land for the purpose of tax assessment or collection. See J. v. Beckerath, "*šmsj-Ḥrw* in der ägyptischen Vor- und Frühzeit," *Mitteilungen des Deutschen Archäologischen Instituts, Abteilung Kairo*, Vol. 14 (1956), pp. 5-6 (whole article, pp. 1-10). See also Gardiner, "Regnal Years," p. 13, where he describes these tours as "royal progresses by river". The emphasis on travel by river arises from the fact that *šmsy-Ḥrw* is determined on the stone by a ship with a falcon on the foredeck and the sign for "following" on the afterdeck.

It may be that these biennial tours were the forerunners of the biennial "countings" of the wealth of the land which we shall mention below.

12. Most scholars believe that the second line of the stone began with the reign of Aha, as I noted above, and continued through the first two years recorded on the right side of the Palermo Stone. If that is so, then the third year on the Palermo Stone should begin the reign of the Horus Djer. This argument appears to be confirmed by evidence drawn from the first Cairo fragment. For, if the various estimates of the distance (in years) between the left end of the Palermo Stone and the right end of the Cairo fragment (estimates that range from 9 to 14 years) are even gross approximations of the true distance, then that distance would not be great enough to contain the end of a reign prior to that included on the Cairo fragment. Hence the years given on the two fragments are parts of the same reign, namely that of Djer, since the superscribed name of the king above line 2 of the Cairo fragment is almost certainly that of Djer (see note 24 below). A brief word should be added to explain my conclusion that the lacuna between the fragments would not have been large enough to have included the end of a preceding reign begun in the Palermo Stone. Studying the various superscribed royal names in the two fragments (those names above lines 2 and 3 of the Cairo fragment and the partial one above line 4 of the Palermo Stone), we can deduce that the superscribed name ordinarily occupied a space of at least seven years. Furthermore, it is evident from the name over line 3 of the Cairo fragment that such superscribed names are so placed that just as many years precede the beginning of the superscription as follow the end (i.e.,

that the name is placed as nearly as possible over the middle years of the reign). But there are nine years of the reign beginning in line 2 of the Palermo Stone and there is no trace of a superscribed name above those nine years. Hence if that reign is said to have ended in the lacuna between the two fragments, the lacuna will have had at least sixteen years recorded in it, that is, the seven years under the name plus nine years to balance at the left end the nine years given at the beginning and recorded on the Palermo Stone. But that 16 years (which is only an estimated minimum) exceeds even the largest estimate of the distance between the fragments. In fact, the lacuna probably would have had to include far more than just 16 years, for we know that Djer's reign must have been a fairly long one. This judgment is based not merely on the untrustworthy figures of 31 or 39 years found in the remains of Manetho's account but rather on the quite extensive archeological remains around the tomb of Djer at Abydos (see W. M. F. Petrie, *A History of Egypt*, Vol. 1 [11th ed. London, 1924], p. 15). Hence the lacuna would have had to have had not only at least 16 years for the end of the reign that began in the Palermo Stone but an additional minimum of at least 11 years for the beginning of the reign of Djer. Hence we seem to have confirmed that the reign of Djer began in the Palermo Stone and continued in both the lacuna between the fragments and in the Cairo fragment.

13. The figure of "four months and thirteen days" was interpreted by Schäfer and Breasted as a date, namely as the thirteenth day of the fourth month. It was Borchardt (*Die Annalen*, pp. 2-5) who correctly saw these and the similar numbers on the Palermo Stone as being time periods which are fractions of civil

(or, as he called them, calendar) years. But notice in this first example that, if we add to the four months and thirteen days of Djer's first year the six months and seven days of the last year served by his predecessor, the total is ten months and 20 days, instead of the full civil year of twelve months and five days (as was the case in the year of transition from the reign of Sahure to that of Neferirkare, where the nine months and 28 days of the end of Sahure's reign may be added to the two months and 7 days of the beginning of Neferirkare's reign to equal 12 months and 5 days [see the Palermo Stone, verso, line 4, Nos. 1 and 2], and apparently also in the year of the change to the reign of Shepseskaf where a plausibly-read [4] months and 24 days may be added to [3+] 4 months and 11 days to equal 12 months and 5 days [ibid., line 1, Nos. 1 and 2]). Now we do not know whether the calendar year of twelve months and five days which existed in the Old Kingdom prevailed in the first dynasty. All we know from the recital of months and days given at the end of Aha's reign and that given at the beginning of Djer's reign is that some kind of calendar with divisions of months and days was in use (assuming that the compositor of the fifth dynasty was using a first-dynasty source). If the 365-day calendar was already being used, then here in the case of the transition to Djer's reign we lack 1 month and 15 days of a full civil year. Hence between the reigns there must have been 1 month and 15 days or that period plus one or more years. It might be argued that, since there was no Nile height recorded for the end of the preceding reign, while we have such a height designated for the first year of Djer's reign, the two time spans are fractions of a single year and thus the two reigns

were separated by only one month and fifteen days.
This argument is considerably weakened by the
previous observation that no Nile height is recorded for
any of the extant years of the reign of Aha, and so it
is not significant that no such height was recorded in
his last year. But an even stronger argument against
there being an interval of only one month and fifteen
days is the fact that the author not only has a straight
line that extends vertically to the bottom of the line
above to indicate the end of the reign but he also
includes to the left of that vertical line the top of the
palm branch that constitutes the hieroglyphic sign for
"year", thus giving us the impression that not only do
we have a new reign to the left of the vertical line but
also a new civil year. However, in Palermo Stone,
recto, line 5, seventh year (see Fig. I.32) there is a clear
case where the civil year is divided up between the
two reigns, for to the left of the vertical line ending
the one reign there is no whole palm branch indicating
a new year, but rather there immediately follow the
signs indicating the initial ceremonies of the beginning
of a new reign. One clear conclusion emerges, namely
that the years being recorded in the annals are
calendar years and not regnal years, as the last case
cited shows. No doubt the cases where the fractions
add up to twelve months and five days also show this
to be so, for it is unreasonable to suppose that the new
king after the lapse of one or more full years would
start his reign on the exact day which would allow his
fraction of the year when added to the fraction of the
preceding king's last year to equal exactly twelve
months and five days. Incidentally in those cases
where the fractions add up to twelve months and five
days the tops of the reign-ending lines are absent either

because the fragment was broken (Palermo Stone, verso, line 1) or because the line was worn away (*ibid.*, line 4). Presumably if the vertical lines were still intact we would find that there were no palm branches indicating new years to the left of the lines ending the reigns.

The Ceremonies of the Union of the Two Lands and the Circuit of the Wall were traditional ceremonies of the coronation of the King. The "Wall" was presumably the "White Wall" of Memphis, perhaps built by the first of the dynastic kings (Menes?) at the founding of Memphis. The procedures of Succession and Coronation as persisting into the New Kingdom are described in lively fashion in H. Frankfort, *Kingship and the Gods* (Chicago, 1948), pp. 101-09.

14. Quite clearly one of the main services of the early annals was the inclusion of the maximum height of the Nile for each year, which no doubt served as a measure of the produce due. It is supposed that the measurements recorded here were taken in the neighborhood of Memphis or of Old Cairo, on a standard Nilometer from some agreed zero point. The numbers were most often in terms of cubits, palms, and fingers with an occasional use of spans. The royal cubit equaled 523 mm. or 20.6 inches = 7 palms = 28 fingers. There is another shorter cubit of 6 palms and two different spans, one of 3 1/2 palms and the other of 3 palms. See F. Ll. Griffith, "Notes on Egyptian Weights and Measures," *Proceedings of the Society of Biblical Archaeology*, Vol. 14 (1891-92), pp. 403-04 (whole article, pp. 403-50). In regard to the Nile height we may conveniently quote the brief account of H. Kees, *Ancient Egypt* (Chicago and London, 1961; Phoenix edition, 1977), pp. 50-52, based on the research of Borchardt and others: "In the annals from the Thinite

ANCIENT EGYPTIAN SCIENCE

Period up to Dynasty V, between about 2950 B.C. and 2500 B.C., the rise of the Nile, probably measured in the neighborhood of Old Cairo, was on an average about 4 cubits (approximately 7 feet); therefore a recorded High Nile of 8 cubits and 3 fingers caused in the following year [according to Sethe's interpretation of Palermo Stone, recto, line 3, No. 4 given below] 'flooding of all the western and eastern nomes' in the Delta. Lists compiled in the reign of Sesostris I, about 1950 B.C., give the following, far higher, figures for the desirable height of the flood-waters at various points on the Nile: Elephantine, 21 cubits, 3 1/3 palms (approximately 39 feet); 'the House of the Inundation' near Old Cairo, 12 cubits, 3 palms and 3 fingers (about 21 1/2 feet); for Diospolis or Tell Balamun, the most northerly town of the Delta, 6 cubits, 3 palms, 3 fingers (about 11 feet). The many measurements recorded up to Roman times reveal a further rise of about 20-30 per cent with the result that the figure for Elephantine reaches 24 cubits, 4 palms (about 42 feet) and, as an idealized figure, 28 cubits (i.e. equal to 4 x 7); for the 'House of the Inundation' at Old Cairo the figure becomes 14-16 cubits, while for the Northern Delta it remains unchanged at 6-7 cubits. This increase can only partly be explained by the rise in the bed of the river which, according to Lyons, amounts to about 4 inches a century; in the course of the centuries separating the reign of Sesostris I from 150 B.C. the rise of the bed would amount to about 6 feet or 3 cubits, 3 palms. The fact that ancient writers made use of theoretical and idealized figures is proved by a completely reliable statement by Strabo (Book XVII, 788) that when Petronius was Prefect 12 cubits, that is about 21 feet, of flood-water assured an abundant harvest. He adds that with only 8 cubits there was still no shortage whereas before his time apparently 14 cubits had to be attained before bountiful crops were assured. This measurement of 12 cubits corresponds almost exactly with the Memphite standard under Sesostris I about 2,000 years earlier. As a check on these figures it should be mentioned that in the Third Century B.C. for the purpose of constructing dykes at Memphis the figure of 12 cubits was taken as the high-water mark whereas the actually recorded heights attained by the flood-waters in the years 259 B.C. and 258 B.C. were respectively 10 cubits, 3 palms, 1 1/6 fingers and 10 cubits, 6 palms, 2 2/3 fingers. An inundation that reached the height of 14 cubits at Memphis must therefore have overflowed the dykes.

"The heights recorded in these figures were not apparently measured from low water, for its level on the monument of Sesostris II is taken to be almost the same throughout the whole length of the Nile with 4 cubits, 2 palms, 3 1/3 fingers for Upper Egypt and 4 cubits, 3 fingers for Lower Egypt. According to Borchardt, however, all the Nilometers in the country measured from a fixed zero at Roda with a theoretically assumed fall from Elephantine to Memphis. The big rise in the measured numbers, especially for Elephantine, cannot be accounted for by the fact that because of the narrowing of the river through the cataract the highwater mark at Elephantine is higher above the low water than is the case at Memphis (Lyons estimates about 23 feet for the former and about 16 feet 6 inches for the latter), or in the Delta where the river broadens to an incomparable extent. The explanation must be sought in the predetermined zero on the gauge."

ANCIENT EGYPTIAN SCIENCE

The classic account of Nilometers is that of L. Borchardt, "Nilmesser und Nilstandsmarken," *Abhandlungen der Königlichen Preussischen Akademie der Wissenschaften*, 1906 (*Abhandlungen nicht zur Akademie gehöriger Gelehrter, Phil.-hist. Abh.* I, pp. 1-55). See also W. Helck, "Nilhöhe und Jubiläumsfest," *Zeitschrift für ägyptische Sprache und Altertumskunde*, Vol. 93 (1966), pp. 74-79; K. W. Butzer, "Die Naturlandschaft Ägyptens während der Vorgeschichte und den dynastischen Zeitalter," *Akademie der Wissenschaften und der Literatur, Mainz, Math.-Naturwiss. Kl., Abhandlung* No. 2 (1959); and B. Bell, "The Oldest Records of the Nile Floods," *The Geographical Journal*, Vol. 136 (1970), pp. 569-73. Bell concludes concerning the Nile heights recorded on the Palermo Stone: "Under either of these assumptions about the zero-point of the scale [i.e. either that the zero-point of the scale rose with the alluvium or that the zero-point was fixed by the nilometer's being built into a quay or wall]...it is clear that the height of the inundation, and thus the amount of the summer monsoon rainfall over East Africa, averaged less from Dynasty II onward than in Dynasty I. The difference between the average flood-height for Dynasty I and for Dynasties II-V is 0.7 metres, under the assumption of a zero-point that rose at a uniform rate with the alluvium. Under the assumption of a fixed zero for the nilometer, the decline in flood height is greater, and is also progressive with time. If the alluvium actually rose more rapidly than the rate I have assumed..., an even larger decline in the flood volume would be indicated."

The connection of the measure of the Nile height with taxes was noted by Seneca in his description of the Nilometer at Aswan (quoted by K. Baedeker, *Egypt and the Sûdân* [Leipzig, 1929] p. 383): "The Nilometer is a well built of regular hewn stones, on the bank of the Nile, in which is recorded the rise of the stream, not only the maximum but also the minimum and average rise, for the water in the well rises and falls with the stream. On the side of the well are marks, measuring the height sufficient for the irrigation and the other water levels. These are observed and published for general information.... This is of importance to the peasants for the management of the water, the embankments, the canals, etc., and to the officials on account of the taxes. For the higher the rise of water, the higher are the taxes." See *ibid.*, p. 113, for an account of the Nilometer on the Isle of Roda.

15. Note that *Desher* has as its hieroglyphic determinative a ship. Compare with this the determinative of the Festival of *Djet* in No. 11 below. Presumably in these festivals movement by ship was involved, so much so that the ship becomes a customary determinative for various festivals. The literal meaning of *desher* is "red" and so we have here a "Festival of the Red". Incidentally the word in the feminine means "the desert land" or "the Red Crown". But here it is given in the masculine.

16. It is unlikely that this is a reference to the birth of twins of the Horus Djer, since it is repeated later (Pr, line 6, Nr. 1). According to Sethe, *Untersuchungen*, Vol. 3, p. 63, the reference is probably to the Birth of Shu and Tefenet, the children of the sun-god. As a ceremony it would symbolize the pouring of life into Shu and Tefenet by Atum (see Frankfort, *Kingship and the Gods*, pp. 134-35) and have significance for the earthly king as well.

17. The event being celebrated here is mysterious but appears to represent the early defeat and subjection of the *Rekhyt*-Folk of the Delta. One thinks of the pile of decapitated people on the Palette of Narmer (see Fig. I.8 right) and similarly of the lap-wings *(rḥyt)* hanging from standards on the mace-head of the Scorpion king (see Fig. I.3). Consult the thorough (but rather speculative) account of the hieroglyphs that describe this event in P. Kaplony, *Kleine Beiträge zu den Inschriften der ägyptischen Frühzeit* (Wiesbaden, 1966), pp. 65-71. The interpretation of the *Rekhyt*-Folk as inhabitants of the Delta is questioned in a long and thoughful account in A. H. Gardiner, *Ancient Egyptian Onomastica*, pp. 98*-108*.

18. Schäfer translates this as follows: "Planen (?) des Hauses *Ḥsf-nṯrw.* Fest des Sokaris (?)." Breasted changes it slightly: "Design (?) of the House (called): 'Mighty of the Gods' *(sḥm-nṯrw)*. Feast of Sokar." My translation differs in two respects. I reject utterly the translation of *ḥ¹* by either *Planen* or *Design*, preferring rather the translation "behind" or "at". It could hardly be the planning or design of the castle that is involved, since the same event is repeated a few years later in the same reign (see C-l r, line 2, Nr. 4), and surely "behind" is the normal rendering of it. Furthermore I read *smr-nṯrw* instead of either Schäfer's or Breasted's readings. Hence I have translated the name of the castle as "Companion of the Gods". Note that Kaplony, *Kleine Beiträge*, p. 69, has read *smr(?)-nṯrw* within a castle outline appearing on the second register of an ivory label bearing Djer's name (which, incidentally, Kaplony always writes as *Sḥtj*). See also his "Gottespalast und Götterfestungen in der ägyptischen Frühzeit," *Zeitschrift für ägyptische Sprache und*

Altertumskunde, Vol. 88 (1962), p. 12, n. 5 (full article pp. 5-16). A conclusive bit of evidence that this was a castle of Djer's is found on a stone vessel in the Berlin Museum (Borchardt, *Annalen*, p. 31 and my Fig. I.45). Kaplony in the article just quoted mentions six castles or fortifications that have *nṯrw* as a part of the name and all are in the first three dynasties. These *Götterfestungen* are distinguished from the *Gottespalast*.

Incidentally I add here "[the First Occurrence of]" so that this year can be adequately distinguished from the later entry in C-1 r, line 2, Nr. 4, where I add "[the Second Occurrence of]".

19. Schäfer remarked that the goddess Yamet was named in the later Pyramid texts. With Giustolisi, I think that it is probable that the sign read by Schäfer as *Yamet* is actually the sign later named *ʾImy-wt* (at least from the sixth dynasty), the well-known symbol of a pelt hanging from a rod, which (probably from the first dynasty and certainly from the third) symbolized the god Anubis. The name *ʾImy-wt* may be translated: "He who is in the bandaging room". Such a symbol from the Middle Kingdom was found in a wooden shrine during the Metropolitan Museum's excavations at Lisht in 1914 (see A. M. Lythgoe, "Excavations at the South Pyramid at Lisht in 1914," *Ancient Egypt*, 1915, pp. 145-53) and in Tutankhamen's tomb (see my Fig. I.46). We see evidence of the appearance of this symbol earlier than its apparent reference here in Djer's reign, in fact in the reign of Aha, who preceded Djer (see Fig. I.16, first register). Furthermore the symbol appears unequivocally two reigns later in Den's reign, still in the first dynasty (see below C-5, line 2, Nr. 3 [Cf. Fig. I.40]). Indeed U. Köhler, *Das Imiut*, 2 vols. (Wiesbaden, 1975), notes (vol. 1, p. 1) that she has found eighteen

representations of the symbol from the first dynasty alone. Her very thorough study points out that the use of the name *Imy-wt* for the cult-symbol cannot be found before the sixth dynasty, though the symbol was probably associated with Anubis as early as the first dynasty and certainly so from the third. I might add that I had already thought of the possibility that the cult-symbol of Anubis instead of *Yamet* should appear here before I had read Giustolisi's articles and so my judgment was quite independent of his earlier suggestion. Incidentally, even if this name "He who is in the bandaging room" were to be found associated with the symbol in the early dynasties (and this is not improbable since the term *wt* was connected with Anubis at least in the third dynasty when, we have seen, the symbol certainly stood for Anubis), it would not, however, ensure that mummification was already practiced by that time. In such a case it would simply reflect the Egyptian routine of bandaging limbs, a routine evident before they took up mummification.

The exact date at which mummification first began in Egypt is unknown, "but the first definite evidence of any attempt is from the beginning of the Fourth Dynasty" (see A. Lucas and J. R. Harris, *Ancient Egyptian Materials and Industries* [London, 1962], p. 271). These authors (p. 270) speak of the early history of burial practices in Egypt as follows: "During the neolithic and predynastic periods, the body was buried in a shallow grave,...generally wrapped in an animal skin or loose folds of linen, but by the early dynastic period the graves of the kings and wealthier classes had become deeper,...: and the previous loose covering on the body had given place to close-fitting linen wrappings, which in some instances eventually became elaborated into the separate wrapping of each limb, with further wrappings for the whole body, examples of which are known from the First, Second, and Third Dynasties respectively, before mummification was introduced." Chapter XII of the Lucas-Harris book is one of the best accounts of Egyptian mummification. See also the recent treatment in Spencer, *Death in Ancient Egypt*, Chaps. 2 and 5.

20. Min, the ithyphallic god of Coptos, was an early, widely venerated god, as is evident from the supposedly predynastic (but probably early dynastic) colossal statues found by Petrie at the Temple at Coptos. See H. Schäfer and W. Andrae, *Die Kunst des alten Orients* (Berlin, 1935), p. 179, for one such statue. Note that the symbol of Min appears not only on this statue but also on the mace-head of the Scorpion king (see above, Fig. I.3, first register, the middle standard). Later Amon-Re absorbs some of his attributes and is often shown in ithyphallic form. Indeed Wainwright believed Amun to be essentially Min (see below, Chap. 2, n. 60).

21. I have added here "[First Occurrence]" because of the repetition of the Festival of the Birth of Anubis in Djer's reign (see below, C-1 r, line 2, Nr. 1).

22. Here it was only necessary to select a single event because for the first time the selector of the year's name introduced the ordinal number of a festival. He tells us that this is the First Occurrence of the Feast of Djet or Wadjet (presumably in Djer's reign) and hence no other identification is necessary. Wadjet, Goddess of Buto, became the tutelary goddess for Lower Egypt.

23. Only the minutest traces of the tenth year are visible, but not enough to be read.

24. Since the preceding years are almost certainly also a part of Djer's reign, and the reading of this superscription is quite generally accepted as being the name of Djer, perhaps it would have been better to introduce this whole name where I simply have Horus Djer above note 12. The reading of the superscription is quite clear except for the Horus name, where one can see the vertical strokes of the sign for Djer but nothing else. But the name as king and Horus of Gold, namely Itet, appears to clinch the identification. I note that the reading "[Horus] of Gold" is somewhat doubtful for all we have on the fragment is *n-nbw*. If this does stand for the Golden Horus name, we would have to conclude that already in the first dynasty (at least by the time of Horus Den, the second king after Horus Djer) four of the five traditional titles for the kings of Egypt were in use: (1) the Horus title in the *serekh* or palace facade (see Fig. I.47a A), (2) the *nebti* or Two-Goddesses title (see Fig. I.47a B), (3) The Golden-Horus title, (4) the *nesu-bit* or King-of-Upper-and-Lower-Egypt title (see Fig. I.47a C). Only the Son-of-Re title was missing, it being used sporadically as an epithet or designation and not as part of a full titulary in the fourth, fifth and sixth dynasties (see Figs. I.47c-d) and formally and regularly as part of the fivefold titulary in the eleventh dynasty (see Figs. I.47e). The reader should note that the *serekh* in which the Horus name appears was also used in the first dynasty by Queens Neithhotep and Meryetneith (see Fig. I.47a D and E), with the crossed arrows of the goddess Neith surmounting the *serekh* and thus appearing where the falcon appeared in the Horus name. Similarly notice that in the second dynasty the Seth animal replaced the Horus falcon in the name of Seth Peribsen (see Fig. I.19a) and that the same animal

was used in conjunction with the Horus falcon in the name of the Horus-Seth Khasekhemuy (also see Fig. I.19a). For a discussion of the fivefold royal titulary, see Gardiner, *Egyptian Grammar*, pp. 71-74, H. Müller, *Die formale Entwicklung der Titulatur der ägyptischen Könige* (Glückstadt/Hamburg/New York, 1938), and J. von Beckerath, *Handbuch der ägyptischen Königsnamen* (Munich and Berlin, 1984), pp. 1-42. All scholars depend on the monumental work of H. Gauthier, *Le Livre des rois d'Égypte* (Cairo, 1907-17).

25. For my addition of "the Second Occurrence", see note 21 above.

26. This is unreadable in the photographs but was suggested "with reserve" by Gauthier (and I tentatively follow him). Incidentally, Daressy also gave this reading. Giustolisi adds the Festival of Sokar as a second event of this year.

27. The suggestion of the reading "Sed" is that of Daressy, who claimed to see the sign for "Birth" and a part of a vertical stroke followed by a standard. This stroke he believed to be part of "s", and on the base of the standard he hypothesized the wolf which stands on the *shedshed* standard crossed by a mace in Pr, line 3, Nr. 11 and is read "Sed" (see Fig. I.32 and *Wb*, Vol. 4, p. 365). Sed seems to have been an early form of the wolf god Wepwawet and is closely tied to the king, his (Sed's) standard accompanying the king in the king's Sed Festival (see below, note 41).

28. See note 18 above.

29. Both Gauthier and Daressy report the hill-country or foreign-land sign on a standard, which is to be read as the desert God Ha (see Gardiner, *Egyptian Grammar*, p. 488, sign N 25).

30. So reads Daressy. Gauthier suggests "Min".

31. Daressy remarks that the "Red Festival" could be a ceremony celebrated in Upper Egypt in connection with an episode in the life of Seth.

32. The sign of the particular God whose birth is being celebrated is, according to Daressy, "unreadable and partially destroyed".

33. The reading "the Great White One" derives from the seated baboon or monkey which remains on the edge of the fragment (see de Cenival, "Un nouveau fragment," p. 16). Another possible interpretation of the baboon is as a sign for Thoth. J. Černý, *Ancient Egyptian Religion* (London, 1952), p. 21, speaks of an early baboon god as follows: "Numerous small statuettes of baboons and a representation of this animal on an ivory label suggest that its cult dates from the beginning of Egyptian history; it may have been practised at Khmun (Hermopolis) where presumably it preceded the cult of the ibis of Thoth. The original reading of the name of this baboon god is uncertain; but later he was called Hedj-wer or Hedjwerew and interpreted as the 'Great White One' or 'Whitest of the Great Ones'." Cf. Petrie, *Royal Tombs*, Part I, plate XVII, 26, for a tablet of Semerkhet which seems to have a similar seated baboon (given above as Fig. I.18, right). It is interpreted tentatively (p. 42) as representing Thoth.

34. Compare the battle against the Wolf People two years hence and also that against the *Intyw* (Pr, line 3, Nr. 2). How these various people are to be distinguished is not clear. The guess that this is Year 19 of Den's reign is based on the not-too-certain assumptions that (1) there are five years between fragments C-5 and Pr, and (2) that the year reported in Pr, line 3, Nr. 3 as the year of a Sed Festival is Year 30

(see note 41 below).

35. Compare note 19. The *Senuty-Shrine*, or Double Shrines of Re, combined the two primitive shrines of Hieraconpolis (the *Pr-wr* Shrine) and Buto (the *Pr-nsr* or *Pr-nw* Shrine). See Fig. I.48 and Gardiner, *Egyptian Grammar*, pp. 494-95, signs O 19 and O 20.

36. I do not know who the enemies were in this engagement, unless they were the followers of Seth. In one of the tablets of Den, a bearded enemy (who apparently is one of the *'Intyw*) has beside him a standard with a wolf symbol on it (Fig. I.18, left). See P. E. Newberry and G. A. Wainwright, "King Udy-mu and the Palermo Stone," *Ancient Egypt*, 1914, pp. 150, 152 (full article, pp. 148-55).

37. Notice that the castle called "Companion of the Gods" is still being mentioned as a place to hold ceremonies in Den's reign, though it was apparently built in Djer's reign.

38. Because the superscribed title covers the space of about seven years, there ought to be about five years between fragments C-5 and Pr.

39. I follow Schäfer in this rendering. But Giustolisi says that we should read the god Ptah and not Saw, which would make this the earliest reference to the cult of Ptah.

40. See note 36 above. The Troglodytes are described as Easterners. Newberry and Wainright call them "People of the Eastern desert and Sinai" (see *op. cit.* in n. 36, p. 152).

41. The Sed Festival is one that renews the King's rule. Tradition (still persisting at the time of the Rosetta Stone) suggests that the first Sed Festival of a king was to be held after thirty years of rule. But this

tradition was often violated, and hence the tentative assumption made here that this is the thirtieth year of Den's reign is anything but certain. For a splendid description and discussion of the Sed Festival, see Frankfort, *Kingship and the Gods*, Chap. 6. There is a fragment of a tablet of King Den (Petrie, *Royal Tombs*, Part I, plate XI, 5, cf. 14) that pictures the double-staired pavilion of the Sed Festival. Frankfort's discussion of the tradition of a 30-year period before the first Sed Festival is worth quoting in full (see *op. cit.*, p. 366, n. 2):

"The texts (e.g., at Abydos, translated by Moret, *Royauté*, [i.e. *Du caractère religieux de la royauté pharaonique* (Paris, 1902)], p. 256) leave no doubt that kingship is renewed at the Sed Festival. The Rosetta Stone, which calls the king κύριον τριακονταετηρίδων [i.e. lord of the thirty-year feasts], proves the existence of a tradition connecting the festival with a thirty-year period; and the practice of several kings (Pepi II, Senusert I, Tuthmosis III, Amenhotep III, Ramses II, Ramses III) suggests that it was normally celebrated for the first time thirty years after the king's accession. But this period was not essential, for, after the first celebration, the Sed Festival was celebrated again at two-, three-, or four-year intervals until the king's death. Moreover, some kings celebrated it before their thirtieth year (e.g., Pepi I in his eighteenth year, Ranebtaui Mentuhotep in his second year), while others, who did not reign for thirty years, celebrated Sed festivals nevertheless (Dedkare, Amenhotep II, Seti II, Psammetichos II). The assumption that the celebration took place thirty years after the admission of the king to coregency (Sethe, *ZÄS*, XXXVI, 64, n. 3) is contradicted by the known facts of the reign of Ramses

II and by the two celebrations of the Sed Festival by Tuthmosis IV, whose mummy shows that he could hardly have been more than twenty or twenty-five years old when he died. See Moret, *Royauté*, pp. 256-61, who also quotes the older literature, and Breasted, in *ZÄS*, XXXIX, 55-61.

"It seems that thirty years, or, in a more general way, 'a generation' (Édouard Naville, in *PSBA*, VII, 135), was the normal time to elapse between a king's accession and the celebration of the Sed Festival but that certain symptoms (the nature of which we cannot guess) might at any time indicate to the ancients that a renewal of kingship was due. It is possible that the king's health may have been one of the symptoms; but the widespread belief that the Sed Festival was a modification of an earlier custom which required that the incarnation of the god be replaced by a more perfect man as soon as the present king showed signs of senility or illness projects into Egypt an East African custom... which may have been adhered to, of course, but for which there is no evidence at all."

In this quotation I have added the phrases in square brackets. To the kings mentioned who celebrated their Sed Festival in the thirtieth year we can add Niuserre of the fifth dynasty (see Kees, *Ancient Egypt*, p. 156). Kees notes that among the extensive sacrificial offerings at that festival were 100,600 meals of bread, beer, and offering-cakes on New Year's Day. C. J. Bleeker, *Egyptian Festivals. Enactments of Religious Renewal* (Leiden, 1967), pp. 96-123, reviews this ceremony. E. Hornung and C. Seeber, *Studien zum Sedfest* (Geneva, 1974), evaluate in excellent fashion the historical evidence for the various Sed Festivals.

42. The event reported here is subject to many

interpretations. I have followed Sethe in interpreting it as the report of the flooding of the Delta after the unusually high Nile noted in the previous year (see note 14 above). But Sethe's full rendering is somewhat different from mine: "Flooding of the western nomes of Lower Egypt, sickness of all people." (See A. H. Gardiner, *Ancient Egyptian Onomastica*, Vol. 1, p. 104*.) Breasted and much later Giustolisi interpreted this event as the taking of a census of "all people of the nomes of the west, north, and east" (Breasted) or "of all the people of the western, northern and eastern regions of the Delta" (Giustolisi).

43. Notice that holding of the Sokar festival has shifted from the castle called "Companion of the Gods" to this castle, now mentioned for the first time. It was this event in connection with that of the next year that caused some scholars to translate *ḥ¹* as the planning or designing of the castle. But I have already established the inappropriateness of such a translation (see above, note 18).

44. Both Schäfer and Breasted simply have "Great Door" as a second item. But I believe that the castle was far enough along so that the Great Door was being added, and that this whole statement speaks of surveying the area for the Great Door. For tablets from Den's reign that show the opening of a door of a castle, see Petrie, *The Royal Tombs*, Part I, plate XI, 14, 15. It is of interest that a priest of the Goddess Seshat (a goddess of writing and learning) should be the one to perform the surveying ceremony. The role of Seshat in the foundation ceremonies of Egyptian temples is well known (see H. Kees, *Götterglaube im alten Ägypten*, 2nd ed. [Berlin, 1956], p. 212, and E. Hornung, *Conceptions of God in Ancient Egypt. The One and the*

Many [Ithaca, N. Y., 1971], p. 282). For the mention of this goddess under the special name *Sfḫt-ꜥbwy*, see W. Murnane, *United with Eternity*, Fig. 62, where she is shown participating with Tuthmosis III in the ceremony of stretching the rope during the foundation ritual of the temple at Djamet (in front of the temple of Ramesses III at Medinet Habu). She is often shown as a companion of Thoth. Such is the case of the portico of the second court in Ramesses III's temple, where north of the portal Seshat says to Ramesses: "I am inscribing for you jubilees in myriads and years like the sands of the sandbank. I am establishing your records in your august mansion eternally, and I am granting that your titulary be established forever as Re rises every day", while Thoth similarly speaks to Ramesses south of the same portal (*ibid.*, p. 40). We should also note Seshat's appearance (again under the name of *Sfḫt-ꜥbwy*) in a relief picturing the foundation ceremony of Seti I's temple at Abydos (see R. David, *A Guide to Religious Ritual at Abydos* [Warminster, England, 1981], pp. 24-25, where we read: "In the Lower Register, the king and Sefkhet-ꜥAbwy measure out the temple building in the presence of Osiris. The 'Lady of Writing, pre-eminent in the Library' says 'It is Ptah who lays out its foundations in person; it is thy father Re who establishes it like his horizon.' Osiris says '(I) cause thy mansion to abide like the sky... It is entitled: 'Stretching the cord in the Mansion of Usirmare-setepenre, near to the necropolis'"). Later (p. 173) David summarizes information on Seshat as follows: "SESHAT *(Sš't)*. A very ancient goddess, the 'Mistress of Builders', whose chief centre was Memphis. Her duties included the recording of the royal names at birth, the writing of the king's name and titulary at his coronation, and the granting to him of sed-festivals. She was an account-keeper, who became associated with Thoth; she is often shown involved in the foundation ritual of temples, where her task was to measure out the ground-plan of the building. Another of her aspects was as a sky-goddess. SEFKHET-cABWY was a name introduced for Seshat by Tuthmosis III, and derived from her symbol, the feathers, which came to be represented as two horns. She came to be regarded as a goddess of Fate, perhaps because of her connection with writing the life-span and with numbers and writing." This should be clarified somewhat. The name Sefkhet-cAbwy literally means: "She of the Seven [Rays of the Star] and the Two Horns". It is then a description of the symbol that Seshat has on her head, which consists of a pole carrying a star with seven rays surmounted by inverted horns (see the use of this name on the east wall of the temple at Edfu: "Seshat, the Great Lady of Writing, of the Seven [rayed-star] and the Two Horns, and Mistress of the House of Books"; cf. E. Chassinat, *Le temple d'Edfou*, Vol. 3 [Cairo, 1928], p. 350, E' e. 2 g. I, lin. 2-3). See samples of Seshat's portrait in Fig. I.31: (a) on a twelfth-dynasty block where she is presiding over the House of Sacred Book(s) and recording the numbers of prisoners, cattle, and other booty; and (b) on the east wall of the Library at the Temple of Edfu. See in Fig. I.31(c) part of a limestone block from the early twelfth dynasty showing a "King of Upper and Lower Egypt" (perhaps Amenemhet I) and a divinity (Seshat?) together stretching the cord.

45. I have translated the first part of this as indicating the opening of the lake of the castle "Seats of Gods". Notice that the rectangle sign for "lake" has

some vertical strokes, and so it could be that we again have here the sign for Door of the Castle. Then we should have to translate the first part as "The Year of the Opening of the Door of the Castle". The second event, namely the shooting of the hippopotamus, may be alluded to in a seal of the time of Den (see Petrie, *The Royal Tombs*, Part II, plate VII, 5, 6) that shows a man (presumably Den) wrestling with a hippopotamus.

46. The use of the expression the "First Occurrence" refers to the first occurrence in the reign of Den. For a possible reference to an Apis Bull, see note 8 above. Apis was the sacred bull of Memphis. Frankfort, *Kingship and the Gods*, p. 167, says concerning the cults of sacred bulls: "It is generally assumed that they represent primitive cults, originally unconnected with the great gods with whom they were related in historic times. But the assumption may well be unfounded. That relationship is expressed by a significant title in the case of both the Apis bull of Memphis and the Mnevis bull of Heliopolis. The bulls are the 'heralds' of the gods. The full titles were 'the living Apis, the Herald of Ptah, who carries the truth upward to Him-with-the-lovely-face (Ptah)' and 'the Herald of Re, who carries the truth upward to Atum.'" See also Kees, *Ancient Egypt*, p. 150, who says the following regarding the report of this event in the Palermo Stone: "So too the sacred bull in the ceremony of the 'running forth of Apis' when it was led forth from the eastern door of its stall near the temple of Ptah to join the ceremonial procession, an event frequently recorded in the annals of the Thinite Period." In the Sed Festival the running of Apis also played a part by symbolizing Lower Egypt and the king's renewing his possession thereof (*ibid.*, p. 62). The bull

accompanied the king as he ran around the ceremonial track four times.

47. Seshat is the Goddess of Writing and she appears on many later monuments (see note 44 above). Concerning the goddess Mafdet, Černý, *Ancient Egyptian Religion*, p. 22, writes: "The cat- or mongoose-goddess Mafdet, 'lady of the Castle [or Mansion] of Life,' attested from the Ist Dynasty, was very early invoked as a protectress against snake bites, as both Egyptian cats and mongooses were fearless snake-killers. The centre of the cult of this goddess is still unknown." Cf. A. H. Gardiner, "The Mansion of Life and the Master of the King's Largesse," *Journal of Egyptian Archaeology*, Vol. 24 (1938), pp. 89-90.

48. I accept Kaiser's conclusion that Adjib's reign preceded the Cairo fragment on the right.

49. There is here a very small fraction of a year which lies on the right side of line 3, and its traces are completely unreadable.

50. I follow Emery in reading the priestly figure given for Semerkhet's *insibya* and *nebti* names as Semenptah (W. B. Emery, *Archaic Egypt* [Harmondsworth, 1961], p. 84). For the reading of his mother's name, see P. Kaplony, *Die Inschriften der ägyptischen Frühzeit*, Vol. I (Wiesbaden, 1963), pp. 473-74.

51. I have mostly followed the readings of Gauthier and Daressy in deciphering the readings in this line of the Cairo fragment. Here in Year 1 one would also expect notice of the Ceremony of the Circuiting of the Wall.

52. Daressy believes that the end of this year has signs for *mst*, *wp*, *r* and he further suggests that the whole reading might be *mst Ḥr wp r nṯrw*, which we

could translate as "Birth of Horus, Opening of the Mouth of the Gods". If this is a correct reading, presumably we have reference to a ceremony of the opening of the mouth of statues of gods (see above, note 6). Otherwise the ceremony of opening of the mouth in respect to gods would make little sense. I note also that this is the last year of the reign and that it must have extended through almost the whole civil year or we would have some indication of the number of months and days of the year that the king reigned before dying.

53. This is almost certainly the reign of Ka^ca, since his reign followed that of Semerkhet. His was the last reign of the first dynasty.

54. This is unreadable but it probably should read "The Year of the Appearance of the king as King of Upper Egypt and [his] Appearance as King of Lower Egypt at the Ceremony of the Union of the Two Lands".

55. I have taken the reading "Son of Nub-" from Schäfer, who credits it to Sethe. I have added "[Nufer]" since the *insibya* name of his father (Nebre) is given as Nubnufer. Presumably this would be followed here in the *Annals* by his mother's name. It could be, however, that the *nwb* sign refers not to a part of his father's name but to the king's Golden Horus name, which may be *Ren*. For the king's Horus name, Ninetjer is the simplest reading of the hieroglyphs. A. H. Gardiner in the Appendix of his *Egypt of the Pharaohs* (Oxford, 1961), p. 432, suggests as an alternate reading Nutjeren, while Sethe (*Untersuchungen*, Vol. 3, p. 40) believes it ought to be read *Ntrj-mw*. Emery, *Archaic Egypt*, p. 93, simply calls him Neteren (Netermu).

56. The bracketed material cannot be read but is

restored on the basis of the readings in Nrs. 3, 5, 7, 9, 11, 13, and 15. This is the first reign reported in the fragments in which we find the biennial "countings" mentioned. As we shall see in the later reigns, the countings are specified as either "the counting of gold and fields" or "the counting of oxen and small cattle". My suggestion that this was the sixth year of Ninetjer's reign is based on the presumed addition of "the Second Occurrence of the Counting" to this year and the conventional view that the first of the biennial countings was undertaken in the second year. It is, of course, not impossible that it took place in the first year of the reign. If so, then this would be the fifth year of Ninetjer's reign.

57. We do not know the nature of the building that is being laid out in this year. It could be a temple or a palace but is unlikely to be his tomb. In fact, Ninetjer's tomb has not yet been discovered, though there are several possible tombs which can be dated to his reign (see Emery, *Archaic Egypt*, p. 94).

58. See note 56.

59. See notes 8 and 46.

60. R. Anthes, "Egyptian Theology in the Third Millennium B.C.," *Journal of Near Eastern Studies*, Vol. 18 (1959), p. 186, translated the first part of this entry as "the first time of the sailing of the heavenly Horus, the star", which does not appear to be in character for entries in the *Annals* that concern repeated events. For the most part they represent festivals and the use of the boat determinative for such festivals is probably because the king and celebrants sailed to such a festival. The star-sign before the boat, which, following Schäfer, I have translated as "Worship" (or Adoration), Anthes translates as "the star". Even if this is correct, I think

that the proper translation would then be "the first occurrence of the festival of the heavenly Horus, the star". But frankly, I doubt even that translation. In the second part of the entry, I follow Breasted's interpretation of this event as the "hacking up" or destruction of these two mansions or cities, though Schäfer believed that the events referred to were the foundings of these two places. P. E. Newberry, "The Set Rebellion of the IInd Dynasty," *Ancient Egypt* (1922), pp. 40-46, and particularly p. 45, n. 8, believed that these events signaled "trouble in the north", prelude to the Seth rebellion of this dynasty. As the result of the civil war we notice that the later king Peribsen had a Seth title rather than a Horus title and that the last king of the dynasty, Khasekhemuy, had a Horus-Seth name. But see H. Te Velde, *Seth, God of Confusion* (Leiden, 1977), pp. 72-80.

61. I am by no means sure of this translation. One would suspect that if both Wadjet and Nekhbet were being celebrated we would have the king appearing both as King of Lower Egypt and as King of Upper Egypt. I also note that I am not sure of the reading of Nekhbet.

62. It was first thought that these two were names of the same king. But this seems unlikely. Still, perhaps the Horus Sekhemib-Perenmaat (his full name) switched allegiance and renamed himself Seth Peribsen. See Gardiner, *Egypt of the Pharaohs*, p. 417.

63. Daressy notes that only scattered signs can be seen in the fourth register, though he claims that in the first year after the division between the reigns of Ninetjer and his successor the reading is clear enough and I translate it: "The Appearance of [the king as] the King of Upper Egypt and [his] Appearance as King of

Lower Egypt....." One would suppose that this would be followed by "at the Ceremony of the Union of the Two Lands". This seems to be the ninth section of the fourth register.

64. There is considerable discussion and difference of opinion regarding the last kings of the second dynasty. See Emery, *Archaic Egypt*, pp. 99-103, and see Helck, "Bemerkungen zum Annalenstein," p. 35. Helck also assumes that there was an anonymous reign of 1 year and 8 months after that of Neferkaseker. Cf. the recent remarks of Barta, "Chronologie der 1. bis 5. Dynastie," pp. 12-13. See also von Beckerath, *Handbuch der ägyptischen Königsnamen*, pp. 48-49.

65. There is division of opinion concerning whether Nebka should be classified as the last king of the second dynasty or the first king of the third dynasty.

66. This is the first time in the *Annals* that the item or items being counted are specified. But notice that it precedes by a year the Second Occurrence of the [General] Counting. This may mean that a first occurrence of the General Counting took place in the year before (i.e. in the second year of the reign). Presumably that General Counting was of the cattle and fields and so is to be distinguished from the Counting of Gold reported here in this year. Notice that later in Years Y + 3 and Y + 5 the different counts are joined together in a General Counting of the Gold and the Fields. Note that Giustolisi assigns these four years of the London fragment and the years in the same line on the Palermo fragment to the reign of Khasekhemuy instead of to that of his successor Nebka. W. M. Stewart, *Egyptian Stelae....* Part II..., p. 6, merely says that the dimensions of the compartments on the recto

of the London fragment "match those of the fifth register of the Palermo Stone, which deals with Dynasty II". This is an incorrect statement if the fifth register of the Palermo Stone commences in the midst of the reign of Nebka, that is the first reign of Dynasty III, and continues with parts of Djoser's reign. I follow Helck in locating the London fragment.

67. Again we have mention of the Rekhyt-Folk, although it is not clear what the ceremony is. See note 17 above. Stewart interprets the birds as jabirus and translates "the souls of Pe".

68. Only the "Third" is visible, but the whole year's report is almost certainly reconstructed from the reports for Years X + 2, Y + 1, Y + 3, and Y + 5.

69. The interest here is that the temple is being built in stone. Stone was used in floors, door jambs, and stairs before the third dynasty, but we know of no large structures entirely of stone earlier than the complex of the Step Pyramid of Djoser, which was constructed in the next reign after Nebka's.

70. See note 66.

71. See K. Sethe, "Hitherto Unnoticed Evidence Regarding Copper Works of Art of the Oldest Period of Egyptian History," *Journal of Egyptian Archaeology*, Vol. 1 (1941), pp. 233-36, and particularly pp. 233-35. Presumably the making and dedication of the statue were pious acts on the part of Khasekhemuy's successor Nebka (perhaps as a step toward legitimizing the change of dynasty). However the older view, that we are concerned here with Khasekhemuy's reign, would of course interpret the builder of this statue as Khasekhemuy himself.

72. I have no idea what the first item is. But I suggest that the whole year might read "The Year of

the Fourth Occurrence of Bringing [the Gods] to the Wall [in which year took place] the Building of the Ship called "The City of Dewadjefa'." Giustolisi suggests: "The Capture (?) of the Wall of Duagefa. The Breaking of the Red Vases."

73. As I remarked above in note 13, this last year of what is almost certainly Nebka's reign is apparently the first year of Djoser's reign, since the perpendicular indicating the end of the reign is not accompanied by a palm branch on its left, which would indicate a separate year for the new king's reign.

74. I follow Kaiser and others in believing that the new reign is that of Djoser.

75. See note 35 above.

76. It is rather interesting that in the five years of Djoser's reign reported here we find no mention of the biennial countings but simply a return to the custom of citing the Following of Horus in which some specific festival took place. Helck believes that the last three and one-half years of the reign of Djoser are on the right side of Fragment C-1r, line 4. They cannot be read. Giustolisi opines that the years we have assigned to Nebka are those of Khasekhemuy, and the years assigned to Djoser are those of Nebka (whom he calls by the Horus name Sanakht), and he (Giustolisi) believes that he can read in the first of the years on the right edge "Festa di Horo" and in the second year "The Appearance of the King of Lower Egypt. The Birth of Anubis."

77. I am assuming, along with Kaiser and Helck, that the reign of Djoser's successor, namely Djoser-tety, is completely included in the unreadable fourth line of Fragment C-1. I follow J.-P. Lauer, *Saqqara* (London, 1976), p. 137, in concluding that Djoser-tety's Horus

name was Sekhemkhet. His incomplete step-pyramid complex was discovered in the 1950s by the Egyptian Egyptologist M.Z. Ghonheim.

78. The full name of Djoser-tety's successor is not known for certain. See Helck, *Untersuchungen*, pp. 20-22.

79. This is the last reign of the third dynasty. Huny's Horus name is not known.

80. The reading of Sneferu's mother's name is that given by Černý. Compare B. Grdseloff, "Notes sur deux monuments inédits de l'ancien empire," *Annales du Service des Antiquités de l'Égypte*, Vol. 42 (1943), p. 118, full article, pp. 107-25.

81. It is possible that this specific counting is in lieu of a more general biennial counting. If this is so, then this is probably the fourth year of Sneferu's reign. From this reign onward, the stone *Annals* become true annals rather than a collection of year names, as they were in the first three dynasties.

82. See note 35 above for these shrines.

83. Compare note 71 above. The statue there appears to have been made by Khasekhemuy's successor while here the statue was made by Sneferu himself.

84. The only signs visible on the lower register of the London fragment are the top of the palm branch standing for *rnpt*, i.e. year, and the "mast" sign (Gardiner, *Egyptian Grammar*, p. 499, sign P 6) meaning "to stand" but which has been used earlier in the *Annals* to mean "a sojourn" or "staying" of the king at some place. See Pr, line 3, Nrs. 1 and 9. Cf. Stewart, *Egyptian Stelae...Part II...*, p. 6.

85. See note 16 above, and compare the vestiges here of the signs there in Pr, line 2, Nr. 5.

86. Added because two years later the *Annals*

report the Seventh Occurrence of the Counting.

87. For this type of ship in the second dynasty, see P. Lacau and J. P. Lauer, *Fouilles à Saqqara; la pyramide à degrés*, Vol. 4, fasc. 2 (Cairo, 1961), pp. 51-52. For a brief account of shipbuilding, consult Kees, *Ancient Egypt*, pp. 106-09. He notes that *meru*-wood was one of the best-known coniferous woods from Lebanon.

88. The word *ꜥš* has often been translated "cedar" in the early literature, though this is not precisely accurate. All that we can be sure of is that the term indicated some kind or kinds of coniferous trees.

89. These are apparently the names of two gates of the royal palace. It is not clear whether these two gates are to be distinguished from the doors of cedar mentioned immediately afterwards.

90. Notice that this eighth counting follows in the next year after the seventh so that the biennial counting seems to have been set aside in favor of an annual counting. But also notice that this does not set a trend, for the biennial counting is still being used in the fifth and sixth dynasties. See Gardiner, "Regnal Years," pp. 13-15.

91. For the reconstruction of the titulary here, see Helck, "Bemerkungen," p. 34.

92. Sethe, *Urkunden*, I, p. 238, has *ḥd* but notes that Borchardt reads *sd*, and I prefer the latter.

93. Or, less likely, the phrase *wp r nṯrw* could be rendered "except the gods". I think that the reference is to the ceremony of the opening of the mouth as applied to statues of gods. See above, note 6.

94. Bastet is the cat goddess, who was perhaps originally a lioness. She was named after her cult town Bast (in Greek Bubastis). See Černý, *Ancient Egyptian*

Religion, p. 25.

95. Only a small part of a single year is left from this line. The only thing discernible is the last part of the reading of the Nile height: "3/4" (perhaps "fingers").

96. Strictly speaking, Shepseskaf's pyramid at South Saqqara is not a pyramid but a mastaba shaped like a sarcophagus.

97. This is the first time that the expression "The Year *after* the First Counting of the Oxen" appears in these fragments. But with it, we now have a system of numbering every year of a reign, and this system appears to have remained thus through the sixth dynasty.

98. For the "Souls of Heliopolis", see Frankfort, *Kingship and the Gods*, p. 94.

99. In this number and the one succeeding it we have evidence of a primitive place-value concept. Briefly we can note that the number 44 preceding the indicated number of unit-*st'wt* or unit-arouras (i.e. 8) must count 10s. Thus 44 in this position counts 44·10 arouras (440 arouras). The same thing is true of the number 23 in the next gift of land (in III), i.e., it counts 23·10 arouras. This is true as well for the number 35 in Pv, line 2, Nr. 2 (sect. I [Schäfer has "36" instead of "35" but I have followed Sethe]). [Similar examples occur later in the Rhind Papyrus: Problems 48 and 50 (see Fig. I.49 for Problem 48, where we see in line 5 the number 8, which counts 10s, followed by the sign for *st't* with one unit below, making a total of 81)]. But where there is no indication of a distinction between 10s and units, as in the following cases: number 24 [Pv line 2, Nr. 2 (II)], 44 [*ibid.* (III)], 54 [*ibid.*, (V)], 2[*ibid.*, (VI)], etc., the numbers appear to count individual *st'wt*. Note that I have everywhere stressed in my bracketed

totals that they are in modern terms because Egyptian fractions (excepting 2/3 and 3/4) were always expressed in terms of the sum of unit fractions, as will be evident in Volume Two, Chapter 4. Note that a *st^{j}t* is about 2/3 of an acre.

100. This may be a reference to inventorying the estate of the pharaoh, which perhaps was carried out in the odd-numbered years, while the general counting of the oxen and other wealth of the land was made in even-numbered years. If this is so, then this surely is the year 5, which would be further confirmed by the succeeding statement that it is the year *after* the second counting. The reading "the House of Horus-Seth" was first suggested by Naville, "La Pierre de Palerme," p. 80. Afterwards Daressy made it a part of the expression I have given here. This reading is by no means certain, for in Sethe's text we find instead of the falcon (for Horus) the owl (for the letter "m"). The sign that is taken to be read as "Seth" is presumably a pig (which indeed occasionally was used for Seth in the *Book of the Dead* and elsewhere) but looks in Sethe's transcription like an anteater. The depiction of what surely must be Seth as a pig in a scene of Osiris's hall of judgment is given between the fifth gate and door in the *Book of Gates* (see Fig. II.43).

101. As Kees has noted, *Ancient Egypt*, p. 32, "a heron ruled over the legendary birthplace of the gods at Buto, corresponding to the Horus falcon of Nekhen....". Djebakherut appears to be the ancient name of Pe or Buto. By the time of the Old Kingdom this god appears as "Horus of Pe" without changing its heron form (see Kees, *Der Götterglaube im alten Ägypten*, p. 51). Hence this gift of land by Weserkaf was no doubt for Horus of Buto.

102. A form of Horus. I have given the reading of Schäfer and Breasted. Another possibility is *Dunanu*, a falcon god which is determined by a falcon on a perch (see *Wb*, Vol. 5, p. 432).

103. In this number no "hundreds" or "tens" are involved. The whole number 1804 counts individual arouras to which are added the indicated fractions. Hence in this number there is no evidence of a place-value concept. Note that only seven signs of 100 are visible, but by their placing it seems very likely that an eighth sign (now worn away) was once present. Similarly only 12 units appear with the cubit sign, but from their placing it seems likely that there was a thirteenth. I remind the reader that an area cubit was a strip 100 cubits long and 1 cubit wide and was thus 1/100 of a *st¹t* or aroura. See Griffith, "Notes on Egyptian Weights and Measures," pp. 410-421 and Gardiner, *Egyptian Grammar*, p. 200, for Egyptian area measurements (see my Fig. I.50).

104. See note 100 above.

105. This is Sethe's guess.

106. Perhaps intended here is the wooden statue of the goddess Mes, for the determinative is a statue. So the literal translation of this might well be "Mes in wood". Similarly the next god may well be a statue of the god Sem and thus the name here would be literally translated "Sem [in] wood".

107. In sections XI-XIV we again see evidence of a place-value concept, but in these cases the first number 2 counts 100s and precedes the signs for 10s. In the first two cases (sects. XI and XII) these signs of 10 *st¹wt* are followed by the sign for a unit *st¹t*, of which eight and six, respectively, are indicated. Each number is completed by fractions. In the other two

cases no units are given but only fractions. There are two further cases of the apparent counting of 100s in Pv, line 4, Nr. 2 (sects. 2 and 4). They respectively produce numbers involving 110·100 and 352·100 which seem to be way out of line in comparison with the other gifts of land, so much so that one wonders if there are not errors of copying or interpretation involved. If the latter, it might mean that here we do not have cases of the use of the place-value concept.

108. I am convinced by J. R. Harris's conclusion that d^cm-gold is indeed "electrum" (*Lexicographical Studies in Ancient Egyptian Minerals* [Berlin, 1961], pp. 48-49), although it is often translated merely as "fine gold" (see Gardiner, *Egyptian Grammar*, p. 509, signs S 40-41).

109. *Sen*-gold appears to be some alloy of gold, perhaps "2/3 gold".

110. Instead of the "Seventh Counting", as Sethe gives it in his text without question, Schäfer reads this as the "6. (?) Mal der Zahlung". Schäfer's reading may have been influenced by the fact that the Turin Canon appears to have assigned to Sahure's reign 12 years and some unknown number of months. The Year of the Sixth Counting would then be the twelfth year, and this particular year, of which Sahure served only nine months and 28 days, would be the year after the Sixth Counting and thus the thirteenth year.

111. Schäfer says that the number of months and days is very uncertain, but Borchardt and Sethe give the reading which I have translated. Indeed it is this reading which combines with that of the first year of Neferirkare's reign to produce the civil year of 12 months and five days.

112. Ceremonies devoted to the gods whose births

were associated with the five epagomenal days, i.e. the five days added to the twelve months to make the civil year of 365 days. These gods were called the children of Nut: Osiris, Horus the Elder, Seth, Isis, and Nephthys. See Kees, *Der Götterglaube*, p. 259. It is evident that since Neferirkare assumed the throne toward the end of the civil year, these ceremonies associated with the Birth of the Gods would have been among his first celebrations.

113. See note 107.

114. This town was a kind of suburb of Heliopolis. Its name means "The Place of the Battle" (i.e. the battle between Horus and Seth). It was called "Babylon" by the Greeks (e.g. Strabo). Its gods are mentioned in later texts.

115. "Great Seer" is the title of the chief priest of Heliopolis. It is evident by this reference that in the fifth dynasty there were two of them.

116. Probably a temple of thirty divine judges. Groups of thirty human judges were part of the administration of justice in the Old Kingdom.

117. Following the arrangement of the fragments proposed by both Giustolisi and Helck, I estimate that the verso of the London Fragment contains part of the second year of Neferirkare's reign. I follow the readings of Reeves, "A Fragment of the Fifth Dynasty Annals," pp. 47-51. I have, however, substituted the letters y and z as the unknown multiples of 10 and 1 arouras respectively.

118. "*Nswt Ḥn*", as this is now usually read. He was the last king of the third dynasty. Formerly read as "Setneh" and not identified. In my opinion, this is by no means certain.

Document I.2: Introduction

The "Biography" of Metjen: Inscriptions from his Tomb

As I pointed out in my general account in Chapter One, the inscriptions from the tomb of Metjen (found by Lepsius in Saqqara and now in Berlin) represent one of the earliest biographical accounts, if not the earliest, appearing on the walls of an official's tomb. Its composition was a significant step in the development of systematic writing. The inscriptions apparently date from the early years of the fourth dynasty, in the reign of Sneferu (the first king of that dynasty), and they refer to a career which in all likelihood began during the reign of Sneferu's predecessor, Huny, the last king of the third dynasty. Breasted summarizes the significance of the biography as follows:[1]

> Apart from the fact that it is our earliest document of the kind..., the biography is especially valuable because it deals with the geography and government of the North, narrating Methen's activity in the Delta, of the administration of which at this early period we otherwise know almost nothing. The narrative tells us of his gradual rise from a beginning as scribe and overseer of a

provision magazine until he governs a considerable number of towns and districts in the Delta. He also obtained in Upper Egypt the rule of the eastern part of the Fayyum[2] and the Anubis nome (Seventeenth). He was liberally awarded with gifts of lands, became master of the hunt, and tells us of the size of his house, with some account of the grounds; all of which, from an age so remote, is of especial interest. He died in the reign of Sneferu; all his affiliations were with the families preceding Sneferu, and he was naturally buried beside the terraced pyramid of Zoser [Djoser], of the earlier part of the Third dynasty.

Though it is not specifically stated in the inscriptions that Metjen began as a scribe, this is surely a correct inference by Breasted, since we are told in the document that he was to be appointed "over [or at the head of] the scribes of a provision bureau" with the title of "Administrator of the Provisionbureau" and that appointment (even if it was hereditary as it probably was) would be inconceivable if he had not himself been trained as a scribe. Such training would certainly have been undertaken or provided by his father, who was himself given the title in the document of "Judge-Scribe" *(s'b-sš)* which, even if it is instead translated "Elder-Scribe" or "Honorary Scribe" to indicate a pensioner,[3] betokens a scribal career before assuming that title, a career in which his son and heir was no doubt raised.

One thing of significance for us in this document is the window it offers us to view the preparation of written royal decrees, for we note quotations from such decrees interspersed in its inscriptions (and singled out below in the translation by bracketed rubrics entitled Royal Documents I-V). These quotations provide support to the belief that the royal chancery was flourishing as a document-producing organization by the end of the third dynasty (see above Chapter 1, n. 12). We can infer that these quotations were taken from copies of such royal decrees on file at either the chancery itself or a local bureau like the one that seems to be referred to in Royal Document III. From the later copies of such decrees we can judge both the physical form and epistolary style of the royal documents produced on papyrus in the chancery (see Fig. I.51). The evidence then that Document I.2 provides of the activity of the royal chancery can help us to explain the third-dynasty relief (Fig. I.29) containing the list of scribal names from the time of Djoser that we mentioned in the general account of Chapter One. The scribes organized in the chancery and elsewhere had clearly become an important element in the burgeoning governmental activity of Djoser's reign.

Though it is of no great importance for the exposition of Egyptian science, the study of the above-mentioned excerpts from the royal documents included among Metjen's inscriptions is very important in giving us some hints of the complex arrangements of private and public land division, ownership and acquisition of land, and the establishment of income-producing personal foundations in the third and fourth dynasties and of the relationships between landholding and public office (see in particular the study of K. B. Gödecken mentioned in the literature below). But surely we may conclude from the evidence of land transfers in our document (as was the case with the land donations listed in Document I.1) that full play was given to measurement, and particularly to land measurement. In this regard I cannot help but note that one of the offices held by Metjen was that of "Crier" (*nḫt-ḥrw*: lit. "Strong of Voice") who apparently evaluated the quantity and value of produce and cried out the figures to the recording scribes. It seems probable that the holder of such an office (if indeed he did anything but draw the yield or income arising from that title) would be particularly conversant with the numerical techniques used and written down by the "reckoners" in mathematical papyri like those described in Volume Two, Chapter Four. In particular, I remind the reader that one of such reckoners was the "Caller" (*nisw*), translated as "accountant" in Document IV.1, Problem 67, in Volume Two. For a pictorial representation of a "Crier" see Lepsius, *Denkmäler*, Abt. II, plate III, bottom register, where we see him depicted in the fifth-dynasty tomb of Reshepses and labeled as "Crier of the Granary" (*nḫt-ḥrw-šnwt*).

Texts and Studies

R. Lepsius, *Denkmäler aus Ägypten und Äthiopien*, Abt. I-VI (Berlin, 1849-58; photographic reprint, Geneva, 1972); see particularly Abt. II, plates III-VII. Contains reproductions of reliefs and inscriptions from Metjen's tomb.

H. Schäfer, *Ägyptische Inschriften aus den Königlichen Museen zu Berlin*, Vol. 1 (Leipzig, 1913), pp. 68, 73-87, a reproduction of the text of the inscriptions (see below, Figs. I.52-I.55).

K. Sethe, *Urkunden des Alten Reichs*, Vol. 1, 2nd ed. (Leipzig, 1933) (abbreviated *Urkunden* I; see Bibliography), pp. 1-7, another reproduction of the text.

J. H. Breasted, *Ancient Records of Egypt*, Vol. 1 (New York, 1906), pp. 76-79, an English translation of some of the inscriptions from Lepsius, Schäfer, and Sethe's first edition.

G. Maspero, "La carrière administrative de deux hauts fonctionnaires égyptiens," *Études égyptiennes*, Vol. 2 (1890), pp. 113-272, a remarkably perceptive study despite its age.

W. Helck, *Untersuchungen zu den Beamtentiteln des ägyptischen Alten Reiches* (Glückstadt / Hamburg / New York, 1954), still the most important single work on early titles, with many references to the titles of Metjen in the context of Old Kingdom titles.

H. Goedicke, "Die Laufbahn des *Mtn*," *MDAIK*, Vol. 21 (1966), pp. 1-71, a very subtle interpretation of Metjen's various titles. It is also very useful for the photographs of almost all the inscriptions. They are added at the end as Tafeln I-IX.

K. B. Gödecken, *Eine Betrachtung der Inschriften des Meten im Rahmen der sozialen und rechtlichen Stellung von Privatleuten im ägyptischen Alten Reich* (Wiesbaden, 1976). A thorough and penetrating study of the inscriptions, with references to further literature.

P. Kaplony, *Die Inschriften der ägyptischen Frühzeit*, 2 vols. (Wiesbaden, 1963), mentioning many titles of the early period; see particularly Vol. 2, pp. 1201-18, and *Supplementband* (1964), pp. 44-46; see also his *Kleine Beiträge zu den Inschriften der ägyptischen Frühzeit* (Wiesbaden, 1966), pp. 225-30.

A. Roccati, *La Littérature historique sous l'ancien*

ANCIENT EGYPTIAN SCIENCE

empire égyptien (Paris, 1982), pp. 83-88.

The English Translation

My English translation often reflects Gödecken's opinions of the nature of the various titles of Metjen. For the text, I have depended principally on the works of Schäfer and Sethe, together with the photographic plates of Goedicke. The bold-faced sectional letters are those used in Schäfer's text (see Figs. I.52-I.55). Fig. I.61 from Gödecken's work includes in its legends references to the printed texts for each of the walls and sections of the cult chamber of Metjen's tomb. I have included in the footnotes the various translations of Breasted, Maspero, Goedicke, and Gödecken because of the diverse interpretations and understanding of this difficult document.

Notes to the Introduction of Document I.2

1. Breasted, *Ancient Records*, Vol. 1, p. 76.

2. This results from Breasted's misidentifying the Crocodile-District (or Nome) with the Fayyum, when it should be identified with the Dendera District. But Metjen did indeed hold a title (apparently inherited from his father) of Palace-governor of the Towns of the Southern Lake (i.e. the Fayyum), as is evident in the document below (D, line 10). By the way, the reference to the eastern part of the Crocodile-District (*ibid.*, E, line 11) is in a title that apparently belongs to the addressee of a royal document rather than to Metjen. Notice also that Breasted recounts the offices listed in what I have called Royal Document V below as if they were successively acquired by Metjen.

3. Gödecken, *Eine Betrachtung*, pp. 69-70.

Document I.2

The "Biography" of Metjen: Inscriptions from his Tomb

A   Magnate of the Ten of Upper Egypt,[1] Palace-governor, Metjen.

B   Magnate.....(of the Ten of Upper Egypt?), Palace-governor of the Great Estate of the towns of Sahu *(S⁣iḥw)*,[2] Palace-governor of Hutihet *(Ḥwt-iḥt)*,[3] Palace-governor of.....

[Royal Document III][4]

C   /1/ [To] the [Land-] Administrator, District-governor, Overseer of the [District-] Commission in the Anubis-District (i.e. the 17th nome of Upper Egypt), and Overseer of the Sinu *(snw* or *sinw)*.[5]

/2/ [Regarding] 4 arouras of arable land in the place [named] Basahu *(B⁣i-s⁣i-ḥw)* in the Fish-District[6] (i.e. the 16th nome of Lower Egypt), with all the [attached] people (i.e. field-workers). [This estate is the concern of] the estate-decree [addressed or pertaining to][7] the Scribe of the Provisionbureau [and its beginning is] "Behold, Give [it] to the single son (Metjen) [etc.]. Execute the royal estate-decree, for he has the [validating] document, /3/ which has been provided to him [by his father] in his lifetime."[8]

[Royal Document IV][9]

[To] the Overseer of the [District-]Commission in the Neith-District and the West-District (i.e. the 4th and 3rd nomes of Lower Egypt).

/4/ There shall be provided for him (i.e. Metjen) 12 "Metjen-Foundations" in the Neith-, Desert-Bull-, and Haunch-Districts (i.e. the 4th, 6th, and 2nd nomes of Lower Egypt)[10] and their yield to him at the [Provision-]Hall, /5/ [for] there has been conveyed to him in return for compensation 200 arouras of arable land by many of the king's people *(nswtyw)*, /6/ and the Columned Hall shall provide 100 rations of bread daily from the Ka-Estate (or Mortuary Temple) of the king's mother Nimaathapi, /7/ a house 200 cubits long, 200 cubits wide, built and provided with good wood, with a very large pond made there, and planted with figs and vines. /8/ The inscription here corresponds to a royal document. Their names accordingly correspond to [those in] the royal document. /9/ Let trees (*or* flowers) and a great number of vines be planted so that a great amount of wine can be made therefrom. /10/ Let the vineyard made for him be arable land 10 plus 2 arouras in area[11] and within the walls [of the estate], and planted with trees (*or* flowers). /11/ Iymers *('Iimrs)* [is] a "Metjen-Foundation" and Iatsebek *('I't-sbk)* [is] a "Metjen-Foundation".

[Royal Document VI][12]

D /1/There shall be conveyed to him (i.e. Metjen) the things of his father the Judge-Scribe (*or* Honorary Scribe) Anubisemankh, without barley and grain, or anything of the House, and with the [attached] people (i.e. field-workers) and small cattle, /2/ [namely,] that he be appointed at the head of (*or* over) the scribes of the Provisionbureau as the Administrator *(iry-iḥt)* of the Provisionbureau,[13] and /3/ that he be appointed Crier *(nḫt-ḥrw*--lit. "strong of voice") among the People of the Grindery (*or as later*, of Offerings),[14] /4/ [as] an Administrator of the Desert-Bull-District (i.e. the 6th or Xoite nome) who accompanies (*or* follows after) the Judge-Land-administrator (or Honorary Land-administrator) of that district. /5/ Let him take [the rank of] Judge-Crier (*or* Honorary Crier) and be appointed Overseer of all the Linen (*or* Flax) of the King. /6/ Let him be appointed as Estate-governor (*or* Palace-governor?) of the Towns of Per-Desu *(Pr-Dsw)* as the one having the staff (*or* in authority).[15] /7/ Let him be appointed Overseer of the People of Dep (part of Buto), /8/ being the Palace-governor of the Towns of Per-Mesedjaut *(Pr-Msd'wt)* and of the Towns of Per-Sepa *(Pr-Sp')* as Administrator of the Neith-District (i.e. the 4th nome of Lower Egypt), /9/ the Governor of the Towns and Fields of the Mortuary Estate of Senet *(Snt)* as the one having the staff (*or* in authority), /10/ the Estate-governor of the Town of Per-Shestjet *(Pr-šstt)*, and Palace-governor of the Towns of the Southern Lake (i.e. the Fayyum). /11/ Let one "Metjen-Foundation" be founded of that which his father Anubisemankh first gave him.

[Metjen's Titles][16]

E /1/ Estate-governor (i.e. Palace-governor?) of the Town of Per-Desu, /2/ Estate-governor (i.e. Palace-governor?) of the Towns of Sahu, /3/ Palace-governor of the Towns of Per-Hesen *(Pr-Hsn)*, Administrator of the Harpoon-District (i.e. the 7th and/or 8th nome of Lower Egypt), /4/ Palace-governor of Sekhemu *(Shmw)*, Administrator of the Desert-Bull-District, /5/ Palace-governor of Dep, Administrator of the People of Dep, /6/ Palace-governor of the Estate of Mesdjawt,

Administrator of the Neith-District, /7/ Palace-governor of Shety *(šty)* (Two Hounds?), Administrator of the Fish-District (i.e. the nome of Mendes, that is the 16th of Lower Egypt), /8/ Palace-governor of the Heswer *(Ḥs-wr)* Canal, Governor of the Fields in the West-District and the Neith-District (i.e. the 3rd and 4th nomes of Lower Egypt), /9/ Palace-governor of Hutihet *(Ḥwt-iḥt)*, Administrator of the Desert,[17] Controller of the [wild-animal] Hunt (*or* Hunters), /10/ Governor of the Fields as one having the staff (i.e. in authority), Administrator of the Haunch-District (i.e. the 2nd nome of Lower Egypt).

[Royal Document I][18]
/11/ [To] District Governor (i.e. the nomarch), Administrator of the Land, and Overseer of the [District-] Commission in the Crocodile-District, Eastern Part (i.e. the nome of Denderah--the 6th of Upper Egypt.[19]
/12/ [Concerning] the Judge-leader (i.e. Honorary Leader) over the Grindery (*or as later*, over Offerings), the Palace-governor of the West- and Neith-Districts, and Controller of the Door of the West *(ꜥꜣ-tyw)* (*or* of the Mercenaries in the West), [Metjen].[20]
/13/ [The Content of the Decree.] There has been conveyed to him (i.e. Metjen) in return for compensation 200 arouras of arable land by many of the king's people *(nswtyw)*. /14/ He has given [for it] 50 arouras of arable land [once] belonging to his mother Nebsent, /15/ she having made a will thereof for her children; /16/ their shares were laid down according to the king's records everywhere.

[Royal Document II][21]

/17/ [To] the Governor of the Mortuary Foundation of the Property of King Huny in the Haunch District (i.e. the 2nd nome of Lower Egypt).

[Contents of the decree.] Let him (i.e. Metjen) with his son be given 12 arouras of arable land, /18/ with the [attached] people (i.e. field workers) and small cattle.

{Here I end the translation, omitting the remaining inscriptions of Schäfer's Sections F-N [see Figs. I.56-I.60], which consist largely of titles repetitious of those already mentioned and of offerings. For the notes to this document, see page 157.}

Notes to Document I.2

1. See Chap. 1, note 11. "Magnate of the Ten of Upper Egypt" *(Wr mḏ šmᶜw)* seems to be a general title of rank for Metjen, having few or no official duties attached to it. As given here, Palace-governor also seems to be a general title of rank; but notice that this title is joined to a number of specific estates and towns in the titles mentioned for Metjen in the remainder of the document below. It clearly was a district (or nome) title in the early dynastic period (see P. Kaplony, *Die Inschriften*, Vol. 1, p. 434; Vol. 2, p. 1048, Anm. 1776, and Gödecken, *Eine Betrachtung*, p. 82).

2 Maspero, "La carrière administrative," p. 169, thinks that this was a kind of feudal estate, acquired by Metjen after that of the estate of Desu (which we shall mention later). Both estates have plural town signs for a determinative, i.e. three of them. Still Maspero believes that each of the estates consisted of only one town.

3. Hutihet was to become the capital of the 3rd nome of Lower Egypt when that nome was formed. Metjen was the first to hold this title, but he also held (presumably earlier) the old title of "Administrator *(ᶜḏ-mr)* of Hutihet". He may have been the last one to hold that older title. See Gödecken, *Eine Betrachtung*, pp. 90-91, 145-46.

4. This is one of the royal decrees inserted in Metjen's inscriptions, the significance of which has been mentioned above in my Introduction. Gödecken, *Eine Betrachtung*, p. 11, calls it "Akte III" and translates it as follows:

(Adressat:) *ššm-tꜣ ḥḳꜣ.t* Vorsteher der

Aufträge im 17. o. ä. Gau, Vorsteher der
sjnw. (Dekrettext:) (Was angeht) den 16. u.
ä. Gau, Ort *Bⁱ-sⁱḥw*, 4 Aruren Feld, Leute
und alle Sachen, die enthalten sind in den
⟨image⟩-Verfügung ⟨an⟩ den Schreiber des
Versorgungsbüros: (Siehe!) 'Gib ⟨sie⟩ einem
einzigen Sohn!'-Veranlasse, dass die
⟨image⟩-Verfägung von ihm geholt werde,
(denn) er hat die Urkunde und er hat (zum
Amt?) verpflichtet (noch) zu seinen
Lebzeiten.

I note here that I have italicized Gödecken's phonetic
renderings of Egyptian words and changed her phonetic
"q" to "*k*" (e.g., for Gödecken's "ḥq3" I have written "*ḥkⁱ*").
The passage has been translated twice before, by
Goedicke and Breasted. Goedicke, "Die Laufbahn des
Mṯn," pp. 64-65, translates this and the succeeding
decree (which, following Gödecken, I have called Royal
Document IV) as a single decree with interpretations of
titles that differ greatly from those later adopted by
Gödecken:

Dekret über Belehnung mit Amt und
Pfründe und Genehmigung zum Hausbau.

(Inschrift C)
Verwalter des Freilandes bei der
Landvergebung, Vorsteher der Teilung als
Vorsteher der Kolonisten aus dem
Kynopolitischen Distrikt von den 4 Aruren
Ackerland, Leuten und allen Sachen in
sⁱḫt-Zr (?) im Mendetischen Distrikt.

Inhalt des Dekrets an den Schreiber
der Speisenverwaltung sowie (an) einen

Sohn. Veranlassen, dass das Königsdekret zu ihm gebracht werde: berantwortet ist ihm auf Lebenszeit (das Amt eines) Vorstehers der Teilung im 4./5. und 3. unterägyptischen Distrikt. Gegrändet werden für ihn 12 *Mtn*-Gründungen im 4./5., 6. und 2. unterägyptischen Distrikt.

Der Diener des (Königshauses), der die Speisenhalle unterhält, gebracht werde ihm die Pacht von 200 Aruren Ackerland von zahlreichen Königsleuten und eine Belieferung der Werkhalle von 100 Rationen täglich von dem Stiftungsgut der Königsmutter *Nj-m'ct-Ḥp*: ein Haus von 200 Ellen Länge, 200 Ellen Breite, gemauert und ausgestattet, belegt mit gutem Holz; angelegt werde ein sehr grosser Teich darin, (sowie) Feigen und Weinstöcke.

Die Inschrift hier entspricht einer königlichen Urkunde, ihre Namen entsprechen vollauf der königlichen Urkunde. Gepflanzt werden Bäume und gar viel Weinstöcke und gar viel Wein werde davon gemacht. bertragen werde ihm ein Weingarten von *1 ḥ' + 2 t'* und 1 Arure umwallt, bepflanzt mit Bäumen: das *Mtn*-Gut von *'Ij-mr.s* und das *Mtn*-Gut von *'Ʈt-*.

Breasted did not cast these two passages as a decree but translated them together as a simple historical narrative (*Ancient Records*, Vol. 1, pp. 77-78):

Honors and Gifts

⌜Administrator,⌝ nomarch, and overseer of commissions in the Anubis nome, overseer of ⌜-⌝ of the Mendesian nome, ⌜--⌝ 4 ⌜stat⌝ of land, (with) people and everything..........There were founded for him the 12 towns of Shet-Methen *(Št-Mtn)* in the Saitic nome, in the Xoite nome, and the Sekhemite nome....., There were conveyed to him as a reward 200 stat of lands by numerous royal ⌜--⌝; a [mortuary] offering of 100 loaves every day from the mortuary temple of the mother of the king's children, Nemathap *(N-m¹ᶜ.t-ḥ¹p)*, a house 200 cubits long and 200 cubits wide, built and equipped; fine trees were set out, a very large lake was made therein, figs and vines were set out. It was recorded therein according to the king's writings; their names were according to the decree *(sr)* of the king's writings. Very plentiful trees and vines were set out, a great quantity of wine was made therein. A vineyard was made for him: 2,000 stat of land within the wall; trees were set out, ([in]) Imeres *(Yy-mrs)*, Sheret-Methen *(Šr-Mtn)*, Yat-Sebek *(I¹t-Sbk)*, Shet-Methen *(Št-Mtn)*.

For Breasted the half-brackets used above did not indicate additions to the text but rather uncertainties of translation. I have eliminated the line numbers and Breasted's notes and their references in the text.

5. The addressee here is the one to bear these titles, if indeed it is a correct conclusion ·that this is part of a royal decree. For the nature of these titles

and particularly that of Overseer of the [Land-] Commission *(mr-wpwt)* see Gödecken, *Eine Betrachtung*, pp. 53-71. Maspero, "La carrière administrative," p. 212, suggests *sounou* and translates it as "coureurs".

6. Kees identifies the deity of this district as the "First-of-the-Fish" in his *Ancient Egypt* (Phoenix ed., Chicago and London, 1977), p. 92. The hieroglyph of this district, the fish with the descending tail *(ḫʾ)*, appears as a representation for marshy or swampy land. Hence we could perhaps translate the name of this district as the "Swampy-District". The principal town in this district was Mendes.

7. My added phrase suggests two possibilities, either the estate-decree was addressed to an unknown Scribe of a Provisionbureau or the decree concerned a specific Scribe of a Provisionbureau, who could then be Metjen's father Anubisemankh. Gödecken believes the latter to be true (see *Eine Betrachtung*, p. 35).

8. My understanding of this sentence differs from that of Gödecken. See her translation in note 4 above.

9. Gödecken labeled this as "Akte IV" and translated it as follows (*Eine Betrachtung*, p. 12):

> (Adressat:) Vorsteher der Aufträge im 3. und 4. u. ä. Gau. (Dekrettext:) Man soll einrichten für ihn 12 "*Grg.t-Mṯn*" in den u. ä. Gauen vier, sechs und zwei und ihre "Belieferung" für ihn in die Speisehalle - (denn/indem) er hat gekauft (geholt) gegen Entgelt 200 Aruren Feld von/bei mehreren *nswtjw* und die Grabvorhalle 'liefert' 100 Portionen Brot (und) aus dem *Ḳs̱*-Gut der Königsmutter *Nj-mꜢ̊t-Ḥꜥpj* täglich - (dazu) ein Haus / Anwesen, 200 Ellenlang, 200

Ellen breit, bedeckt mit gutem Holz, indem
ein Teich darin angelegt ist, sehr gut, und
mit Feigen und Weinstöcken angepflanzt.
Es sei eine Urkunde darüber in der
königlichen Akte und ihre Namen dazu in
der königlichen Akte. Man pflanze die
Bäume und Weinstöcke sehr ordentlich,
damit daraus sehr guter Wein gemacht
wird. Man mache für ihn einen
Weingarten von 1 $\underline{h}^j$ und 2 t^j innerhalb der
Mauer, bepflanzt mit Bäumen. *'Ij-mr.s̱* (ist
ein) *"Grg.t-Mṯn"*, *i̯.t-S̱bk* (ist ein)
"Grg.t-Mṯn".

This should be compared with the translations made by
Goedicke and Breasted given in note 4 above. In
Goedicke's translation the title given here as belonging
to the addressee of the decree was assigned to Metjen
for his lifetime. In Breasted's translation, the part
containing this title was omitted.

10. For an illuminating treatment of the various
kinds of personal foundations in this period, see
Gödecken, *Eine Betrachtung*, pp. 95-113. She
distinguishes three kinds of "Metjen-Foundations". The
first kind was put together from property that he had
personal ownership of, property that he developed from
the property that his mother left him, as is the case
with the 12 "Metjen-Foundations" mentioned here in
Royal Document IV. The second kind is that provided
for him during a specified time he held some royal
office, the foundation supplying a yield from the land
or towns that were bound to the office which he
occupied. Perhaps this kind of "Metjen-Foundation" is
represented by Iymers and Iatsebek mentioned in C, line
11. A third kind is somewhat like the second but in fact

results from the property bound to the offices he inherited from his father and therefore retained during his lifetime. The last kind is illustrated in Royal Document V below.

11. All of the translations of the size of the land provided for the vineyard are incorrect in some respect. The correct figure is 1 $ḥ^j$ (= 10 st^jwt) + 2 st^jwt, making a total of 12 st^jwt, or in translation 12 arouras. For the correct rendering of areas, see Document I.1: Introduction, paragraph (9). See also Fig. I.50.

12. Gödecken does not number this document but notes that it is a new Decree-text (*Eine Betrachtung*, pp. 12-13):

> (Neuer Dekrettext:) Man soll ihm geben die Dinge des Vaters, des $s^jb-sš$ *Inpw-m-$^cnḫ$* ohne *it* und *bd.t.* irgendwelche Sachen des Hauses, (Feldarbeiter-)Leute und Kleinvieh! Indem man ..
>
> ... ihn einsetzt über die Schreiber des Versorgungsbüros als *irj-iḫt* des Versorgungsbüros.
>
> ... ihn einsetzt als *nḫt-ḥrw* unter den Leuten der Reibanlage (?), indem der ^cḏ-mr des 6. u. ä. Gaues hinter dem $s^jb-ḫrj-sḳr$ des 6. u. ä. Gaues ist (nachfolgt), welcher das Amt eines $s^jb-nḫt-ḥrw$ einnehemen wird!
>
> ... ihn einsetzt als Vorsteher aller Flachsarbeiten des Königs!
>
> ... ihn einsetzt als $ḥḳ^j$ von $pr-dšw$ als einer mit Amtsstab! ... ihn einsetzt als ^cḏ-mr der Leute von Dep, $ḥḳ^j-ḥw.t-^c^jt$ von $pr-mšd^jwt$ und $pr-šp^j$ und (gleichzeitig) ^cḏ-mr vom Neithgau, als $ḥḳ^j$ der Döfer und Äcker des

Totenstiftungsgutes der *Šn.t*, die unter dem
Amtsstab sind, als *ḥḳ¹* von *pr-ššṭt* und als
ḥḳ¹ der Dörfer des Gutes vom Fajjum!
Man gründe ein *˙Grg.t-Mṭn* von dem, was
ihm der Vater *'Inpw-m-ᶜnḫ* gegeben hat.

Goedicke had translated the same passage as follows
("Die Laufbahn des *Mṭn*," pp. 65-66):

Dekret für Übertragung des väterlichen
Besitzes

(Inschrift D)
Gegeben werde ihm das
Eigentum seines Vaters, des *Z'b-Zš*
'Inpw-m-ᶜnḫ, und nicht falle Weizen und
Emmer oder irgendwelche Sachen an das
Königshaus, oder Leute unde Kleinvieh.

Er erlasse Vorschriften der
Speisenverwaltung als Verantwortlicher
der Speisenverwaltung.

Er lege fest den Beuteanteil unter
den Militärkolonisten, die im Kataster des
6. unterägyptischen Distrikts sind, und
das-für-das-Opfer-des-verstorbenen-Königs-
Bestimmte der Leute des 6.
unterägyptischen Distrikts.

Er nehme die Beute des Königs in
Empfang.

Er bestimme ⟨als⟩ Vorsteher allen
Flachses (?) des Königs.

Er bestimme als Verwalter der
Gründungen des Königs der
(Königs)verwaltung im unterworfenen
(Gebiet).

Er bestimme im Gebiet von Dep (und

als) Verwalter des *msd'wt*-Landes der
Königstotenstiftung und des
Königsverwaltung und des Freilandes der
Königsverwaltung im 4./5.
unterägyptischen Distrikt (und als)
Verwalter der Liegenschaften und des
unterworfenen Freilandes der
Grenzverwaltung (und als) Verwalter der
Liegenschaften der Deltaresidenz der
Königsverwaltung (und als) Verwalter der
Liegenschaften der Königstotenstiftung der
südlichen Residenz.

    Mtn gründe ein Gut von dem, was
ihm sein Vater *'Inpw-m-ᶜnḫ* gegeben hat.
Breasted's translation again was narrative in form since
he did not consider the inscriptions to be part of or
paraphrasing decrees (*Ancient Records of Egypt*, Vol. 1,
p. 77):

Death of Methen's Father

    171. There were presented to him the
things of his father, the judge and scribe
Anubisemonekh; there was no grain or
anything of the house, (⌐but⌐) there were
people and small cattle.

Methen's Career

    172. He was made chief scribe of the
provision magazine, and overseer of the
things of the provision magazine. He was
made ⌐____⌐ becoming local governor of
Xois (Ox-nome), and inferior field-judge of
Xois. He was appointed ⌐_____⌐ judge, he
was made overseer of all flax of the king,

he was made ruler of Southern Perked
(Pr-ḵd), and ⌐deputy⌐, he was made local
governor of the people of Dep, palace-ruler
of Miper (⌐My⌐-pr) and Persepa (Pr-spⁱ),
and local governor of the Saitic nome, ruler
of the stronghold of Sent (Snt), [deputy] of
nomes, ruler of Pershesthet (Pr-šsṯt), ruler
of the towns of the palace, of the Southern
Lake.   Sheret-Methen (Šrt-Mṯn) was
founded, [and the domain which] his father
Anubisemonekh presented to him.

I should remark that Maspero, "La carrière
administrative," p. 167, gives a translation of the first
part of this passage that is also in narrative form:

Le maitre-scribe Anoupoumânkhi lui
donna sur son bien les choses nécessaires à
sa vie, quand il n'avait encore ni blé, ni
orge, ni aucune autre chose, ni maison, ni
domestiques mâles et femelles, ni
troupeaux, ânes et porcs.   Il fut mis
premièrement scribe de place
d'approvisionnement, directeur responsable
des biens d'une place d'approvisionnement.
Il fut mis crieur, taxateur des colons; puis,
étant détaché à la suite du curateur du
nome Xote et matre-chef des chaouiches du
nome Xote, il fut pris comme matre-crieur.
Il fut mis directeur de tout le lin du roi. Il
fut mis régent de Pidosou et eut le droit de
porter la canne de commandement.

13.   I have followed Gödecken in translating this
paragraph, containing this and other titles, as if these
various offices were a part of Metjen's inheritance from
his father.   If this is correct, they indeed did constitute

the "things of his father" mentioned in line 1. And, as I said in my introduction to this document, I believe that even though Metjen does not list "Scribe" as one of his titles in the various inscriptions devoted entirely to his titles, we almost certainly should assume that his father taught him the craft, knowing that he would take over his office of Administrator of the Provisionbureau, where presumably he would have to supervise the scribes.

14. The nature of this office was discussed at length by Maspero, "La carrière administrative," pp. 134-39. He concluded that the Crier is one who oversees operations of measurements and evaluation and reports the results to the scribes. We know of cases where the Crier is called a Crier of the Granary (see p. 146 above). Presumably his duties were not unlike Metjen's. It is of course possible that the title already entailed supervising offerings as it apparently did later or even that it was a title without any duties. Further on in the list we note that Metjen was to be appointed "Judge-Crier" (i.e. Honorary Crier?), which title is immediately followed by that of Overseer of all the Linen (or Flax) of the King. It seems likely that the title of Judge-Crier would not pass to him until he was ready to hand over the "things of his father" to his son. Or it could be that Breasted and Maspero were correct in simply reporting this list of offices as if they were offices successively acquired by Metjen. Where I have "People of the Grindery" Jonckheere, *Les médecins de l'Égypt pharaonique* (Brussels, 1958), pp. 117-18, reads *sinw grgt* and he seems to approve the translation "Médecin des colons". He also discusses the various other possibilities. I note that there is no other indication among the titles that Metjen was a physician.

15. I am not sure of the translation of the phrase *ḫr mḏ* in this context. Presumably it refers to Metjen's position as "having authority" and thus is specifying that this is not merely an income-producing position without official duties. Notice that the phrase is repeated in sect. D, line 9, and sect. E, line 10 below.

16. It will be noticed that the first ten lines of sect. E contain titles that are clearly those belonging to Metjen, and for the most part they are repeated in other sections of the inscriptions where they are followed by Metjen's name. With line 11 we find titles which appear to be those of the addressee of Royal Document I.

17. This title and the one succeeding it give us hints of the role that Metjen probably played in the Libyan border region.

18. Gödecken labels this document as "Akte I" and translates it as follows (*Eine Betrachtung*, p. 11):

> (Adressat:) *Ḥḳꜣ-spꜣt šš m-tꜣ* Vorsteher der Aufträge im 6. u. ä. Gau, Ostteil. (Betrifft:) (Den) *sꜣb-ḫrj-sḳr, ḥḳꜣ-ḥw.t-ꜥꜣt* im 3. und 4. u. ä. Gau, Leiter der *ꜥꜣtjw*. (Dekrettext:) Er hat gekauft (geholt) ⟨gegen⟩ Entgelt 200 Aruren Feld bei mehreren *nswtjw*. Er hat (dafür) gegeben die 50 Aruren Feld der Mutter *Nb.s*-Neith, da sie ein *imj.t-pr* darüber gemacht hat für die Kinder, indem ihre Anteile gelegt wurden (d. h. eingetragen wurden) zu den königlichen Akten eines jeden Büros.

Goedicke's earlier translation ("Die Laufbahn des *Mṯn*," pp. 66-67) includes the titles I have given in the preceding paragraph (often differently interpreted) and the texts of both Royal Documents I and II without any break (thus not distinguishing Metjen's titles from those of the addressees of both royal documents):

Dekret für die bertragung des mütterlichen Besitzes

(Inschrift E)

Verwalter der Gründungen des Königs der Königsverwaltung, Verwalter der Lehen des oberägyptischen Reichsheiligtums, Verwalter der Liegenschaften der Grenzverwaltung im Küstendistrikt, Verwalter der Opferstiftungen der Königstotenstiftung im Distrikt von Xois, Verwalter (der Liegenschaften von) Dep der Königstotenstiftung im Distrikt von Dep, Verwalter des *msd'wt*-Landes der Königstotenstiftung im Satischen Distrikt, Verwalter (der Liegenschaften) der Königstotenstiftung von *štwj* im Mendetischen Distrikt, Verwalter (der Liegenschaften des) Marschlandes der Königstotenstiftung, Verwalter des *cht*-Besitzes der Königstotenstiftung im West- und Satischen Distrikt, Verwalter (der Liegenschaften der) Königstotenstiftung von *Hwt-iht*, Leiter der Jäger des Wüsten-distrikts, Verwalter des *cht*-Besitzes des unterworfenen Landes im Mendetischen Distrikt, Verwalter des Freilandes in der Landvergebung und Vorsteher der Verteilung (für die Leute) aus dem Distrikt von Dendera (im) Ostland, Verwalter der Ablebensgabe der Königstotenstiftung im West- und

Satischen Distrikt, Leiter der Söldner des
Westens, dem die Pacht von 200 Aruren
Ackerland durch viele Königsleute gebracht
wird:

Gegeben werde ihm die 50 Aruren
Ackerland der Mutter *Nbsnt*, indem sie ein
Testament darüber machte fr die Kinder.
Festgelegt werde ihr (pl.) Besitz durch
Königsurkunde ⟨in⟩ allen Stellen.

Der Verwalter des Stiftungsgutes
(Königs) *Nj-swtḥ* im Letopolitischen
Distrikt gebe ihm 12 Aruren Ackerland mit
seinen Kindern, sowie Leute und Vieh.

Breasted (like Goedicke later) has the two documents
together as a single section, which he precedes by
Metjen's titles (*Ancient Records of Egypt*, Vol. 1, pp.
78-79):

Methen's Offices

174. Ruler of Southern Perked
(*Pr-ḳd*); Ruler of Perwersah (*Pr-wr-sꜣḥ*);
Ruler and local governor of the stronghold,
Hesen (*Ḥsn*); in the Harpoon nome;
Palace-ruler and local governor in Sekhemu
(*Sḫmw*) of Xois (Ox-nome); Palace-ruler
and local governor in Dep (Buto);
Palace-ruler and local governor in Miper
(ꜥ*Myꜣ-pr*), of the Saite nome; Palace-ruler
and local governor in Two hounds, of the
Mendesian nome; Palace-ruler in Heswer
(*Ḥs-wr*); Ruler of fields in the west of the
Saitic nome; Palace-ruler of the Cow
stronghold; local governor in the desert,
and master of the hunt; Ruler of fields,

[deputy] and local governor in the Sekhemite nome; Nomarch, [administrator], and deputy in the eastern Fayum; Field-judge, palace-ruler of the west of the Saitic nome, leader of ˹___˺.

Gifts of Land
175. There were conveyed to him, as a reward, 200 stat of land by the numerous royal ˹___˺.

There were conveyed to him 50 stat of land by (his) mother Nebsent *(Nb-snt)*; she made a will thereof to (her) children; it was placed in their possession by the king's writings (in) every place.

Ruler of ˹___˺ of the Sekhemite nome. There were given to him 12 stat of land, with his children; there were people and small cattle.

As before, I have eliminated Breasted's inclusion of the line numbers and notes.

19. Presumably these titles were all held by a single person who was expected to see that the decree was executed. It is not clear why the chief official of the Denderah nome should be addressed in connection with this decree unless the original 50 arouras of Metjen's mother were located in that district.

20. This last title is further evidence of the role played by Metjen in the Libyan border area. The translation "Controller of the Door of the West" is found in Kees, *Ancient Egypt*, p. 185; for the alternate translation, see Goedicke's translation given in note 18.

21. This document was entitled "Akte II" by Gödecken, who translates it as follows (*Eine*

Betrachtung, p. 11):

> (Adressat:) Leiter des Totenstiftungsgutes des Königs Hwnj im 2. u. ä. Gau. (Dekrettext:) Man gebe ihm 12 Aruren Feld zusammen mit seinem Sohn, dazu Leute und Kleinvieh.

Translations of it by Goedicke and Breasted are included in note 18.

Document I.3: Introduction

Inscriptions from the Tomb of Niankhsekhmet

This short document contained on a well-produced stela illustrates one of the ways in which the king could reward the successful career of one of his physicians (in fact any one of his competent courtiers), namely by providing the stone for and the decoration of the false door of the retainer's tomb. The ruler in question in this document is Sahure, the second king of the fifth dynasty, while the honored physician is Niankhsekhmet. The tomb is at Saqqara. It was excavated and first published by Mariette (see the literature below).

Note that Niankhsekhmet is designated as a Chief of Physicians. Jonckheere claims that we can detect three further ranks above this one in the Old Kingdom: Inspector of Physicians, Overseer of Physicians, and Controller of Physicians.[1] Hence it appears that when the false door was decorated Niankhsekhmet was on the lowest rung of the senior medical personnel. Jonckheere points out that Niankhsekhmet also bore the titles of Chief of Physicians of the Palace and Chief of Dentists of the Palace, that he had a brother named Anubishotep (*'Inpw-htp*) and two daughters named Hetepheres (*Htp-hr-s*) and Mersankh (*Mr.s-ʿnh*), and that the names of his wife and sons were effaced.[2]

ANCIENT EGYPTIAN SCIENCE

Texts and Studies

For a bibliography on Niankhsekhmet's tomb, see the second edition of B. Porter and R. L. B. Moss, *Topographical Bibliography of Ancient Egyptian Hieroglyphic Texts, Reliefs, and Paintings*, Vol. 3, Part 2 (Oxford, 1978), pp. 482-83.

The texts which I have used for my translation are those of K. Sethe, *Urkunden des Alten Reichs*, Vol. 1, 2nd ed. (Leipzig, 1933) (=*Urkunden* I), pp. 38-40, and L. Borchardt, *Denkmäler des Alten Reiches* (=*Catalogue général des antiquités égyptiennes du Musée du Caire*), Vol. 1 (Berlin, 1937), pp. 169-73 (and particularly pp. 172-73 for my extracts), and plate 39 (No. 1482).

I note also the early publication by A. Mariette, *Les mastabas de l'Ancien Empire*, edited by G. Maspero (Paris, 1882-89), pp. 203-05 (and especially pp. 204-05).

Consult the English translation of J. H. Breasted, *Ancient Records of Egypt*, Vol. 1 (New York, 1906), pp. 108-09, and the French translation of A. Roccati, *La Littérature historique sous l'ancien empire égyptien* (Paris, 1982), pp. 96-98.

See also P. Ghalioungui, *The House of Life: Per Ankh. Magic and Medical Science in Ancient Egypt* (Amsterdam, 1963; 2nd ed. 1973), plate 8; F. Jonckheere, *Les médecins de l'Égypte pharaonique* (Brussels, 1958), pp. 49-50; and P. Ghalioungui, *The Physicians of Pharaonic Egypt* (Cairo, 1983), pp. 19-20.

See further the still useful article of G. Maspero (with Mariette's text and a French translation), "De quelques termes d'architecture égyptienne," *PSBA*, Vol. 11 (1889), pp. 304-17.

The English Translation

In my translation I have kept a close eye on Breasted's translation, though I have several times departed from it. I have translated only the passages to the left and right of the false door which recount the exchanges between Niankhsekhmet and the king.

Notes to the Introduction of Document I.3

1. Jonckheere, *Les médecins*, p. 96.
2. *Ibid.*, pp. 49-50.

Document I.3

Inscriptions from the Tomb of Niankhsekhmet

[A. Inscriptions on the Left]

/1/ The Chief of Physicians Niankhsekhmet spoke before his majesty: "Would that your ka (*k¹* =double or vital force), which is beloved of Re, would command that there be given to [me] a false door of stone /2/ for the tomb of mine in the cemetery." Thereupon his majesty caused that there be brought for him two false doors[1] of limestone from Tura[2] /3/ to be set [initially] in the audience-hall of the House [called] "The White Crown of Sahure shines forth"[3] and that there be appointed for them the two chief-controllers of the crafts (i.e. the two high priests of Memphis?)[4] /4/ and *wᶜb*-priests of the House of the Craftsmen so that the work might be done on them in the presence of the king himself. The stone cutting continued every day. That which was done on them was inspected in the court daily. His majesty had inscriptions (?)[5] placed on them and had them painted in lapis lazuli.[6]

[B. Inscriptions on the Right]

/1/ His majesty said to the Chief of Physicians Niankhsekhmet: "As these nostrils are healthy, as the gods love me, may you depart to the cemetery /2/ at a very old age as one honored." I gave adoration to the king greatly and worshiped every god for the sake of

Sahure, /3/ for he knows the desires of [his] entire Following. Now as for everything which goes forth from the mouth of his majesty (i.e. when he commands or expresses something), it immediately happens, /4/ for the god has given him knowledge of everything which is in the body [i.e. in the heart of anyone else][7] because he is more august than every god. If you love Re, you will worship every god for the sake of Sahure who did this [for me]. I was one honored by him; never did I do anything evil to any person.[8]

Notes to Document I.3

1. According to Breasted, Erman has suggested that the text is concerned here with a double false door rather than with two false doors. As P. Spencer, *The Egyptian Temple: A Lexicographical Study* (London, 1984), p. 223 (nn. 210, 211), notes, the forms of the determinatives in this and the succeeding writing of *rwt* in line 2 given by Sethe in his text were incorrectly drawn. She gives the corrected forms on page 196.

2. The limestone quarries opposite Memphis.

3. This may well have been the name of one of Sahure's residences. We are not to conclude that the false doors were to be set up there permanently but only while the work on them proceeded. They were transported to Niankhsekhmet's tomb after their completion.

4. The title *wr ḥrp ḥm(wt)*, translated here as "chief-controller of the crafts", is the title of the High Priest of Ptah at Memphis, but it may be that at this early period it simply meant "master-craftsman". See R. O. Faulkner, *A Concise Dictionary of Middle Egyptian* (Oxford, 1962), p. 170. M. S. Holmberg, *The God Ptah* (Lund and Copenhagen, 1946), pp. 51-55, translates it as "supreme leader of handicraft" and says (p. 53): "not until we come to texts from the fifth and sixth dynasties, all of which may be of a later date than the above cited text in the tomb of Niankhsekhmet, is the *wr ḥrp ḥmw.t*-title regularly associated with the designations 'the hem-priest of Ptah, the hem-priest of Sokaris'." That he was also the High-Priest of Ptah cannot be surely affirmed before the sixth dynasty (*ibid.*, p. 54). See also A. H. Gardiner, *Ancient Egyptian*

ANCIENT EGYPTIAN SCIENCE

Onomastica Vol. 1 (Oxford, 1947), p. 38*, for other examples where architects were assisted by the two high priests of Memphis. This matches the existence of two high priests of Heliopolis (see Document I.1, n. 115). At any rate the king appointed two such master-craftsmen to this project of preparing the false doors along with a number of lesser craftsmen.

5.   The word here questioned ("inscriptions") seems appropriate to the context, but I confess that I have not seen *ḫr-ᶜw* (followed by three circles) so translated. I have written *ᶜw* by assuming that the three circles are signs of the plural. See Gardiner, *Egyptian Grammar*, p. 490, sign N 33. Breasted suggested "color" as a possible translation, and Maspero, "De quelques termes," p. 309, wrote "sculptures (?)" following it by "dont la peinture est bleue".

6.   Breasted and Maspero translated this as "blue". With the determinative of three circles (which Sethe's text omits but which are clearly present in Ghalioungui, *The House of Life*, pl. 8, and are given by Borchardt, *Denkmäler*, Vol. 1, p. 172), we would expect the rendering "lapis lazuli" (perhaps not the stone itself or its pigment but some blue pigment imitating it; see J. R. Harris, *Lexicographical Studies in Ancient Egyptian Minerals* [Berlin, 1961], pp. 148-49).

7.   My bracketed phrase follows the remarks given by Breasted in *Ancient Records*, Vol. 1, p. 109, n. e.   This is another way of saying that the king was sensitive to the desires of his followers.

8.   This is a commonplace refrain in tomb biographies throughout Egyptian history, a refrain that becomes expanded as Spell 125 of the *Book of the Dead* (see Doc. II.3, Spell 125, S 2).

Document I.4: Introduction

Inscriptions from the Entrance to the Tomb of Washptah

The two inscriptions here translated give us an intimate picture of the relationship between a favorite official and the king, i.e. between the Architect, Judge, and Vizier Washptah and his king Neferirkare (the third king of the fifth dynasty). The most interesting feature of these extracts from the point of view of this volume is the reference to the sudden illness and death of Washptah in the course of his showing the king a new building for which he (Washptah) was responsible and the facts recorded in connection with this event, namely that the king summoned not only the chief of physicians but also a king's companion (or companions) and a lector-priest, and that he caused to be brought to the scene a chest of writings. We can guess with some confidence that this chest contained one or more medical papyri like those which we have included in Volume Three, Chapter Five, and that it perhaps also contained collections of spells which the lector-priest would have found useful if the illness turned out to be a case of possession. Alas, neither books nor attendants were of any use. Washptah's condition was hopeless, as the attendants informed the king, no doubt delivering their prognosis in the customary way, "an ailment not to be treated" (e.g. see Volume Three, Document V.1, case 8). Following this prognostication the king prayed mightily to Re promising to do everything for the deceased that he had desired. In fact he ordered the event recorded in writing in Washptah's tomb, demonstrating once more the favors that kings were wont to bestow on favorite officials in connection with their burials.

Texts and Studies

See the bibliography concerning Washptah's tomb in Porter and Moss, *Topographical Bibliography* Vol. 3, Part 2, p. 456. To this add P. Ghalioungui, *The Physicians of Pharaonic Egypt* (Cairo, 1983), pp. 73-74. The full titles of all the works cited below are given in the Introduction to Document I.3.

For the early publication of the tomb, see Mariette, *Mastabas*, pp. 267-71. The later texts, which I have used in my translation, are Sethe, *Urkunden* I (2nd ed.), pp. 40-45, and Borchardt, *Denkmäler*, Vol. 2, pp. 40-42, 129, 144, and plates 69, 70.

See Breasted's *Ancient Records*, Vol. 1, pp. 111-13, for an English translation and Roccati's *La Littérature historique*, pp. 108-111, for a French translation.

The English Translation

My English translation demands little comment except to note that the text is in poor shape. I have, of course, paid special attention to the translation by Breasted, as well as that of Roccati. I have included Breasted's headings since they are helpful to the reader. As usual my footnotes constitute a brief commentary on the text and my translation.

Document I.4

Inscriptions from the Entrance to the Tomb of Washptah

[Inscription A]
Erection of the Tomb by His Son

/1/ It was his eldest son, King's First Deputy[1] having the Authority over the Rekhyt-Folk,[2] Merneternesut, who made [this] for him (Washptah) when he was in his tomb in the cemetery.

The King Visits a New Building

/2/..... Neferirkare came to see the beauty of ....[3] /3/.... when his majesty came forth upon them.... /4/ His majesty caused that it be.... /5/....the royal children saw /6/ ....and they marveled greatly /7/ ....Then, lo, his majesty praised him (i.e. his architect Washptah) because of it.

Washptah's Sudden Illness

But his majesty saw him [Washptah] [attempting to?] kiss the ground [in obeisance] (i.e. to respond properly).[4] /8/ His majesty said "..... Do not kiss the ground (i.e. do not bend over so far?), but rather kiss my foot." When the royal children and the companions [of the King] [who] were in the court [heard this exchange], fear beyond everything (i.e. exceedingly great fear) was in their hearts.

[Inscription B]
He is Conveyed to Court and Dies

ANCIENT EGYPTIAN SCIENCE

/1/ [His majesty had him taken] to rest in the residence. His majesty had the royal children, the companions [of the king], the lector-priest, and the chief of the physicians[5] go [into Washptah]. /2/ They said before his majesty....His majesty had brought for him (the sick Washptah) a chest of writings.[6] /3/ ....They said before his majesty that he (Washptah) was helpless.[7] [Then his majesty] prayed to Re on the [sacred?] lake ....exceedingly beyond everything. His majesty said that he would do everything according to [Washptah's] desire. Then he returned to the Residence. /5/ ....Never was it done for the likes of him before.[8] He prayed to Re every day /6 / ....in the heart of his majesty beyond anything (i.e. exceedingly). Then his majesty ordered [this event] put into writing on his (Washptah's) tomb.[9]

Notes to Document I.4

1.   By the time of Washptah this had become a title of rank. See W. Helck, *Untersuchungen zu den Beamtentiteln des ägyptischen Alten Reiches* (Glückstadt / Hamburg/New York, 1954), p. 60.

2.   For the *Ṛhyt*-Folk, see Document I.1, n. 17.

3.   This was obviously a reference to the building which Washptah was erecting for the king.

4.   That is, Washptah, stricken with his illness, had not made the expected deferential response. In giving this sense to the passage, namely that Washptah did not acknowledge the king because he was stricken, I am following what I believe to be the sense of the passage. I suppose the king urges Washptah not to kiss the ground but rather his foot because that would be easier if the king were on a raised chair or throne. The state of the text is particularly bad here and an examination of Borchhardt's plates is not very fruitful.

5.   As I pointed out in the introduction, it is of interest that the lector-priest was called in along with the chief of physicians so that there could be magical as well as medical help for the stricken architect.

6.   No doubt this chest contained medical papyri. It could well be that both Documents V.1 and V.2, in Volume Three, were based on medical papyri that themselves went back to the Old Kingdom. At any rate, it appears certain that some such books were kept in the chest of writings made accessible to the physician at Washptah's bedside. A similar case can be made for the existence of some collection of spells, which would be useful to the lector-priest. Examples of each type of book to be consulted are evident in the

two parts of the Edwin Smith Papyrus given as Document V.1.

7. As I have suggested in the Introduction to this document, this is probably a reference to a negative diagnosis like that found occasionally in Document V.1.

8. This is a standard phrase that appears in tomb biographies throughout Egyptian history, to the effect that the king has treated the deceased in an exceptional way.

9. We have already noticed in the preceding biography that it was the custom of the king to assist in the preparation of the tombs of his favored officials. Here we are specifically told that this event was to be recorded in the deceased's tomb.

Document I.5: Introduction

Inscriptions from the Entrance of the Tomb of Senedjemib

These extracts from inscriptions that appear on Senedjemib's tomb at Giza near the Great Pyramid of Cheops (Khufu) present a warm picture of King Djedkare Issy (*or* Izezi), the penultimate king of the fifth dynasty, and of his talented architect and official Senedjemib. It will not surprise readers of Chapter One above that Senedjemib was one of the royal officials who was trained as a scribe and rose through a series of extremely important positions, not the least of which for our interest was his position as Overseer of the Scribes of the King's Records (*or* Writings). The first list of positions in Inscription A records his titles of Overseer of the Treasury, Overseer of the King's Palace, Overseer of the Armory, and Overseer of the Southern Estates. In the course of the events described in that inscription it is noted that Senedjemib was honored before the king as a Master of the Secret Things of the King and as a Favorite of the King. Then we are told that when the king was at the Place of Records (presumably to examine the plans drawn up by Senedjemib for some building or other), he sent a servant to fetch his official so that he might honor Senedjemib by having him anointed and washed in the presence of the king. The king's visit to the Place of Records reminds us that a large number of papyrus fragments (among the earliest discovered) date from the reign of Issy.[1]

This inscription contains a striking bit of evidence of the king's education and interest in writing. It is said here that the king wrote "with his own fingers in order to praise" Senedjemib. Presumably Issy learned to write as a boy along with other princes and sons of officials.

Another point of interest in these inscriptions is the inclusion among them of two royal letters (one in each inscription) addressed to Senedjemib, continuing a tradition that began at least as early as the time of Metjen (see Document I.2). It should be realized that royal letters (particularly if they confirmed the reception of honors bestowed by the king) represented high points in the lives of the officials. One of the most celebrated of such letters is that inscribed on the entrance of the tomb of the Overseer of Scouts or Caravan-leader Harkhuf in the cliffs across the river from Aswan, a letter written by King Pepi II (end of the sixth dynasty) when he first came to the throne as a boy. The letter expresses the young king's enthusiasm concerning the news that Harkhuf was bringing back a dancing pygmy (or dwarf) from his fourth voyage. It is worth quoting almost in full:[2]

> Seal of the king himself: Year of the second occurrence (=Year 4, *or perhaps* Year 2),[3] Month 3 of Akhet (i.e. the first season), Day 15. Decree of the king to the Sole Companion, Lector-priest, Overseer of Scouts, Harkhuf.
>
> I have taken notice of the matter of this letter of yours, which

you sent to the king at the palace to inform him that you had come down safely from Yam with the army that was with you. You said in this letter of yours that you have brought all kinds of great and good gifts, which Hathor, Lady of Immaau, has given to the *ka* of the King of Upper and Lower Egypt Neferkare, living forever unto eternity. You said in this letter of yours that you have brought a pygmy of the god's dances (i.e. one who dances) from the Land of the Horizon-dwellers like the pygmy brought back by the God's Seal-bearer Bawerded from Punt in the time of Issy. You have said to my majesty that no one like him (the pygmy) has ever been brought by anyone else who has visited Yam before. Truly you know how to do everything your lord desires and praises. Truly you spend days and nights planning and executing what your lord desires, praises, and commands. His majesty will provide you with many excellent honors for the benefit of the son of your son forever, so that all the people will say when they hear what [his] majesty has done for you: "Is there anything like what was done for the Sole Companion Harkhuf when he came down from Yam on account of the vigilance he showed in doing what his lord desired, praised, and commanded?" Come north to the [royal] Residence immediately. Hurry and bring with you this pygmy which you have brought back from the Land of the Horizon-dwellers, live, hale, and healthy, for the dances of the god, to rejoice and gladden the heart of the King of Upper and Lower Egypt Neferkare, living forever. When he (the pygmy) embarks on the ship with you, station trustworthy people around him on both sides of the ship to prevent him from falling into the water. When he lies down at night, have trustworthy people lie down around him in his tent and have them make an inspection ten times in the night. [His] majesty wishes to see this pygmy more than all the gifts (*or* tribute) from the Mine-Land and Punt. If you arrive at the Residence with this pygmy in tow, live, hale, and healthy, [his] majesty will do great things for you, more than was done for the God's Seal-bearer Bawerded in the days of Issy....

The first inscription (A) comes from the early period of Issy's reign, i.e. the fifth year (or less likely, the tenth year; see note 3 to the document below). And the second inscription (B) is dated from the end of the reign, i.e. the sixteenth occurrence [of the counting] (equal to Year 32 of the reign; see note 12 to the document below). If the transcriptions of these numbers are correct, and this seems likely, then we have in these two inscriptions evidence of the use of two different systems of numbering regnal years, one annual and the other biennial.

The second inscription (B), in its letter from the king, lists the following titles for Senedjemib: Chief Judge, Vizier, and Overseer of the Scribes of the King's Records (*or* Writings), all very important positions, showing us how far Senedjemib had advanced in the years between the dates of the composing of the two inscriptions. This second letter from the king has a remarkably sophisticated conclusion for a document arising from ca. 2356 B.C.: "O, Senedjemib, what you desire, you have, for assuredly knowledge is what you desire." This statement epitomizes the best attitude of a society confident in the power of its scribal activity and of the knowledge it would bring.

In a third inscription (Sethe's C), which I have not included in my translation below, Sebedjemib-Inty's son, Senedjem-Mehy, remarks that a royal command was issued to assemble the princes and other people (Sethe, *Urkunden* I, 2nd ed., Vol. 1, p. 64, line 1), no doubt for the purpose of endowing his father's tomb. He notes that the tomb was completed in one year and three months (*ibid.*, line 5). The king appointed mortuary priests (lit. "servants of the *ka*") for the endowment (*ibid.*, p. 65, line 3), and the endowment was put in writing and engraved on the tomb by sculptors (*ibid.*, line 4). All of this tells us once more what an important part writing played in the burial procedures of the Old Kingdom.

Texts and Studies

Porter and Moss, *Topographical Bibliography*, Vol. 3 (2nd ed.), pp. 85-87. Lepsius, *Denkmäler*, Abt. II, plate 76c-f (especially 76d and f) was the earliest publication of the tomb.

For my extracts I have used Sethe, *Urkunden* I (2nd ed.), pp. 59-63 (which includes the additional fragments found by G. A. Reisner and C. S. Fischer, "Preliminary Report on the Work of the Harvard-Boston Expedition in 1911-13," *Annales du Service des Antiquités de l'Égypte*, Vol. 13 [1914], pp. 227-52, and particularly p. 248, nos. 1 and 2).

Breasted, *Ancient Records*, Vol. 1, pp. 121-25, has an English translation based on the first edition of Sethe's *Urkunden* and hence without the additions discovered by Reisner and Fischer. See also the French translation of Roccati, *La Littérature historique*, pp. 122-28.

The English Translation

My English translation has been made from the revised text of Sethe's (see literature above) and thus is more extensive than Breasted's translation. Again observe that the headings are taken from either Breasted or Sethe (or both). The line numbers are those indicated in the text of Sethe.

Notes to the Introduction of Document I.5

1.   A. H. Gardiner, *Egypt of the Pharaohs* (Oxford, 1961), pp. 86-87.

2.   Sethe, *Urkunden* I (2nd ed.), pp. 128-31. Compare the translations of Breasted, *Ancient Records*, Vol. 1, pp. 160-61; Gardiner, *Egypt of the Pharaohs*, pp. 58-59; and M. Lichtheim, *Ancient Egyptian Literature*, Vol. 1 (Berkeley/Los Angeles/London, 1975), pp. 26-27.

3.   We have here the question of whether the Egyptian expression *rnpt-sp* 2 (perhaps to be read as *ḥ'ṯ-sp* 2) at this time still meant "The second occurrence" of the counting of the cattle that took place biennially or whether it simply meant "Regnal year 2", the biennial counting having been abandoned or changed to an annual affair.   In this translation I prefer to assume that the biennial count is referred to.

Document I.5

Inscriptions from the Entrance of the Tomb of Senedjemib, Pet-named Inty[1]

[Inscription A]

/1/....Overseer of the Two Houses of Silver (i.e. the Treasury),[2] /2/ Overseer of the King's Palace, /3/ Overseer of the Armory, Overseer of Every Seat of the Residence, Overseer of the Southern Estates, [Senedjemib]. /4/ Year 5, Month 4, Day 3, today, under Issy,[3] /5/ when I was honored before Issy, more than anyone like me, as the Master of the Secret Things of the King [and] as a Favorite of His Majesty. /6/ What his majesty wished was done. Indeed, his majesty praised me for every project which his majesty commanded to be done. I did it according to the desire of his majesty. /7/ ...his majesty while he was in the Place of Records (or Writings).[4] When it came to pass /8/ that a servant was standing by the house and his majesty had ordered him to hasten....and his majesty caused that I be anointed with unguent /9/ and that my flesh be cleaned in the presence of his majesty.[5] Never before was the like done in the presence of the king for anybody, /10/ because of [my being] more noble, trustworthy, and loved before Issy than the likes of anybody.... His majesty himself wrote with his own fingers[6] in order to praise [me] /11/ because I did everything.... (he commanded?) well and excellently and according to the desire of his majesty concerning it.

Letter by the King's Own Hand[7]

/12/ Royal Command [to] the Chief Judge, Vizier, and Overseer of the Scribes of the King's Records (i.e. Writings), /13/ Overseer of All the Works of the King, Senedjemib. /14/ My majesty has seen this letter of yours which you have written to inform his majesty of everything you have done....the writing (i.e. plan) /15/ for the building [called] "Beloved of Issy" which will be upon the lake of the palace.... Now you are rightly [named] Senedjemib (i.e. He Who Makes the Heart Rejoice), for can it not be said: "Lo, it is a thing (i.e. a building) /16/ which makes the heart of Issy rejoice?" His majesty causes the truth to be known immediately. For assuredly it happens that you /17/ say what Issy loves (i.e. desires) better than any noble who is in the land.... /18/ when indeed his majesty knows that every vessel (?) is for his good.....of the statement (i.e. plan) of the Overseer of All the Works of the King. /19/ It rejoices the heart of Issy very truly....

[Inscription B]
A Statement of the King to Senedjemib[8]

/1/ ....was brought to you a command of the king....[when] you said to his majesty that you would make a lake according to what was said in the palace.... /2/ when you said that you would .... Sed Festival[9], my majesty greatly liked these words of yours ....

A Further Letter from the King[10]

/7/ Royal Command [to] Chief Judge, Vizier, Overseer of All the Works of the King, /8/ Overseer of the Scribes of the King's Records (i.e. Writings), Senedjemib.

/9/ His majesty has seen that plan which you caused to be brought forth according to the understanding [reached] in the palace, [the plan] for the lake of the mansion Hutmehwasekh *(Ḥwt-mḫ-wsḫ)* [11] and /10/ for the Palace of Issy [called] "Nekhbet...." .....of the [Sed?] Festival; when you said to his majesty: "This will be done for you.... /11/ [the building or lake?] 1220 cubits [in length] and 220+.... cubits [in width] as was commanded of you in the palace.... /12/ Assuredly the god has made you a favorite of Issy. His majesty knows that you are more skilled than the overseers of all the works /13/ produced in the whole land. I will do for you more than I have done for anybody. /14/ For I shall give you the strength unto eternity.... You will be made [forever?] Overseer of All the Works of the King. /15/ O, Senedjemib, what you desire (*or* love) you [already] have, for assuredly knowledge *(rḫt)* is what you desire (*or* love)." /16/ Year[12] of the sixteenth occurrence [of the counting?], Month 4 of Shemu (i.e. the third season), Day 28.

{For the notes to this document, see page 199.}

Notes to Document I.5

1. Pet-name" is my translation of Egyptian *rn-nfr* (lit. "good name"). It is often a shortened name that contrasts with *rn-wr* ("great name"), which latter in this case is Senedjemib. *Wb.*, Vol. 2, p. 428, suggests that *rn-nfr* is equivalent to the rarer expression *rn-nds* ("small name"). For a full discussion of these various name forms, see H. Ranke, *Die ägyptischen Personennamen*, Vol. 2 (Glückstadt/Hamburg/New York, 1952), pp. 6-8. See also "Name,"*Lexikon der Ägyptologie*, Vol. 4, c. 322.

2. For this title, see A. H. Gardiner, *Ancient Egyptian Onomastica*, Vol. 1, p. 26˙ (and for the next title, see *ibid.*, p. 27˙).

3. As the text stands we have ⫶ (i.e. ⫶) (=*rnpt 5*), which literally is "Year 5". If this is the correct reading, it surely looks as if the biennial system was not being employed. Of course it could simply be that the glyph ☉ (meaning "time" or "occurrence") has accidentally been omitted from the text. If so, the text should have been ⦙☉⫶ (i.e. ⦙☉⫶). If this glyph was meant to be there, as it was in the later date discussed in note 12, the meaning would then be the "Year of the fifth occurrence" [of the counting]. Consequently this would imply that the date given here was "Year 10", as would be expected if the biennial count was being followed. See Document I.1: Introduction, notes 6 and 7.

4. My guess is that Senedjemib's description and plans of the building whose construction was agreed upon in the palace were deposited in the Place of Records, where the king went to study them.

5.   It was apparently a signal honor to be anointed and washed in the presence of the king.

6.   I have already mentioned the significance of this statement for estimating the king's education.

7.   This is Breasted's heading, which he took substantially from Sethe, and of course it is not in the document (nor indeed are any of the headings in my translations).

8.   This heading comes from Sethe's text.

9.   Senedjemib seems to have been entrusted with some part of the preparations for the king's Sed Festival (see Document I.1, n. 41). If this was to be celebrated in the thirtieth year of the king's reign, as it often was, then it means that the events described in this paragraph took place at least two years before the apparent end of Issy's reign (see note 12 below).

10.   Again I reproduce Sethe's heading.

11.   This is a sheer guess since I have not located any reference to this mansion.

12.   The date given here seems to be in the biennial system of dating and should be understood as the "Year of the sixteenth occurrence" [of the counting], which would accordingly be Year 32 of Issy's reign. Of course it is not impossible that the *rnpt sp* used by itself here simply means regnal year, which, if Sethe's reading of the number is correct, would then mean "Regnal year 16". There is one further complication. According to Sethe, Reisner believed "26" instead of "16" was a possible reading. In this case, if the biennial system was being employed, we would then have at least 52 years in Issy's reign, but there is no other evidence that he reigned so long. In fact, the latest year to which we have a reference (aside from the date here) is also the "year of the sixteenth time (*or*

occurrence)" found in one of the papyrus fragments mentioned in my introduction to this document (see Gardiner, *Egypt of the Pharaohs*, p. 435, n. 2), which, in the biennial system, would again be a reference to the Year 32. Needless to say, "Year 26" (interpreting the expression "26th occurrence" as simply meaning "Regnal year 26") would present no problem of falling outside of Issy's probable reign length of 32 years. On the whole, I prefer accepting the number "16" and referring it to a biennial system.

Document I.6: Introduction

Tales of Wonder at the Court of King Cheops (Khufu)

I have included as this document two from a series of at least five tales of magical acts performed by lector-priests. These tales are present in a single manuscript, the Westcar Papyrus (Papyrus 3033 of Berlin), which appears to date from the time of the Hyksos (ca. 1640-1532 B.C.). They were probably composed in the twelfth dynasty (1991-1783 B.C.). The form of the narrative is that of tales told by the sons of King Cheops (the second king of the fourth dynasty) to their father.

Of the five tales, the third and the fourth have been given here. The entire narrative part of the first tale is missing and only the end persists, but enough to show that it concerned a magical wonder supposedly performed in the third dynasty during the reign of King Djoser, the builder of the great complex of the step pyramid at Saqqara.

The second story, a tale of adultery, was laid in the reign of Djoser's predecessor, King Nebka. The magical act performed by the lector-priest named Webaoner was the manufacture of a wax crocodile seven fingers(?)[1] long, over which the lector-priest spoke magical words, so that when thrown in the water in which the adulterer was bathing it grew to life-size (seven cubits in length) and seized the offender, holding him for seven days until the lector-priest (accompanied by the king) told the crocodile to bring him forth. At this point the lector-priest picked up the crocodile and once more it became a crocodile of wax in his hand.

The third and fourth stories may be entitled "The Boating Party" and "The Deeds of the Magician Djedi". The third takes place in the time of King Sneferu, the father of Cheops and the first king of the fourth dynasty, while the fourth is laid in the reign of Cheops himself.

Toward the end of the fourth tale there is a reference (which I have not translated here) that provides a connecting link with the fifth tale. After the account of Djedi's magical acts of restoring the severed heads of two birds (a goose and a water fowl) and that of an ox, there is an omission, followed by a passage in which Djedi is asked by the king whether he knows the number of the sacred chambers of the sanctuary of Thoth. Djedi says that he does not but that he does know where that number may be found, namely in a chest of flint in a building or chamber in On (Heliopolis) called "Inventory". When the king then asks Djedi to fetch the chest for him, Djedi replies that he cannot. But, he says, the eldest of the three children who are in the womb of Ruddedet, the wife of a priest of Re, will be able to get it for him. The children in the pregnant woman's womb are declared to be the sons of Re, Lord of Sakhbu, and they are further said to be the future rulers of the land after the reigns of Cheops's son and grandson, the eldest serving first as the "Great Seer" (i.e. High Priest) of Heliopolis. This then is a fanciful, historical account of the origin of the first three kings of the fifth dynasty: Weserkaf, Sahure, and Neferirkare.

As I said earlier, this dynasty was particularly devoted to Re and the priests of Heliopolis must have exerted considerable influence on it. This story, then, telescopes the rest of the fourth dynasty into two reigns, those of Cheops's son (Chephren) and his grandson (Mycerinus), to be followed by those of the first three kings of the fifth dynasty.

The fifth and last story, which I have not included here, concerns the birth of the three children of Ruddedet and the assistance provided by the goddesses Isis, Nephthys, Meskhenet, and Heket as midwives. Re says to the goddesses and to Khnum, who accompanies them as a porter and the one who "gave health" to the bodies of each child as he was born: "Please go and deliver Ruddedet of the three children who are in her womb, those who will exercise this potent office (i.e. the kingship) in this entire land." One last point to be made in connection with this bit of fictional history is that it had Re as the father of the future kings. This was no doubt a rationalization of the fact that "Son of Re" occasionally appears as a title of the king in the fifth dynasty (see Document I.1, n. 24). In a way this prefigures the accounts by Hatshepsut and Amenhotep III in the eighteenth dynasty of the impregnation of their mothers by Amon-Re. I can mention finally that this last story is replete with magical tricks and devices.

The main reason I have included the third and fourth tales here is to indicate how strongly in the minds of literary artists (and presumably the people at large) the lector-priest was connected with magic. As I said in Chapter One, the title itself, *ẖry-ḥbt*, literally means "the one who carries the festival roll". Indeed the tomb reliefs often depict a lector-priest as one carrying a papyrus roll. He was primarily concerned with the magical spells and incantations uttered during the funeral and temple rites. One tomb of the Old Kingdom describes the role which the lector-priest was to play in the burial of the deceased: "Beloved of the King and of Anubis is the lector-priest ...who shall perform for me [the deceased] the things beneficial to a blessed spirit according to that secret writing of the lector-priest's craft."[2] Furthermore, he is shown on occasion as "reciting incantations with upraised arms".[3] In view of these activities it should not surprise the reader that he was called in on medical cases where possession or the like was suspected, as we saw happen at the illness of Washptah recounted in Document I.4, or that the lector-priest was described as the discoverer of semi-magical spells,[4] or indeed that the author of these tales of the fantastic took him as the magician par excellence. We shall see in the next chapter how magic pervaded the whole religious fabric of Egyptian society (at least at its top), how Egyptians attempted to achieve afterlife by means that were fundamentally magical, and further how they were concerned with preserving the cosmic order by those means.

Texts and Studies

For the text, see A. Erman, *Die Märchen des Papyrus Westcar (Mittheilungen aus den orientalischen Sammlungen*, Hefte 5-6) (Berlin, 1890). This includes not only plates of the hieratic text, but also a hieroglyphic transcription of the text, a transcription of it in the Roman alphabet, and a German translation.

The hieroglyphic transcription was also published by K. Sethe, *Ägyptische Lesestücke* (Leipzig, 1924; 2nd ed. 1928), pp. 26-36.

See also Sethe, *Erläuterungen zu den ägyptischen Lesestücken* (Leipzig, 1927), pp. 32-45.

Of the many translations of these tales available, that of M. Lichtheim, *Ancient Egyptian Literature*, Vol. 1 (Berkeley / Los Angeles / London, 1975), pp. 215-22, appeals to me the most because it is highly accurate and literary at the same time. Also useful is the translation of W. K. Simpson in *The Literature of Ancient Egypt*, new ed. (New Haven and London, 1973), pp. 15-30. For some of the extensive literature, see the introductions to the translations of Lichtheim and Simpson.

The English Translation

My English translation follows closely (but not literally) the translations of Simpson and Lichtheim. The plate and line numbers are those of Erman's edition.

Notes to the Introduction of Document I.6

1. The measure "fingers" is a guess, since it is not readable in the hieratic text. All that we know is that the wax crocodile was small enough to fit in the magician's hand. In Blackman's translation of A. Erman, *The Ancient Egyptians: A Sourcebook of their Writings*, Torchbook ed. (New York, 1966), p. 37, we find "spans" instead of "fingers", reflecting Erman's German translation in the original (*Die Märchen*, p. 8) where he wrote "sieben Spannen". But in the commentary in the same volume (p. 25) we find "7 [Zollen lang war?]", that is to say "7 fingers" rather than "7 spans".

2. Gardiner, *Ancient Egyptian Onomastica*, Vol. 1,

p. 55*.

3. *Ibid.*, p. 56*.

4. *Ibid.*, cites the London Medical Papyrus (8,12).

Document I.6

Tales of Wonder at the Court of King Cheops (Khufu)[1]

The Boating Party

/4,17/ Baufre stood up to speak and said: I shall let your majesty hear of a wonder produced in the time of your father Sneferu, justified (i.e. deceased), by the chief lector-priest (*ḥry-ḥbt* /4,20/ *ḥry-tp*) Djadjaemankh, [an event] that illuminates the past.....[2]

[One day King Sneferu wandered through] all [the rooms] of the palace, L.P.H.,[3] in search of [something refreshing (i.e. a diversion) without finding it. Thereupon he said:][4] "Go and bring me the chief lector-priest and scribe of books, Djadjaemankh." He was brought to him immediately. Then his majesty said: "I have [wandered through all the rooms] of the palace, L.P.H., in search of something /5,1/ refreshing without finding it." Then Djadjaemankh said: "May your majesty proceed to the lake of the palace, L.P.H., and equip a boat with [a crew of] all the beautiful girls from inside your palace. Then your majesty's heart will be refreshed by seeing them row up and down /5,5/ [the lake]. As you see the beautiful nestings of your lake and you see its beautiful fields and banks, your heart will be refreshed on account of it." [His majesty said:] "Indeed I shall go boating. Let there be brought to me twenty oars of ebony worked (i.e. decorated) with gold, their grips of sandal-wood (?) worked with electrum. Let there be brought to me twenty women /5,10/ with the most beautiful bodies, breasts, and braids who have not yet given birth (i.e., who are virgins). Also let there be brought to me twenty nets and let these nets be given to those women in place of their clothes." Everything was done as his majesty commanded.

Thereupon they rowed up and down. The heart of his majesty was happy on account of /5,15/ seeing them rowing. Then the one who was at the stroke-oar became entangled with her braids, and a fish-shaped pendant of new turquoise fell into the water. Then she stopped rowing and so the rowers on her side stopped rowing. His majesty said: "Why are you not rowing?" Then they said: "The stroke-rower /5,20/ has stopped rowing." So his majesty said to her: "Why are you not rowing?" She replied: "My fish-shaped pendant of new turquoise has fallen into the water." [Then his majesty said to her: "Row! I shall replace it for you."] Then she said: "I like my own better than one like it." Then his majesty said: "Go and bring me the chief lector-priest Djadjaemankh." and he was brought immediately. Then said /6,1/ his majesty: "Djadjaemankh, my brother. I did what you said. The heart of his majesty was refreshed by seeing them row. Then the fish-shaped pendant of new turquoise belonging to one of the lead-rowers fell into the water. So she stopped rowing, upsetting the rowers on her side. Then I said to her /6,5/ 'Why are you not rowing?' And she said to me: 'My fish-shaped pendant of new turquoise has fallen into the water.' So I said to her: 'Row! I shall replace it.' Then she said: 'I like my own better than one like it.'"

Then the chief lector-priest Djadjaemankh spoke his words of magic. Then he put one side of the lake's water on top of the other, and he found the fish-shaped pendant /6,10/ lying on a potsherd. Then he brought it and gave it to its owner. But now the water which had been 12 cubits deep in the middle[5] ended up as 24 cubits deep when folded back. So he spoke his words of magic and brought back the waters of the lake to their [former] position. His majesty spent the day in holiday with his entire palace, L.P.H. Then he rewarded the chief lector-priest /6,15/ Djadjaemankh with all good things.

Behold the wonderous act that took place in the time of your father, the King of Upper and Lower Egypt Sneferu, justified, by the chief lector-priest and scribe of books Djadjaemankh.[6]

His majesty, the King of Upper and Lower Egypt, Cheops, justified, said: "Go and give an offering of 1000 loaves of bread, 100 jugs of beer, an ox, two measures of incense to the majesty of the King of Upper and Lower Egypt Sneferu, justified. Further let there be given one loaf, one jug of beer, one measure of incense to the chief lector-priest and scribe of books Djadjaemankh, for I have seen his deed (sp) of knowledge (or skill: rḫt)."[7] Everything his majesty commanded was done.

The Deeds of the Magician Djedi

Then Hardedef, the son of the king, stood up to speak. He said: "[So far you have heard] about the knowledge (rḫt) of those who have passed away and [so] one does not know the things that are true from those that are false [concerning them].[8] [But there is a subject][9] of your majesty in your very own time /6,25/ whom you do not know [but who is a great magician]." His majesty said: "What is this, Hardedef, my son?" Then Prince Hardedef said: "There is a man /7,1/ whose name is Djedi. He resides in Djed-Sneferu, justified. He is a man 110 years old. He eats 500 loaves of bread and one half an oxen as his meat, and drinks 100 jugs of beer up to this very day. He knows how to join a severed head, and he knows how to make a lion /7,5/ follow behind him, even though its leash is on the ground. He knows the number of secret chambers of the Sanctuary of Thoth." Now his majesty, King of Upper and Lower Egypt Khufu, justified,[10] had spent a long time seeking the secret chambers of the Sanctuary of Thoth in order to make a copy for his tomb.[11]

Then his majesty said: "You yourself Hardedef, my son, bring him to me." Then ships were readied for Prince Hardedef. He traveled /7,10/ upstream to Djed-Sneferu, justified. Then after the ships were moored to the riverbank, he traveled by land, seated in a sedan chair of ebony, its poles being of *sesenedjem*-wood mounted with gold. When he reached Djedi, the sedan chair was set down. He (Hardedef) got up to greet him (Djedi) and found /7,15/ him lying on a mat on the threshold[12] of his house. One person beside his head was smearing him [with unguent?] and another was wiping his feet.

Then Prince Hardedef said: "Your condition is like one who has not reached old age (for old age is the state [lit. 'seat'] of dying, the state of [being ready for] burial, and the state of [being ready for] interment)--you who sleeps until dawn free of illness and without a hacking cough. Greetings /7,20/ to an honored man. I have come here commissioned by my father, Khufu, justified,[13] to summon you [to the court where] you will eat delicacies which the king gives, the food of those who are in his service. He will convey you in good time to your fathers (i.e. ancestors) who are in the necropolis."[14] Then Djedi said: "In peace! In peace! Hardedef, king's son, beloved of his father. May your father Khufu, justified,[15] praise you.   May he advance /7,25/ your rank [to be] among the elders. May your ka[16] prevail over your enemy.   May your ba[17] know the ways leading to the portal which conceals the weary-ones (i.e. the dead).[18]   So greetings /8,1/ to a prince."

Then Prince Hardedef extended his hands to him, helped him stand up, and proceeded with him to the riverbank, while giving him his arm.   Then Djedi said: "Let me have one of [those] transport barges *(k'k'w)* to transport for me my children[19] and my books."   Then two ships with their crews were given to him.   /8,5/ Djedi journeyed downstream in the ship in which Prince Hardedef was.   After he arrived at the Residence, Prince Hardedef entered to report to the majesty of the King of Upper and Lower Egypt Khufu, justified.[20]   Prince Hardedef then said: "O, sovereign, L.P.H., my lord, I have brought Djedi."   Thereupon his majesty said: "Go and bring him to me."   His majesty proceeded to the Forecourt (*or* Great Hall) of the /8,10/ palace, L.P.H. Djedi was ushered into him.

His majesty then said: "What is this, Djedi, I have not seen you [before]?" Djedi then said: "It is [only] he who is summoned who comes, O sovereign. I have been summoned, and, lo, I have come."   His majesty then said: "Is it true, what they say, that you know how to join a severed head?"   Djedi said: "Yes, I know how, O sovereign, L.P.H., my lord."   /8,15/ Then his majesty said: "Let there be brought to me a prisoner who is in the prison that he may be executed."   Then Djedi said: "But not to people (i.e. this may not be done to a human being), O sovereign, my lord, for surely the doing of such a thing to the august cattle (i.e. man) is forbidden." Then a goose was brought to him and his head was cut off. Then the goose was placed on the west side of the Forecourt, its head on the east /8,20/ side of the Forecourt. Thereupon Djedi spoke his words of magic and the goose stood and waddled, as did also his head. When the one part reached the other [and they became joined], the goose stood up cackling. Then he had a waterfowl[21] brought to him, and the same thing was done to it. Then his majesty caused an ox to be brought to him /8,25/ and his head was felled to the ground. Then Djedi spoke his words of magic and the ox arose[22]....

Notes to Document I.6

1. The title is approximately that given by Sethe in his edition.

2. The narrative is announced as if it were an historical account. Literally the fragment says: "illuminating its yesterday".

3. "L.P.H." stands for "Life, Prosperity, and Health" (or perhaps better, "May it live, prosper, and be in health"). By the time of the Middle Kingdom it became conventional to speak of the palace as a kind of royal entity and hence to append to it "L.P.H.". Then later the palace came to be more directly identified with the king. Thus "palace", when translating "pr-^{c_l}", i.e. "pharaoh", became the designation of the king himself, perhaps for the first time in the reign of Amenhotep IV (i.e. Akhenaten) in the eighteenth dynasty and more often from the nineteenth dynasty onwards (see Gardiner, *Egyptian Grammar*, 3rd ed., p. 75). Notice also that throughout this document the epithet "justified" follows the mention of a king's name. This literally may be rendered "true of voice", with the probable meaning "acclaimed as right or true". It was an epithet applied to dead kings (and later, from the Middle Kingdom, to dead persons of any rank). As Gardiner (*ibid.*, pp. 50-51) says, it was "often practically equivalent to our 'deceased'. Originally applied to Osiris with reference to the occasion when his legal rights, being disputed by Seth, were vindicated before the divine tribunal in Heliopolis. The same epithet is also used in connexion with Horus as the 'triumphant' avenger of the wrongs done to Osiris". Indeed it was apparently Horus rather than Osiris who first bore the

epithet. For its applications to Horus and Osiris, and its various interpretations, see R. Anthes, "The Original Meaning of *m^jc-ḥrw*," *Journal of Near Eastern Studies*, Vol. 13 (1954), pp. 21-51. Cf. J. G. Griffiths, *The Origins of Osiris and his Cult*, 2nd ed. (Leiden, 1980), pp. 178-80. The point to notice here in my translation is that the epithet is applied not only to Kings Djoser, Nebka, and Sneferu, all of whom are already deceased when the tales are being told to King Cheops, but also to King Cheops himself, even though he is alive as the stories are being told. The reason for this apparent inconsistency seems to be that the author (or scribe) who was composing (or writing down) these tales lived long after all of these kings were dead and by his time it was conventional, when any dead king was mentioned, to describe him as "justified" regardless of the period depicted in the story.

4. The reconstructions in brackets are possible because of the narrative practice in these stories of having the speaker who has been engaged in some activity which the author has just recounted repeat aloud his action when he encounters the person to whom he is speaking. Notice that the phrase *st ḵbt*, which I have translated "something refreshing", literally means "a place of refreshment". In the context here it surely means that the king was seeking some sort of diversion.

5. Lichtheim prefers "across" to "in the middle".

6. Throughout the document these two titles are linked to guarantee that the person producing the magical deeds has the best possible credentials for such deeds.

7. Again notice the stress on *rḫt*, i.e. knowledge, learning, or skill, so admired in Egyptian society.

8. The prince is emphasizing here the difficulty of determining the truth of past events when the participants are no longer alive. He offers in contrast examples of deeds in Cheops's time.

9. So Lichtheim would restore the text of the lacuna. The next bracketed phrase also comes from Lichtheim's translation. Something like these restorations was no doubt in the original text.

10. See note 3 for an explanation of why the term "justified" is used here even though Cheops is still alive in the narrative.

11. The Egyptian term is *ʒḫt*, meaning "horizon" or "tomb". Lichtheim translates it as "temple", which is probably more specific than is intended in this place. Of course, in the time of Cheops, the interior burial chambers of the king's pyramid were not decorated or engraved, though the mortuary temple on the east side of the pyramid had some decoration.

12. Lichtheim translated this as "forecourt of his house".

13. See note 3.

14. That is to say, the king will give a good burial to Djedi when his time comes. This would have been a decisive argument to an Egyptian, namely, to be guaranteed a good burial among his ancestors.

15. See note 3.

16. That is, may your ka (vital force), which is in control after you are deceased, overcome opposition in the afterlife. For the concept of the ka, see S. Morenz, *Egyptian Religion* (Ithaca, 1973), p. 170, and U. Schweitzer, *Das Wesen des Ka im Diesseits und Jenseits der alten Ägypter (Ägyptologische Forschungen*, Heft 19) (Glückstadt/Hamburg/New York, 1956).

17. The ba as early as the Old Kingdom was

analogized as a bird in flight and later in the New Kingdom was depicted as a bird with a human head (and sometimes with human hands) which sat near the coffin of the deceased. It is usually translated as "soul", but not without ambiguity since by the time of this document the ba seems to have developed from a manifestation of the power of the deceased (as it was in the Old Kingdom) into a personified agent of the deceased. At any rate, it was thought to be able to fly back to the living world left by the deceased. The wish, then, in this passage is that Hardedef's ba may know the correct way to fly. See Morenz, *Egyptian Religion*, pp. 157-58, and, above all, L. V. Žabkar, *A Study of the Ba Concept in Ancient Egyptian Texts* (Chicago, 1968), particularly p. 116.

18. Blackman-Erman and Simpson would associate this portal with an individual, Blackman-Erman translating the phrase as "the portal of Him-that-hideth-weakness" who is identified as "a door-keeper in the underworld" and Simpson translating it as "the Portal of One Who Clothes the Weary One", whom he identifies in a note as the embalmer.

19. So the text says, but in the Blackman-Erman translation a note is added: "His pupils?". Simpson simply says in the text "students" instead of "children".

20. See note 3.

21. So Simpson translates $ht\text{-}c_I$. The Blackman-Erman translation has "duck" and Lichtheim "'long-leg'-bird" for it.

22. At this point in the text the scribe has omitted the account of how Djedi was able to make a lion follow him without a leash and only the very last words of that account are here added directly after the story of the severed ox's head.

Document I.7: Introduction

Scribal Immortality

As I have remarked in the first chapter, Egyptian scribes were unusually conscious of the significance and desirability of their profession. This document from the end of the nineteenth dynasty or a little later stresses that the scribe's writing was a more efficacious route to immortality than his building of tombs. His books are read and remembered long after his tombs and those of others have crumbled and are forgotten. In brief, "their names are [still] proclaimed on account of the books which they produced since they (the books) were good, and the memory of him who made them will last to the limits of eternity". The document goes on to cite the names of a number of sages whose fame was still present in the author's time: Hardedef, Imhotep, Nefry, Khety, Ptahemdjehuty, Khakheper[re]seneb, Ptahhotep, and Kaires (four of whom are pictured on the wall of a Ramesside tomb at Saqqara; see Fig. I.62).

Gardiner, the editor of this piece, has suggested that the rather skeptical tone of its author toward the efficacy of building tombs may well reflect the famous Song of the Harper that probably goes back to an original carved in a Middle Kingdom royal tomb:[1]

Song which is in the tomb of King Intef, the
justified, in front of the singer with the harp.
He is happy, this good prince!
Death is a kindly fate.

ANCIENT EGYPTIAN SCIENCE

A generation passes,
Another stays,
Since the time of the ancestors.
The gods who were before
      rest in their tombs,
Blessed nobles too are
      buried in their tombs.
(Yet) those who built tombs,
Their places are gone.
What has become of them?
I have heard the words of
      Imhotep and Hardedef,
Whose sayings are recited whole.
What of their places?
Their walls have crumbled,
Their places are gone,
As though they had never been!
None comes from there,
To tell of their state,
To tell of their needs,
To calm our hearts,
Until we go where they have gone!
Hence rejoice in your heart!
Forgetfulness profits you,
Follow your heart as long
      as you live!
Put myrrh on your head,
Dress in fine linen,
Anoint yourself with oils
      fit for a god.
Heap up your joys,
Let your heart not sink!
Follow your heart and your
      happiness,

Do your things on earth as
    your heart commands!
When there comes to you that
    day of mourning,
The Weary-hearted (i.e. Osiris)
    hears not their mourning.
Wailing saves no man from the pit!
Make holiday
Do not weary of it!
Lo, none is allowed to take his
    goods with him,
Lo, none who departs comes back
    again!

It was the mention of the sages Hardedef and Imhotep here that alerted Gardiner to the possibility that this was a source for our document.

This document is a part of a student's miscellany in Beatty Papyrus IV (Brit. Mus. 10684). According to its editor, it is "one of those miscellanies or collections of edifying writings which formed the staple of a Ramesside scribe's education".[2] Another section of this miscellany tells the student: "Be a scribe. It will save you from taxation and will protect you from all labours",[3] a theme stressed in our next document, which is concerned with the value of the profession in this world. This section goes on to compare the scribal profession to menial occupations, to the great benefit of the scribe's. Not only are the physical conditions of his work better than those of other occupations, since he is ordinarily in charge of those engaged in hard work, but the scribe may rise "step by step until he has reached (the position of) magistrate".[4]

Another motive for assuming the scribal calling to set beside those already mentioned is the contribution that writing makes to the furtherance of knowledge. This is briefly alluded to in a well-known instruction-book, ostensibly (but not certainly) composed by a king of the First Intermediate Period (ca. 2075 B.C.) for his son Merikare:[5]

> Copy your fathers who were before you; (achievement?) is determined by knowing. See! Their words are made lasting in writing. Open (the writings) that you may read and emulate what is known. So the expert becomes the one who is instructed.

Text and Study

The Egyptian text and English translation of this document are included in A. H. Gardiner, *Hieratic Papyri in the British Museum. Third Series. Chester Beatty Gift*, 2 vols. (London, 1935). For the text see Vol. 2, plates 18-19. The translation occupies Vol. 1, pp. 38-39, and a discussion of Beatty Papyrus IV appears on pp. 28-44 of the same volume. See also the English translation given by John A. Wilson in J. Pritchard, ed., *Ancient Near Eastern Texts Relating to the Old Testament* (Princeton, 1950), pp. 431-32.

The English Translation

I have kept an eye on both of the earlier translations but have not followed them slavishly. The italicized beginnings of the paragraphs represent words that are rubricated. The references in parentheses are to Gardiner's text as noted above.

Notes to the Introduction of Document I.7

1. Gardiner, *Hieratic Papyri...Third Series*, Vol. 1, p. 41, cites the well-known Song of the Harper in A. Erman's version as translated by A. M. Blackman, *The Literature of the Ancient Egyptians*, p. 133. I have given here the fine and sensitive translation of Lichtheim, *Ancient Egyptian Literature*, Vol. 1, pp. 196-97.

2. Gardiner, *op. cit.*, p. 37. Compare the various texts edited by Gardiner, *Late-Egyptian Miscellanies* (Brussels, 1937), and translated and commented upon by Caminos, *Late-Egyptian Miscellanies* (London, 1954).

3. Gardiner, *Hieratic Papyri....Third Series*, Vol. 1, p. 41. I have changed Gardiner's "thee" forms to "you" to make them conform with the procedure I have followed in translating this document.

4. *Ibid.*

5. T.G.H. James, *Pharaoh's People. Scenes from Life in Imperial Egypt* (Chicago and London, 1984), p. 25.

Document I.7

Scribal Immortality

(18, vers. 2,5) *Now then if you do these things, you are* skilled in [what is needed to produce] writings. As for knowledgeable scribes since the age of those who came after the gods--those who were able to foretell the future--their names have endured for eternity though they have completed their life-spans and all their relatives are forgotten.

They did not make for themselves pyramids of copper[1] and tombstones of iron.[2] They did not know how to leave heirs in the form of children [who would] pronounce their names. [Rather] they made for themselves heirs in the form of the writings and instruction-books they produced.

They gave to themselves [the papyrus-roll] as a lector-priest, the writing-board as a son-whom-he-loves, [the instruction-books] as their pyramids; /10/ the reed-pen was a child, and the back of a stone as a wife. They transformed the great and the humble into their children by being their [writing] supervisor.

Though [mortuary] doors and houses were made [for them], these are [now] in ruins, their *ka*-servants are [gone],[3] their tombstones are covered with dirt, and their tombs are forgotten. [But] their names are [still] proclaimed (*lit.* pronounced) on account of the books which they produced since they (the books) were good, and the memory of him who made them (i.e. the books) will last to the limits of eternity.

Be a scribe; put it in your heart that your name /19,1/ may fare similarly. A book is more effective [in keeping your name alive] than is a decorated tombstone or an established tomb-wall. These houses (i.e. chapels) and pyramids are made so that their names (i.e. those of the deceased) will be pronounced. Surely it is beneficial in the necropolis for a name to be on people's lips (*lit.* in the mouth of men).

A man has perished and his corpse has become dirt. All of his kindred have gone to ground (i.e. crumbled into dust). But it is writing which causes him to be remembered in the mouth of the reciter. More effective is a book than the builder's house (i.e. chapel?) or tombs in the West. It is better than an established estate or than a stela /5/ in a temple.

Is there anyone here like Hardedef?[4] Is there another like Imhotep?[5] No one has appeared among our kin like Nefry[6] and Khety,[7] that chief one of them. I recall to you (*lit.* I cause you to know) the name of Ptahemdjehuty[8] and of Khakheper[re]seneb.[9] Is there another like Ptahhotep[10] or similary Kaires?[11]

Those learned men who foretold what was to come. That which came forth from their mouths happened, [for] it is found as a pronouncement written in their (*lit.* his) books. The children of other people are given to them to be heirs like their own children. Though their magic [in practice] was concealed /10/ from the whole world, it can be read in an instruction-book. Though they are gone and their names forgotten, yet writing causes them to be remembered.

Notes to Document I.7

1. J. R. Harris, *Lexicographical Studies in Ancient Egyptian Minerals* (Berlin, 1961), pp. 50-62.

2. *Ibid.*, pp. 59-60. Though *biꜣ n pt* may originally have been meteoric iron (i.e. "iron of heaven"), it became the general expression for iron however obtained.

3. That is to say, the mortuary services established in the will of the deceased had by this time petered out and the attendants ("the *ka*-servants") were no longer taking care of the deserted tomb, an inevitable eventuality sooner or later.

4. We have already seen that Hardedef was the son of Cheops who recounted the tale of the magician Djedi (see above, Document I.6). He was also considered to be the author of an instruction-book, of which a fragment is extant (see the English translation in Simpson, ed., *The Literature of Ancient Egypt*, p. 340).

5. Imhotep, the vizier and architect of King Djoser, was responsible for building the mortuary complex of the Step Pyramid at Saqqara. He was not only considered as a traditional sage and patron of scribes during the course of Egyptian history (see M. Weber, *Beiträge zur Kenntnis des Schrift- und Buchwesens der alten Ägypter* [Cologne, 1969], p. 43), but in the later period was deified and became a god of healing like Asclepius (see J. B. Hurry, *Imhotep, the Vizier and Physician of King Zoser and afterwards the Egyptian God of Medicine* [Oxford, 1926], passim, and particularly pp. 27-28 and 29-73).

6. Nefry is written here as *Nfrii*. But Gardiner

thinks that the first *i* may well be *t*. If so, perhaps the Neferti who results is the author of a well-known book, the *Prophecies of Neferti*, an account of future events supposedly told to King Sneferu but undoubtedly written in the reign of Amenemhet I (1991-1962 B.C.) since it is propaganda for that monarch. For a translation of the *Prophecies*, see Lichtheim, *Ancient Egyptian Literature*, Vol. 1, pp. 139-45.

7. Khety is read by Gardiner as Akhthoy, whom he believes to be not only the author of the *The Satire of the Trades* (Document I.8 below) but probably as well the author of *The Instruction-book of Amenemes I for his Son Sesostris I*, since a later section of Beatty Papyrus IV mentions an Akhthoy who is said to be the author of this instruction-book (see *Hieratic Papyri....Third Series*, Vol. 1, p. 43).

8. Ptahemdjehuty is not known. Gardiner suggests the possibility that it was an error for Djedihuty, mentioned in another papyrus, apparently as an ancient sage (*ibid.*, p. 40).

9. Khakheperreseneb was, according to Gardiner (*ibid.*), "the native poetaster who, on a writing-board in the British Museum, bemoans the difficulty of finding new things to say".

10. This is apparently the author of the so-called *Instruction-book of Ptahhotep*, a work of which the oldest copy (the Papyrus Prisse) dates from the Middle Kingdom. It has been much studied and often translated (see Lichtheim, *Ancient Egyptian Literature*, Vol. 1, pp. 61-80).

11. Kaires was a sage whose work is not extant. See Fig. I.62, where he is pictured along with other famous men.

Document I.8: Introduction

The Satire of the Trades

The tone of praise for scribal life in this document is more worldly than that found in the preceding document. The setting for this instruction-book, which was composed by one Dua-Khety (*or perhaps*, Khety, son of Duauf), is a boat plying upstream to the Royal Residence, where the author's son Pepy is to be enrolled in the writing-school. Composed in the Middle Kingdom, the document has two purposes: (1) to "sell" the scribal profession to the young Pepy (and presumably to all of the readers of the work) by comparing it very favorably with other callings, and (2) to offer the boy the usual advice concerning manners and conduct found in many instruction-books. It is quite evident that the father and son are not themselves of the "upper class", which in the Old Kingdom probably supplied the overwhelming majority of students to scribal schools, for otherwise the father would scarcely have compared the scribal profession with so many menial professions like those of washermen, reed-cutters, cobblers, and the like. Furthermore, in the very first paragraph the author seems to be stressing that many of the students at the schools were the sons of the elite officials of the Residence. A similar remark was made (no doubt by another father to his son) later in Ramesside times:[1]

I have placed you at school (*tꜣ-ꜥt sbꜣ*)

> along with the children of dignitaries
> (*or* nobles; *srw*) in order to instruct
> you concerning this surpassing office
> (i.e. scribedom).

He goes on to give the usual advice as to how to get ahead:

> See! I shall tell you the way of the
> scribe in his [saying] "Early to your
> place! Write in front of your
> associates! Tidy (*lit.* put your hand
> on) your clothes and attend to your
> sandals!" You bring your papyrus
> roll daily with good intention. Do
> not be idle. They [say: recite
> diligently the tables like?] "Three
> plus three" .... You will make your
> calculations quickly .... Write with
> your hand, recite with your mouth,
> and take advice. Be not weary, pass
> no day in idleness, or woe to your
> limbs (i.e. expect a beating). Fall in
> with the ways of your instructor and
> obey his teachings. Be a scribe!

Many other Ramesside documents offer similar advice, and still others indicate to us that often the advice was not taken, as the students were reported to be inattentive or spending their time carousing.[2] The same Ramesside documents also follow *The Satire of the Trades* in comparing the scribal calling favorably to other trades.[3]

Texts and Studies

For the text of *The Satire of the Trades*, see

especially H. Brunner, *Die Lehre des Cheti, Sohnes des Duauf (Ägyptologische Forschungen*, 13) (Glückstadt and Hamburg, 1944), and W. Helck, *Die Lehre des Dwᵓ-Ḥtjj* (Wiesbaden, 1970). Both have extensive notes and German translations.

The three most recent English translations are worth perusing: those of J. Wilson in J. B. Pritchard, *Ancient Near Eastern Texts Relating to the Old Testament* (Princeton, 1950), pp. 432-34; W. K. Simpson, in *The Literature of Ancient Egypt* (New Haven and London, 1973), pp. 329-36; and M. Lichtheim, *Ancient Egyptian Literature*, Vol. 1 (Berkeley/Los Angeles/London, 1975), pp. 184-92. Lichtheim cites earlier translations and several key works of commentary.

English Translation

The two German and three English translations mentioned above have all proved useful to me in making my own translation. On the whole I have followed Helck's text closely. The use of italics at the beginning of some of the paragraphs represents rubricated phrases.

Notes to the Introduction of Document I.8

1. Gardiner, *Late-Egyptian Miscellanies*, pp. 68-69. Cf. Caminos, *Late-Egyptian Miscellanies*, pp. 262-63.

2. Gardiner, *ibid.*, pp. 3-4, 23-24, 36-37, 47-48, 59-60, 64-65, 65-66, 83, 101, 102, 106-07; Caminos, *ibid.*, pp. 13, 83, 131-32, 182, 231-32, 247, 250-51, 315-16,

377-78, 381-82, 395-96.

      3.   Gardiner, *ibid.*, pp. 44-45, 60-61, 64-65, 83, 84-85, 103-04, 107-09; Caminos, *ibid.*, pp. 169, 235, 247, 315-16, 317-18, 384-85, 400-02.

Document I.8

The Satire of the Trades

Beginning of the Instruction-book which a man from Sile named Dua-Khety[1] made for his son named Pepy when he was traveling upstream to the Residence to put him into writing-school *(pr sbⁱ nt sśw)* among the sons of the officials belonging to (*or* of the elite of) the Residence.[2]

Then he said to him: I have seen beatings [applied to laborers]. Set your heart on books. I have observed those who were taken off for [forced] labors (i.e. the corvée). There is nothing better than books. It (i.e. the profession of writing) is like a boat in water (i.e., it keeps one safe from the dangerous or uncomfortable depths ever present in other professions).

Read then the end of the *Kemyt*-Book[3] where you will find this sentence: "As for a scribe in any position in the Residence City, he will not be poor in it (i.e., he will be well-off in such a position)." For he fulfills the need of another man and [hence] he will not go forth [dis]contented (i.e. unrewarded).[4] I do not see [another] office of which this statement could be made.

I shall cause you to love writing more than your mother. I shall cause its beauty to come before you. For it is the greatest of all professions. There is none like it in this land. When he (i.e. the young scribe) grows sturdy but still is a child he is hailed [with respect] and sent to perform missions. On returning, he is begowned [like an adult].

I have never seen a sculptor [acting as] an envoy, nor a goldsmith sent [on a mission]. But I have seen a smith at work before his furnace door, his fingers like [the claws of] crocodiles. He stinks more than fish roe....

{Then follows a description of sixteen trades with their difficulties and hardships described, presumably for a comparison with the less physically demanding profession of the scribe: carpenter, jewel-maker, barber, reed-cutter, potter, mason, gardener, farmhand, weaver, arrow-maker, courier, furnace-tender, cobbler, washerman, bird-catcher, and fisherman.}

Behold, there is no profession free of a boss except [that of] a scribe; he is the boss.

If you know writing, it will be better for you than [it is] in the professions which I have set out before you.... Behold what I do [for you] when traveling upstream to the Residence. Behold I do it for the love of you. Advantageous to you is a day in school (*lit.* House of Instruction). Its works last forever [like?] a mountain....

Let me speak to you of other matters which you should know...[5]

Behold I have set you on the way of god. The fortune (i.e. the Goddess Renenet)[6] of a scribe is on his shoulder from the day of his birth. [By this] he reaches the Halls of Magistrates, which the people have made. Behold there is no scribe who lacks food and provisions from the King's Palace, L.P.H. The destiny (*lit.* The Goddess Meskhenet)[7] allotted to the scribe promotes him before the Magistrates. Praise god for your father and your mother who set you on the way of life. This is what I set before you and the children of your children.

It (i.e. this work) has come to an end in peace (i.e., it has come to a happy conclusion).

Notes to Document I.8

1. Note that Brunner's earlier text of this work used the reading of three manuscripts to give the names as Khety, son of Duauf, while Helck in his later edition followed Seibert in adopting the reading given in two manuscripts: Dua-Khety.

2. See Weber, *Beiträge zur Kenntnis des Schrift- und Buchwesens*, pp. 92-93, where he notes that the relative exclusiveness of the schools in the Old Kingdom began to break down by the time of this document and seems to have disappeared in the New Kingdom. It seems to have returned in the Late Period, when a reformer of the House of Life notes he has staffed the Houses with "persons of rank, not a poor man's son among them" (see the text above note 31 in Chapter One). Cf. also the discussion given by Weber.

3. The *Kemyt*-Book was an apparently popular instruction-book, of which many fragments remain (G. Posener, *Ostraca hiératiques littéraires de Deir el Medineh*, Vol. 2 [Cairo, 1951], Plates 1-25). T. G. H. James believes that the *Kemyt's* simple, prosaic text, written in vertical columns, played a role in the elementary teaching of scribes (*Pharaoh's People*, pp. 148-50).

4. I have given a translation which embraces the meaning suggested for this passage by Lichtheim in her translation but not her exact words.

5. The passages that follow, and which I have

omitted, include the kind of wisdom and advice found in many of the instruction-books, e.g. be respectful, cautious, dignified, accurate in delivering messages, truthful in what you say about your mother, restrained in the quantity of bread and beer you ingest, and so on.

6. Renenet ("She-Who-Nourishes") is a goddess personifying fate and fortune. See Hornung, *Conceptions of God in Ancient Egypt*, p. 281. The statement in our document means that a scribe's fate is fixed at the time of his birth.

7. Meskhenet is another goddess personifying destiny. She was one of the goddesses acting as a midwife at the birth of the three kings in the last story told in the Westcar Papyrus (see above, Document I.6: Introduction). In the birth-scenes at Hatshepsut's temple at Deir el-Bahri, Meskhenet makes a speech, ordaining many good things for Hatshepsut (see Sethe, *Urkunden* IV, p. 227).

Document I.9: Introduction

The Onomasticon of Amenope

In my general account of the House of Life in Chapter One above, I mentioned the encyclopedic name-list probably prepared at the very end of the twentieth dynasty by one Amenope, son of Amenope, "scribe of the sacred books in the House of Life". Following its editor, I have characterized it as a list of things that exist rather than as a dictionary or collection of glosses. The author himself makes this abundantly clear at the beginning of the treatise when he says that this instruction-book is for "learning all things that exist: what Ptah has created and what Thoth copied down". This was of course not a unique composition. Gardiner discussed and edited two further onomastica in his edition of Amenope's work. The first is the Ramesseum Onomasticon, which is a list probably from the thirteenth dynasty (i.e. ca. 1750 B.C.).[1] It contains the remains of a list that ought to have included 343 items but in fact has only 323 items, among which are names of liquids, oils, plants, birds, fish, quadrupeds, southern fortresses, cities of Upper Egypt (a list of twenty-nine that extends from Elephantine in the south to a little beyond Akhmin in the north), loaves or cakes, cereals, a butcher's list of parts of oxen (as sacrificial joints or parts), condiments, and fruits (or products of trees). To this list is added a supplement that contains varieties of cattle (the distinction being based on color). The second onomasticon is a very short piece found on a University College Writing-Board, which is to be dated about the twenty-first or twenty-second dynasty.[2] On the recto read:

> (1) I acquaint you with the occupations that are in a temple: guardian of the Treasury, guardian of the Granary, (2) maker of *bit*-loaves, baker, maker of *K'w*-cakes(?), baker of *s^c y(t)*-cakes, butcher, confectioner, (3) maker of *psn*-loaves, shaper of incense, basket-weaver (?), dyer of red cloth, (4) maker of rush mats (?), bouquet-maker, gardener, bearer of floral offerings....

On the verso we find:

> (1) I acquaint you with the work of a (wood-)carver *(gnwty?)* and initiate you into what he makes: chapel, (2) divine bark, carrying stands for gods, sanctuary, ..., doors, poles, poles (3) for uraei, statue in its chapel, beds, palanquins, footstools (4) (for the) feet, boxes,..., coffers, chests, receptacles, coffins (*the rest is lost*).

Much the earliest list of this kind is found on a writing-board of the fifth or sixth dynasty which contains (in those parts which are readable) the names of kings, gods, and cities.[3]

In fact, the listing of words, signs, and names for other purposes was widespread. For example, there are more than twelve lists of the parts of human bodies.[4]

Notice also the list of parts of ships found in the *Coffin Texts* (not included in my extracts in Doc. II.2; but see Spell 398 in the edition and translation mentioned there). Of interest likewise is the list of seventy-five names of Re appearing in the *Litany of Re* (see Document II.5), a list which reflects the varied mythological accounts surrounding the creator god, his passage at night, his activities, and his attributes.

I should stress that often lists of names are given with the express purpose of bringing into existence that which is named. Surely this is the purpose of the lists of offerings that appear on the walls of the tombs, the names of which were to be pronounced in the "invocation-offerings" there specified, and particularly the food offerings arranged before the deceased as he sits at an offering table in the stela above the false door in his tomb. We shall have occasion in Chapter Two to see how important for magical spells and acts (both those connected with funerary practices and those involved in ordinary human relations) was the pronouncing or writing of names.

This ancient view, that to name is to bring forth what is named, is reflected in the remarkable Memphite Theology (see Document II.9), where it is said that the gods of the Ennead and indeed everything were conceived in the heart of Ptah and came into being by spoken command. Even if this document was the product of the scribes of King Shabaka in the twenty-fifth dynasty rather than being of great antiquity as used to be assumed,[5] a similar kind of creation by the word appears in numerous other documents from all periods, e.g. in the *Book of the Divine Cow*, in the cosmogonic part of which Re brings into being by spoken command: Nut as the sky (i.e. the raising of Nut to become the sky), the celebrated Field of Offering and Field of Rushes, the Heh gods, which support the sky, and the stars themselves (see Document II.6).

Though the Onomasticon of Amenope (and works like it) are not dictionaries since they contain no definitions, there are evidences of some quite late dictionary-like lists of words. For example, we note the fragments of a hieroglyphic dictionary in the Papyrus Carlsberg No. VII (copied in the first century A.D. and probably composed not too long before that time).

On its first page it includes words that begin with the hieroglyph ☐ *(=h)*, of which I give the first two along with the introduction:[6]

> Explanation of the employment of signs, explanation of the difficulties. Disclosure of the things hidden, explanation of the obscure passages.... by their noble protection.

> Explanation of what emanated from the Gods, the noble ancestors, the sacred traditions from the nomes of Upper and Lower Egypt ....(found on) a leather-roll in the temple of Osiris, the first of the Westerners, the great God, Lord of Abydos, in....

> I.e. An Ibis *(hbw)*. I.e. "A heart descends", in accordance with what Re said about it: "It descended from the body." I.e. A Ba descends. I.e.... Everything is perceived[7] through him. It is a *hjn*, 5....? It is the

ancient one, who emerged from the box. It is the palette.... Everything in this land is perceived through the treatises and the utensils, which came into existence through him. It is his finger .... Thot[h], the chief of the marvels in the house of clothing, who regulates the entire land, the ...? comes into existence through him.

I.e. The Day (hrw), I.e. Re in his rise in the morning, by means of whom everything is perceived. I.e. the Ennead.... The eye is called the Ennead. The sun-disk came into existence from the right eye of Re. It is the Vulture-Goddess who binds(?) the bows, and who binds.... It is Ta-tenen, the male one. The Uraeus came into existence from the right eye of Re; it is the crown of Lower Egypt who unites it with her body.

It is evident in this treatise that we have no ordinary dictionary but rather a mythological exposition, which shows in part how the mythological associations originated in punning or the use of words that sounded like those being discussed. Returning to the Onomasticon of Amenope, we can observe that it is not complete in its longest copy (G) but breaks off at the 610th item. It seems probable, in view of the additional items in the copy known as the Ramesseum fragments and in other such lists, that plants, fish, birds, and quadrupeds were included or were projected for inclusion.

One section of our document is particularly noteworthy, the list of towns in Egypt. The principal copy contains 80 place-names, of which I have included only a few of the more important to illustrate this category. This is surely one of the most useful extant lists of Egyptian towns and the editor in his commentary has made splendid use of it, integrating it into other lists of towns, such as that given in the Ramesseum Onomasticon mentioned above.[8] The table of towns which Gardiner included in his last volume, the one devoted to plates, is of great interest to the student of ancient Egyptian geography.

Text and Study

A. H. Gardiner, *Ancient Egyptian Onomastica*, 3 vols. (Oxford, 1947).

English Translation

I have followed Gardiner's translations of the names except in a few instances. Needless to say, I have omitted the long commentary by which Gardiner so expertly explicates the text, but the interested reader will certainly want to examine some of the long essays included therein. I have occasionally modified the modern spellings of the transliterations of the Arabic names of towns included in Section V (for example, I use Dendera instead of Denderah in No. 313). The item numbers given in brackets are those assigned by the editor, for there are no such numbers in the Egyptian text. When an "A" appears with a number this indicates that the item is not in the principal manuscript (G) but comes from one or more of the copies.

The reader interested in the town-list of Section V will want to consult the very useful sketch-maps included by the editor but which I have omitted here since I used so few of the town names. In this connection the reader will also find of use the maps in Baedecker's guide, *Egypt and the Sûdân*, as well as those in J. Baines and J. Málek, *Atlas of Ancient Egypt*, (Oxford, 1980).

I have included some of Gardiner's comments in the notes to the document without page references but always within quotation marks. The reader will readily locate them in Gardiner's text by turning to the item numbers to which they apply. If a quotation appears in a footnote without attribution, the source is Gardiner.

Notes to the Introduction of Document I.9

1. Gardiner, *Ancient Egyptian Onomastica*, Vol. 1, pp. 6-24.

2. *Ibid.*, pp. 64-68.

3. G. A. Reisner, "A Scribe's Tablet found by the Hearst Expedition at Giza," *ZÄS*, Vol. 48 (1911), pp. 113-14. Reisner dates it as fifth dynasty, but Grapow and Westendorf date it "6. Dynastie" in the *Handbuch der Orientalistik*, Abt. 1, Vol. 1, Part 2 (Leiden, 1970), p. 221.

4. H. Grapow, *Grundriss der Medizin der alten Ägypter*, Vol. 1: *Anatomie und Physiologie* (Berlin, 1934), p. 12, and G. Lefebvre, *Tableau des parties du corps humain mentionnées par les égyptiens*

(*Supplément aux Annales du Service des Antiquités*, 17) (Cairo, 1952).

    5.   See F. Junge, "Zur Fehldatierung des sog. Denkmals memphitischer Theologie oder Der Beitrag der ägyptischen Theologie zur Geistesgeschichte der Spätzeit," *MDAIK*, Vol. 29 (1973), pp. 195-204 and S. Morenz, *Egyptian Religion* (Ithaca, 1973), pp. 154-55. See the remarks reflecting the view that "das Hervorbringen ist die Konsequenz des Sprechens" in Weber, *Beiträge zur Kenntnis des Schrift- und Buchwesens der alten Ägypter*, pp. 68-70, and the rich collection of passages on the creating word from all periods of Egyptian history in J. Zandee, "Das Schöperwort im alten Ägypten," *Verbum: Essays on Some Aspects of the Religious Function of Words dedicated to Dr. H. W. Obbink* (*Rheno-Traiectina*, Vol. 6, 1964), pp. 33-66. I find reasonable the view of H. A. Schlögel, *Der Gott Tatenen* (Freiburg, Switz., and Göttingen, 1980), pp. 110-17, that the Memphite Theology had a more probable origin in the nineteenth dynasty. This concept of the creative word is treated quite thoroughly in Chapter Two below.

    6.   E. Iversen, *Papyrus Carlsberg Nr. VII: Fragments of a Hieroglyphic Dictionary* (Copenhagen, 1958), pp. 13-21. Iversen's general comment (p. 7) on the nature of this dictionary is of interest: "If, however, we consider the commentary, it will be seen that the first page contains 11 separate catch-words [introduced in the right margin by hieroglyphic signs, as is evident in Fig. I.63], the reading of which are given respectively *h'bw hrw hnm.t h h'b hmhm hn hmj hnw*. As it is obvious that these words have nothing in common except the fact that they all begin with *h* [*hnm.t* being no exception owing to the assimilation of *ḥ* by *h*], it

would be most tempting to conclude that alphabetic considerations had determined their arrangement. Unfortunately, the condition of the second page does not permit us to draw any definite conclusions as to the arrangement of the signs there, as all the original readings of the catch-words except *kʲ* have been lost, and the phonetic value of several signs is ambiguous and doubtful. The sequence of the first readable signs of the page does not speak for an alphabetic arrangement but the four last ones...might probably be read as *kʲ kʲrj km*, and *km*." For the much longer sign-list (arranged in no discernible order) edited by F. Ll. Griffith, see *Two Hieroglyphic Papyri from Tanis* (London, 1889).

7. I have usually translated *rḫ* as "known" in this volume, but I let stand Iversen's "perceived" both times that he uses it.

8. See the neat summary of other kinds of geographic lists and inventories in Sauneron, *Les prêtres de l'ancienne Égypte*, pp. 144-50.

Document I.9

The Onomasticon of Amenope

I. *Introductory Heading*

Beginning of the teaching for clearing the mind, for instruction of the ignorant, and for learning all things that exist: what Ptah created, what Thoth copied down, heaven with its designs, earth and what is in it, what the mountains belch forth, what is watered by the flood, all things upon which Re has shone, all that is grown on the back of earth, excogitated by the scribe of the sacred books in the House of Life, Amenope, son of Amenope. He said:

II. *Sky, Water, Earth*

[1] sky [2] sun [3] moon [4] star [5] Orion [6] the Foreleg (i.e. the Great Bear or Big Dipper) [7] Cynocephalus Ape (a constellation) [8] the Strong one (a constellation) [9] the Hippopotamus (also a constellation) [10] storm-cloud, storm [11] tempest [12] dawn [13] darkness [14] sun, light *(šw)* [15] shade, shadow [16] sunlight *(kḫ)* [17] rays of the sun *(sty itn)* [17A] storm-cloud *(šnꜥ)* [18] dew [19] *ꜣwdt* (? meaning unknown) [20] snow (?) [21] *srmt* (meaning unknown, but perhaps something to do with a body of water) [22] primeval waters, Nile *(nw)* [23] flood (i.e. Nile) *(mtr)* [24] river, Nile *(itrw)* [25] sea *(ym)* [26] wave [27] swampy lake *(ḥnw)* [28] pond, lake *(š)* [29] well (in desert; *ẖnm(t)*) [30] *ẖnm(t)* (basin?) [31] *ẖnini* (some

sort of irrigation basin or canal?) [32] waters *(nwy(t))*
[33] pool *(brkt)* [34] frontier, front *or* southern part
[35] back *or* northern part [36] well *(šdt)* [37] cleft?
(for water) [38] *ḏ¹bb* (meaning unknown) [39] river
bank (?) [40] *šgr* (meaning unknown) [41] watercourse
[42] place for drawing water [43] runnel [44] flood
(wḏnw) [45] *wg¹(t)* (meaning unknown) [46] current
[47] water-hole [48] *db(w)* (meaning uncertain) [49]
shores, river-banks [50] *rn(w?)* (meaning doubtful) [51]
standing water? *(isḫm* or *sḫm(t))* [52] island [53] fresh
land *(nḫb)* [1] [54] tired land *(tni)* [55] (normal)
agricultural land *(lit.* "high land"; *k¹yt)* [56] mud, clay,
mud-flat [57] low-lying shoal *(lit.* "it comes in the
return of the year"?; *iw.f-n*(or *-m)-nri)* [58] woodland
[59] sand [60] new land [61] *pᶜt* (some kind of land)
[62] riparian land *(idb)*.

III. *Persons, Court, Offices, Occupations*
[63] god *(nṯr)* [64] goddess *(nṯrt)* [65] (male)
spirit *(¹ḫ)* (*or* Blessed Dead) [66] female spirit *(¹ḫt)* [67]
King *(nsw)* [2] [68] Queen *(nsyt)* [69] King's Wife
(ḥmt-nsw) [70] King's Mother *(mwt-nsw)* [71] King's
Child *(msw-nsw)* [72] Crown-prince *(r-pᶜt* or *iry-pᶜt)*
[73] Vizier [74] Sole Friend (the commonest title of
courtiers) [75] Eldest King's Son [76], [77] Overseer of
the Hosts, (even) the Greatest Ones of the Courtiers (*or
possibly two separate titles:* Overseer of the Army *and*
the Great Ones of the Courtiers)[3] [78] Dispatch-writer
of Horus, Mighty Bull (i.e. of the King) *(sš šᶜt n Ḥr k¹
nḫt)* [79] Chief of Department of the Good God (i.e. of
the King) [80] First King's Herald of His Majesty [81]
Fan-bearer on the Right of the King [82] performing
excellent works for the Lord of the Two Lands (an
epithet not a title) [83] Superintendent of the

Chamberlains of the Victorious King [84] Chief of Bureau (Diwân) of his Lord [85] Royal Scribe within the Palace [86] the Vizier and Overseer of the Cities of Egypt[4] [87] General (*lit.* Overseer of a [Military] Expedition) [88] Scribe of the Infantry [89] Lieutenant Commander of the Army [90] Overseer of the Treasury of Silver and Gold [91] King's Envoy to Every Foreign Land [92] Overseer of cattle [93] Overseer of the Palace (lit. "King's House") [94] Overseer of Horses[5] [95] Lieutenant Commander of Chariotry [96] Charioteer [97] Chariot-warrior [98] Standard-bearer[6] [99] Chief (or Chiefs?) of the Scribe(s) Who Place(s) Offerings before All the Gods [100] Overseer (or Overseers?) of the Prophets of Upper and Lower Egypt [101] The Mayors of the Towns and Villages [102] the Great Controllers of His Majesty [103] in command of the secrets of the palace (less a title than an epithet) [104] at the head of the entire land (an epithet of viziers) [105] Deputy of the Fortress-Commander of the Sea [106] Intendant(s) of Foreign Lands of Syria and Cush [107] Scribe of Distribution [108] Scribe of Assemblage [109] Overseer of the River-mouths of the Hinterland [110] Chief taxing-master of the Entire Land [111] Major-domo of the ruler of Egypt [112] Chief of Scribes of the Mat (?) of the Great Court [113] Chief of the Record-keepers of the House of the Sea [114] Royal Scribe and Lector-priest as? (*or* of?) Horus[7] [115] Scribe of the House of Life, skilled in his profession [116] Lector-priest of the Royal Couch [117] First Prophet of Amun in Thebes [118] Greatest of Seers of Re-Atum (high priest of Heliopolis) [119] Greatest of Artificers of Him who is South of His Wall (i.e. of Ptah, title of the high priest of Memphis) [120] *Setem*-priest of Kindly of Face (i.e. Ptah, a second title of the high-priest of

Memphis) [121] Overseer of the Granaries of Upper and Lower Egypt [122] King's Butler in the Palace [123] Chamberlain of the Palace [124] Great Steward of the Lord of the Two Lands [125] Scribe Who Places Offerings before All the Gods[8] [126] Prophets (*lit.* "god's servants") [127] God's Fathers [128] (ordinary) Priests (*lit.* "pure or clean ones"; *w^cbw*) [129] Lector-priests [130] Temple scribe [131] Scribe of the God's Book (*or* Books) [132] Porter (*lit.* [Builder's] "Workman") [133] Elder of the Portal (*or* Forecourt)[9] [134] Hour-watcher, Astronomer [135] Bringer of Offerings [136] Bearer of the Wine-jar Stand [137] Maker of Runners(?) (*lit.* "a way of rushes?") [138] Milker (of cows) [139] Butcher [140] Preparer (*lit.* "turner over") of Tripe [141] Baker of *š^cy(t)*-cakes [142] Baker of *rḫs*-cakes [143] Maker of *bit*-loaves [144] Maker of *psn*-loaves [145] Brewer [146] Baker (*rtḫty*) [147] Shaper of Incense [148] Cooker (*or* baker) of *ṯrr* [149] Maker of Sweetmeats [150] Confectioner (*bnrty*) [151] Maker of Baskets of Date-cakes (?) [152] Herdsman [153] Milkman (*lit.* "carrier of milk-jug(s)") [154] Carpenter (*ḥmw*) [155] Sculptor (*gnwty?*) [156] Carpenter, shipbuilder (*mdḥw*) [157] (Portrait) sculptor (*lit.* "He who makes to live"; *s^cnḫ(y)*) [158] Coppersmith [159] Goldsmith [160] Worker in Precious Stones [161] Maker of Faience [162] Purveyor of Precious Stones [163] King's Sandal-maker [164] Corselet-maker [165] Chariot-maker [166] Arrow-maker [167] Bow-maker [168] Maker of Necklaces (?) [169] Stone-worker (*lit.* "Wielder of the *bš'* -tool") [170] Bead-maker [171] Rope-maker [172] Fan-maker [173] Barber (*sš'*) [174] Barber (*ḫ^cḳw*) [175] Hairdresser (? *nbdy*) [176] Maker of....(?) (*irw ṯkt*) [177] Miner[10] [178] Quarryman (*lit.* "necropolis-man") [179] Demolisher [180] Draughtsman (*lit.* "the scribe of contours"; *sš ḳdy*) [181] Sculptor (in relief; *lit.* "wielder of the chisel"; *ṯ'y mḏ'ṯ*) [182] Gypsum-worker [183] Bringer of Stone [184] Potter [185] Potter of *hin*-measures (i.e. of the measures containing about 1/2 litre) [186] Builder of Walls [187] Patcher(?) of Stonework [188] Medjay (here, desert police or troops) [189] hunter [190] diver [190A] Harpooners (of hippopotami) [191] Master of the ....(?) (*ḥry mi'ṯ*) [192] Master of the Cowhouse [193] Doorkeeper [194] Guardian [195], [196] Measurer; land-administrator (?) (*ḥ'y n(?) rmnyw*) [11] [197] (Military) Herald[12] [198] an officer of some sort (*sk(i)*) [199] Transport Officer (?) (*mškb*) [200] Policeman, Guard (*sᶜšᴵ*) [201] Bearer of Weapons [202] *sw(??)* (meaning unknown) [203] Groom [204] Guardian of Crops [205] Sailor[13] [206] Pilot (*lit.* "he who is in the front") [207] Steersman [208] Bird-catcher [209] Fisherman [210] Trader, Merchant [211] Buyer(?) (*mḥr*) [212] Seller [213] Female (?) Singer[14] [214], [215] Male and Female Musicians [216] *ᶜwy* (?) (reading and meaning unknown) [217] Libyan Dancers [218] *kmr* (dancers of some kind) [219] *ḏpk* (or *ḏpg*) (dancers of some kind) [219A] Dancers (*ksks*) [220] Leader [221] Servants (*lit.* "hearers") [222] Porter of....(*kᴵ(w))ty....* [223] Clothes-porter [224] Vineyard-keeper[15] [225] Gardener [226] Tenant Farmers, Agricultural laborers (*iḥty (ᶜḫwty)*) [227] Vegetable-dealer(?) [228] Herdsman (repeated from No. 152 above) [229] Keeper of Cattle-pens.

IV. *Classes, Tribes, and Types of Human Beings*
[230] men (*or* man *rmṯ*) [231] patricians (*or* mankind; *pᶜṯ*) [232] plebeians (*or* mankind; *rḫyṯ*)[16] [233] sun-folk (*hmm(t)*) [234], [235] commander(s) of

troops (probably a single category but could be two)
[236] infantry [237] chariotry [238] _Tmḥw_-people (i.e.
Libyans)[17] [239] _Ṯḥnw_-people (Libyan people to the
west of the N. W. corner of the Delta?) [240]
Meswesh (Libyan of some sort) [241] Libu (Libyan, i.e.
North African land and people).... [243] Keshkesh (one
of the peoples of the Hittite confederacy at Kadesh....
[245] Khatti (the land of the Hittites) .... [250]
Carchemish (the well-known city on the Upper
Euphrates).... [252] Kadesh (on the Orontes).... [257]
Byblus.... [260] Nahrin (the country near and mainly to
the east of the Euphrates).... [262] Ascalon (on the coast
N. of Gaza).... [264] Gaza [265] Isr (Assyria or, perhaps
less probably, Asher) .... [268] Sherden (a Mediterranean
people, whose identity is much discussed) [269] Tjekker
(one of the Sea Peoples who attacked Syria and Egypt
in the reign of Ramesses III) [270] Pelesti, Philistines....
[276] _Ḥ'(i)w-nbw(t)_ (Mediterranean islanders or perhaps
occasionally the islands themselves).... [286] Sangar
(quite probably Babylonia).... [294] Begrek (an unknown
locality) [295] man _(s)_ [18] [296] stripling _(mnḥ)_ [297]
old man _(i'w)_ [298] woman _(st)_ [299] young woman
(_lit._ "beautiful one"; _nfr(t)_) [300] various person(s) _(tp
šbn)_ [301] boy [302] child (from infancy upwards;
nḥn) [303] lad _(rnn)_ [304] maiden _(rnnt)_ [305] weaver
[306] subordinate, assistant [307] _ḏ'ty_ (meaning
unknown) [308] Overseer of Carpenters (_var._ Carpenter
of the Dockyard) [309] sailor (_or_ ship's hand) [309A]
boat-builder [310] a goer forth _(pry)_ (meaning not
clear) [311] (male) slaves _(ḥmy(w))_ [312] female slaves
(ḥmy(wt)).

V. _The Towns of Egypt_
[313] town _(dmi)_ [19] [314] Senmet (i.e. the island

of Bigga or Biga opposite Philae) [315] Elephantine (the
island opposite Aswan) [316] Ombi (the modern Kom
Ombo) [317] Silsilis [318], [319] Edfu [320]
Hieraconpolis (here *Mḥn*, but the older form is *Nḥn*, i.e.
Nekhen) [321] Nekheb (the modern El-Kab on the right
bank opposite Nekhen).... [323] Inyet (modern Esna)....
[332], [333] Hermonthis.... [335], [336] Waset, the City of
Amun, mistress of every town (Thebes, the main town
being between the modern Luxor and Karnak).... [340]
Coptus [341] Ombos (Nebet, included the modern town
of Naqada).... [343] Dendera.... [350] Abydos.... [353] This
(*Tn(i)*, old writing *Ṯni*) (exact site unknown) [354]
Akhmin.... [371] Asyut.... [374] Kos (*Ḳis*).... [377] Khmun
(Hermopolis, the modern Ashmunein).... [389]
Heracleopolis (*Nn-nsw*) (modern Ihnasya el-Medina)....
[392] Mi-wer (Moeris; Kom Medinet Ghurab).... [394]
Memphis (normal writing is *Mn-nfr* but here *Mnf*)[20]
[395] Troia (the Modern Tura, the site of the limestone
quarries).... [400] On of Re (Heliopolis; *Iwnw-R*ꜥ)....
[404] Mendes (*lit.* "House of the Soul" [earlier and better
'Ram'], lord of Djedet; *Pr-Bʾ-nb-Ḏdt*).... [410] Piramesses
(*lit.* "The House of Ramesses, Beloved of Amun,
L.P.H.")[21].... [412] Andjet (alternate name for Djedw, i.e.
Busiris).... [415] Buto (*lit.* "House of the goddess Edjo")....
[417] Tanis.... [419] The fortress of Tjel (*Ṯʾrw*)....

VI. *Buildings, Their Parts, and Types of Land* [22]
    [420] castle, pylon (*bḥn*) [23] [421] settlement,
village (*wḥt*) [422] house (*pr*) [423] room (later "house";
ꜥt) [424] living room [425] side room (?) [426] lower
part, basement [427] outer chamber [428] broad hall
(*wsḥt*) [24] [429] alcove, moon-shaped recess [430]
ergastulum, storehouse [431] colonnade(?) (*wḫʾy(t)*) [25]
[432] upper chamber [433] ceiling, roof (*ḥʾtyw*) [26]

[434] stairway   [435] tomb chapel, tomb *(ḥwt-kʾ)* [27]
[436] a hidden place   [437] magazine (for corn and the
like; *mḫr*)   [438] container (for corn; *šʾꜤ*)[28]   [439] garner
(šꜤyt)   [440] chest   [441] storehouse *(wḏʾ)*   [442]
window   [443] chink, crack   [444] surrounding wall
(sbty) [29]   [445] rampart (*or* bastion)[30]   [446] wall
(inb)[31]   [447] street   [448] ankh (meaning unknown)
[449] guard-house   [450] fortress   [451] prison *(kri)*
[452] corner, angle *(ḳnbt)*   [453] angle *(ḳꜤḥ)*   [454]
granary *(šnwt)* [cf. No. 439]   [455] Treasury (*lit.* "White
House")   [456] *miʾt* (probably a farm building)   [457]
cow-house   [458] vineyard   [459] a name whose
meaning and reading are unknown   [460] garden (*ḥsb*, a
late writing of *ḥsp*)   [461] avenue, promenade   [462]
roof *(tp-ḥwt)*   [463] ground-floor   [464] *rbn* (unknown)
[465] *...rt* (unidentified)   [465A] demolition (? or
rubble?)   [466] *mḫn* (exact meaning unknown)   [467]
hillock   [468] vein (? of rock; *ḏꜤt)*   [469] *srft* (perhaps a
kind of earth or building material)   [470] caves (?
mgrt?)   [471] (destroyed)   [472] ruin (? *sḫnn*)   [473]
flood *(nt)*.[32]

VII. *Agricultural Land, Cereals, and their Products*[33]
      [474] ploughland *(skʾ(t))*   [475] hillock (a
repetition of No. 467 above)   [476] mud-flat (repeated
from No. 56)   [477], [478] (lost words)   [479] *pḫʾ*
(meaning unknown)   [480] *Ꜥmd* (meaning unknown)
[481] *wḥm-tmḥ* (meaning unknown)   [482] vegetables
[483] cucumber-beds   [484] *Ꜥmt* (?) (unrecognizable
word) [485-7] (several words that cannot be divided)
[488] *ḥt* (doubtless for *ḥtiw*, "threshing-floor")   [489] *npt*
(?) (doubtful, perhaps to be taken with the next word)
[490] winnow   [491] *pis* (a cereal of some kind)   [492]
sꜤḳ (a cereal)   [493] (lost)[34]   [494] white emmer   [495]

black emmer [496] red emmer [497] ....emmer [498] *ḥrnt* emmer [499] orange-red(?) emmer [500] emmer of Ptah [501] (unknown) [502] (unknown) [503] country wheat (? *stwt? n sẖt*) [504] *bšì* (a cereal of some kind) [505] dates (*bnr* or *bnrt*?) [506] flour [507] *mᶜr* (probably grain from a plant *mᶜr*)....[35]

VIII. *Beverages*[36]

[556] beer (*ḥ(n)ḳt*) [557] *trk* (some kind of intoxicating drink) [558] *ḥmt* (apparently a kind of beer in the process of making).... [565] new wine, must [566] wine of Egypt (*irp n Ḳmt*) [567] wine of Khor (i.e. Syria) [568] wine of the Oases [569] wine of Andjet (i.e. Busiris) [570] wine of Hardai (i.e. Cynopolis) [571] wine of Amor [572] vinegar [573] *ipwr* (a wine of second-rate quality) [574] ᶜ*d* (? fat?, but this does not fit in with this series) [575] ᶜ*wìy(t)* (a fermented beverage of some sort, found in medical sources) [576] dregs (or lees; *gìš*) [577] *dnd* (meaning unknown) [578] *brbs* (meaning unknown).

IX. *Parts of an Ox and Kinds of Meat*[37]

[579] meat (*iwf* or *if*) [580] head (*d̠ìd̠ì*) [581] neck.... [584] breast [585] vertebra of the upper back [586] back [587] rib [588] thighs (or the like; *mìst*).... [590] foreleg [591] under part of foreleg (? *ḥry-ḥpš*?) [592] loins [593] *ḳnḳn* (a joint of some sort) [594] tail [595] *d̠nḥ* (perhaps the upper part of the foreleg).... [598] liver [599] lungs [600] spleen [601] heart (*ḥ̣ìty*) [602] intestines (*imy-ẖt*, apparently identical with *mẖtw*) [603] *rᶜm* (meaning unknown) [604] *sẖn* (? perhaps the pancreas or alternatively the mass of fat in which the kidneys are enveloped) [605] flank, side [606] *t̠rst* (an undetermined part of an ox) [607] *ḳbḥ*

(part of the leg or foot) [608] raw meat [609] cooked meat [610] spiced meat.

Notes to Document I.9

1. Gardiner comments on this and the next two kinds of land: "The three next kinds of land appear from P. Wilbour, Text B, to have formed a recognized administrative classification and to have been assessed (for what purpose is obscure) in the proportions 10: 7 1/2: 5 respectively."

2. I am capitalizing all the titles; Gardiner is somewhat inconsistent in his capitalization.

3. Gardiner finally settles for this title as equivalent to Generalissimo of the Army.

4. Here Gardiner writes: "At this point we pass from the grandees in the immediate entourage of the King to his military staff, but after only three essential members of this have been named (Nos. 87-9) five high administrative officials intervene (Nos. 90-4). Then the list of army officers is continued in Nos. 95-8. Nos. 107-8 deal with scribes concerned with the organization of the army, and later on there are some references to soldiers of lower rank (Nos. 197-8, 201, 202(?), 234-5) and to certain types of troop (Nos. 236-7)...."

5. Again Gardiner says: "The mention of horses recalled to Amenope's mind several other titles connected with chariots; of these all are exclusively military except No. 96, so that the way is paved for the 'standard bearer' of No. 98...."

6. "The next two entries, which name personages concerned with executing the duties owed by the King to the gods, were possibly meant to link up with Nos. 90-4, after which the military or semi-military titles Nos. 95-8 formed a digression."

7. Gardiner adds: "Nos. 114-20 refer to priestly

persons of relatively high rank, the lower ones being dealt with below in Nos. 125ff. In Nos. 117-20 we have the designations of the high priests of the three great cities of Thebes, Heliopolis and Memphis respectively."

8. At this point Gardiner says: "Priests and temple-employments follow, starting with the general terms, in the plural, for the higher grades and thence descending to menial occupations like those of confectioners and the like (Nos. 148-51).... The remaining employments from Nos. 152-229 may also have belonged to the temple service."

9. To the references given by Gardiner, add Spencer, *The Egyptian Temple*, p. 157.

10. Gardiner says here: "Nos. 177-87 enumerate quarry-workers, builders and the like."

11. Combined because of the presence of the *n*, so that it might be a measurer of an administrative domain instead of the two titles given. As usual, Gardiner's commentary is instructive.

12. Gardiner writes here: "Some degree of homogeneity is discernible in Nos. 197-204, all of them implying coercion of one kind or another. Nos. 197, 198, 201, 202 are military titles, of which the higher ones occurred earlier, Nos. 77, 87-9, 95-8."

13. This and the next two "deal with boatmen, after which Nos. 208-9 refer to callings in which a boat would naturally be used...."

14. "Nos. 213-9 are all concerned with music and dancing."

15. "Nos. 224-9, concluding the enumeration of human employments, deal exclusively with agricultural callings. A few more military terms occur near the beginning of the next section (Nos. 234-7) and two nautical occupations near the end of it (Nos. 308-9), the

the latter very much out of place."

16. This entry had these meanings at this time, but see Document I.1, n. 17. Gardiner's long commentary (pp. 98*-112*) on the meaning of Nos. 230-33 is of particular interest.

17. "The list of foreign peoples that follows (Nos. 238-94), insofar as the names are identifiable, mostly reflects the external relations of Egypt in the Ramesside period." I have given here some of the more important names that can be identified. The reader should consult Gardiner's rich commentary for these various names.

18. "The remaining words of this section (Nos. 295-312) refer to differences of age, sex and status among human beings, though Nos. 305, 307, 308, 309 do not fit well with this formulation; the last three, together with No. 309A, are all connected with boats; a sailor's life was perhaps regarded as a slavish occupation."

19. After this general word for town "follows a long enumeration of the towns of Upper Egypt, followed by a shorter one of those in Lower Egypt." This list is "arranged consecutively from South to North." Important as this list (Nos. 314-419) is for students of ancient Egyptian geography, I have included only some of the better known towns to illustrate this category of entities in Amenope's work.

20. This name has its origin in the name of the pyramid of Pepi I (or the settlement around it), i.e. *Mn-nfr-Ppii* (or more commonly *Mn-nfr-Mryr⁹*). As usual, Gardiner gives an interesting essay on this entry. He notes that this name for the city around modern Mit Rahina "occurs perhaps no earlier than Dyn. XVIII (cf. *Urk.* IV,3)...." (p. 123*).

21.   The following conclusion of Gardiner's (pp. 174*-75*) is no longer accepted: "My final verdict must, accordingly, be that the case for Tanis [being identical to Pi-Ramesse], though of considerable strength, is not yet strong enough to put Kantir entirely out of court; much confirmatory evidence in favour of Tanis is required to outweigh completely the separate mentions of Pi-Ramesse and Tanis in On[omasticon] Am[enope]; for Kantir, on the other hand, there is as yet little to be said except that the place possessed a palace of Ramesses II of some magnificence." This opinion is further modified in the succeeding postscript. Gardiner in an earlier article had identified Pi-Ramesse not only with Tanis but also with the Hyksos town of Avaris. Baines and Málek, *Atlas of Ancient Egypt*, distinguish the three cities, suggesting that Avaris may have been located at Tell el-Dabᶜa and Pi-Ramesse at Qantir. See also E. P. Uphill, *The Temple of Per Ramesses* (Warminster, Eng., 1984), pp. 1-3. The key to the recent rejection of the identification of Tanis with either Avaris or Pi-Ramesse is that none of the foundations of Tanis appears to be earlier than the twenty-first dynasty.

22.   "The new category that begins here shows both a continuity and a discontinuity with what precedes: we are still concerned with habitations of men, these passing into parts of buildings and types of land where human occupations were carried on; on the other hand specific localities distinguished by proper names are at an end, and we now embark on a series of general names."

23.   For a discussion of the feminine form (the normal word for a pylon), see Spencer, *The Egyptian Temple*, pp. 192-96. She notes that *bḫn* "was used, from

the New Kingdom onwards, for a large estate, and the two terms can be written in virtually identical ways."

24. *Ibid.*, pp. 71-80. Spencer concludes that both "hall" and "court" are sound translations and we need to know the context before we can decide which translation is appropriate.

25. *Ibid.*, pp. 243-47, discussing *wḫ⸍*.

26. *Ibid.*, pp. 155-61.

27. *Ibid.*, p. 23. It was used as a "funerary temple" as early as the first dynasty.

28. For our purposes Gardiner's comment is worth quoting: "in the Rhind mathematical papyrus either 'circular'.... or 'square'....; in *JEA* XII, 131 Gunn cites an example where the sense may be more abstractly 'volume'."

29. Spencer, *The Egyptian Temple*, pp. 270-78.

30. *Ibid.*, pp. 281-83.

31. *Ibid.*, pp. 260-64.

32. "This entry seems incongruous here, and would have been more in place in Section II, where there are two more or less closely synonymous and possibly related words, namely No. 22 *...nw* and No. 32 *...nwy(t?)*."

33. "In spite of the rubric there is but little distinction of subject between Nos. 474ff. and the preceding numbers. The new section might more fitly have opened with the cereals, Nos. 491ff."

34. Gardiner suggests that perhaps "barley" *(it)* was given here. Then follow seven kinds of *bty* (emmer or spelt).

35. The next forty-seven items "doubtless name all kinds of pastry, bread or cake made from the previously mentioned cereals."

36. Gardiner indicates that there is no rubric

here and he gives the heading as a convenience for his edition.

37. "That parts of an ox are here intended, not parts of the human body, seems clear...."

Section Two

Order

Chapter Two

The World and Its Creation: Cosmogony and Cosmology

Even the briefest glance at the extraordinary volume of ancient Egyptian literature that touches upon the nature of the world and how it came into being will convince a reader that during the three thousand years of Pharaonic Egypt there was no natural philosophy or physics that was separate from religion, myth, and magic.[1]   Hence it is not very surprising that the documents presented in this section for the description of the various cosmogonies and the general view of the world are all religious documents.   They include: (1) funerary spells, like those that make up the first three documents: the *Pyramid Texts*, the *Coffin Texts*, and the *Book of the Dead*; (2) books concerned with one or more gods that are partly mythological and partly ritualistic, like the *Litany of Re* (Doc. II.5) and the *Book of the Divine Cow* (Doc. II.6); (3) books that describe the underworld and the passage of the solar bark through it at night, e.g. the *Book of Amduat (Amdat)* (Doc. II.4); (4) hymns (Doc. II.7) and ritual accounts, like the *Destruction of Apep* (Doc. II.8); (5) miscellaneous inscriptions and texts found in graves, temples, and on separate stones, like the *Memphite Theology* (Doc. II.9); and (6) magical papyri filled with references to the gods (Docs. II.10-II.11).

A convenient manner of arranging the various

cosmogonies is by the locales in which they developed:[2] (1) the Heliopolitan scheme composed at On (i.e. Heliopolis) in the Old Kingdom, a scheme emphasizing creation by the solar deity; (2) the Hermopolitan system, a creation devised in Khmun (i.e. Hermopolis), in which eight primitive gods (the so-called Ogdoad) play the important role; they were the chaos-gods who in later times were thought of as the associates of Thoth, the patron of learning, who himself became a creator god; (3) a Memphite cosmogony, arising at the original capital city of ancient Egypt, Memphis, and featuring Ptah the Creator identified with the creative activity of Tatenen ("the land which rises"). From these three important systems arose synthetic accounts which make use of various features of the three main accounts: (4) a synthesis of (1) and (2) featuring the solar creator assisted by the Ogdoad; (5) a synthesis of (1) and (3) in which Ptah is differentiated into eight hypostases, of which the principal one is Atum (the sun-god), and sunlight is introduced with the emerging land; (6) the Theban cosmogony (or cosmogonies) centering on the syncretic god Amon-Re, "King of the Gods", as the demiurge; (7) later systems like those of Edfu (featuring Horus) and Esna (with Khnum and Neith as the creators).

Before giving the details of the main systems and referring to the various documents for supporting authority, we should note that some features of the diverse systems are shared. For example, all of the systems reject initial creation *ex nihilo* by indicating that before creation at the "first time" the amorphous "Abyss" or "Deep" existed under the name of Nun, described everywhere as the primitive waters in which the creator god, whoever he was, lay formless, and no land, no sky, no light, no life, no time, no birth, no death or strife existed (in fact no entity that formed a part of the later created cosmos existed) (see Doc. II.1, Sects. 1040, 1463, 1466; Doc. II.7d, D.6-7). So creation was to be a process of giving form to the primitive Abyss and separating entities from it. In all systems the creator god, whether he was the sun, or the primitive Eight (the Ogdoad), or Ptah, or any other god or goddess, created himself, that is, first fashioned his form. As the one being who was alone after his self-creation, the creator god is often called the Sole One or the Unique One, with no monotheistic implications. In many systems the creator god after creating himself created an emerging mound of land or an island on which to stand and many a city later held that the original mound emerged in its area (indeed each of the great temples was conceived as being built on the original mound and included the original sanctuary of the creator god).[3] After creating himself and the emerging land he created other gods and then (with or without the help of one or more of these gods) he formed man and the rest of the entities of the ordered world. In all of the systems a principal creation was that of world order or maat $(m^{j\zeta}t)$, which so much of Egyptian religious ritual and magic strove to preserve from the ever-present danger of the still-existing waters of Nun lying on the edges of and even within the ordered world. Accordingly the creator god is usually called Lord of Maat. We see that early spells tended to identify the king with the creator god (see Doc. II.1, Sect. 1040).

The various cosmogonical systems also share, no doubt, their primitive origins in the two pervading natural features of Egypt: the overwhelming importance of the Nile and its annual flooding and the ever-present sun as a continuing source of light and heat. The first surely accounts for both the insistence on the origin of everything in the primitive waters of Nun and the emergence of the mound or island of creation. The second stimulated, in all likelihood, the appearance of the sun as either the creator god or his principal creature in the Egyptian cosmogonic schemes. Nor is it surprising that an early stage in some of the schemes is the appearance of a lotus on the emerged mound as the first step toward life. For each year when the flood subsided and land rose from the receding waters plant shoots were evident on the emerging land. And it is equally understandable how the Egyptian cosmologist might have conceived the daily rising of the sun following the cold, dark night as repeating again and again the creation of the "first time", the time when the sun first appeared shining at creation. A naturalistic analogy for the rising sun is found in the mythological-cosmological accounts which show a great dung-beetle pushing the sun along (Fig. II.1) like its earthly counterpart, which can be seen to roll the ball of dung and mud in which it lays its egg (and it may be that the beetlelike image for the rising sun resulted from the fact that the words for "beetle" and "become" or "come into being" share the same consonants, namely *h*, *p*, and *r*, the word for "beetle" originally being distinguished by having double *r*).

Though the various systems diverge on the procedure of creation by the demiurge (e.g. by spitting, by masturbation, by conception in the heart and speech on the tongue, or by craftwork), as we shall see, they all fail to elucidate the mysterious autogenesis in the Abyss of the creator god. The word *kheper* used for this generation on the "first day of time" corresponds, as Sauneron remarks,[4]

> to a notion very difficult to grasp....According to the contexts in which the word figures, a modern occidental is forced to translate it in very different ways: "to be born", "to come into existence", or "to exist", "to be in existence" or better "to become", "to be transformed (into)", "to be manifest (under such a form)", while the noun *kheperu* will sometimes be rendered by "(mode of) existence" and sometimes by "transformation". Only a better appreciation of the fundamental sense of this root, *kheper*, now static and now dynamic depending on the case, will permit us to determine the true metaphysical conceptions of the Egyptians on the matter of the genesis of the creator (and of the creatures).

Finally, as we shall see later when discussing general cosmological features, all systems of cosmogony emphasized the creation of many gods except that of Akhenaten (Doc. II.7c). This last was the lone example of a truly monotheistic system, as Hornung has convincingly shown in refuting the efforts of many scholars to find a single great god of which the many gods are mere manifestations.[5] To put it briefly, all Egyptian schemes but Akhenaten's were polytheistic.

Before passing to the first of the cosmogonic systems, I should remark on the bewildering complexity and the contradictory nature of many of the concepts accepted without apparent demur or concern on the part of the ancient Egyptians. A syncretic god like Amon-Re could exist as an entity with a life and a history of its own without the component gods losing their individualities. Bonnet has put it nicely:[6]

> The formula Amon-Re does not signify that Amun is subsumed in Re or Re in Amun. Nor does it establish that they are identical; Amun does not equal Re. It observes that Re is in Amun in such a way that he is not lost in Amun, but remains himself just as much as Amun does, so that both gods can again be manifest separately or in other combinations.

Syncretism was carried to more than two gods: e.g. in Ptah-Sokar-Osiris, Osiris-Apis-Atum-Horus, Amon-Re-Harakhti-Atum, and Harmachis-Khepri-Re-Atum. It is of interest that the Great Sphinx at Giza was considered by the Egyptians to be a representation of the last of these syncretic gods, consisting of the three daily forms of the solar god (Khepri in the morning, Re at midday, and Atum in the evening) present in Harmachis (the sphinx itself).

In one scheme Re may be the creator, while in another he is created by Ptah, and both schemes seem to prevail simultaneously. Similarly the Ogdoad may be the primary creator or simply an agent of another creator, again without apparent concern for consistency (see below, note 66). Older scholars like Erman expressed disdain or even horror at the wholesale inconsistency:[7]

A strange curse lay on the Egyptians: they could not forget. At the earliest period writing had been discovered by them and had placed them in the front rank of the nations, but the price of this remained to be paid. Every fresh epoch of their long existence brought them new ideas, but the earlier ideas did not disappear in consequence. It is possible that the latter might fall into temporary neglect, but they were still treasured as sacred possessions, and in another century would once more assume a prominent position. Or again, a book that lay dormant in some temple library would one day become a living influence. In this way the confusion of ideas, national and local, old and new, increased with every successive period, and added to the mass of religious details that rejoiced the Egyptian theologians, but which we regard with horror.

More recent scholars reject the use of modern rational evaluation of Egyptian thought and attempt to explain its rather different approach to reality. Thus Frankfort tells us:[8]

We find, then, in Egyptian religion a number of doctrines which strike us as contradictory; but it is sheer presumption to accuse the ancients of muddleheadedness on this score .... The ancients did not attempt to solve the ultimate problems confronting man by a single and coherent theory; that has been the method of approach since the time of the Greeks. Ancient thought ―mythopoeic, "myth-making" thought― admitted side by side certain *limited* insights, which were held to be *simultaneously* valid, each in its own proper context, each corresponding to a definite avenue of approach. I have called this "multiplicity of approaches," and we shall find many examples of it as we proceed. At the moment I want to point out that this habit of thought agrees with the basic experience of polytheism.

Rudolf Anthes takes an approach somewhat different from that of Frankfort's.[9] He holds that the Egyptians possessed common sense and so did not take varying pictorial representations of concepts (and indeed the concepts themselves) at their face value but only as symbols. As an example he discusses the various concepts of the sky represented in the *Book of the Divine Cow* (our Doc. II.6) and its picture of the cow (see Fig. II.2a). In it the goddess Nut is shown as the Divine Cow, whose underbelly constitutes the sky as seen, and on it appear some stars that lie in the line of the underbelly. It is also quite obviously considered as a waterway on which the solar bark is seen twice (once at the beginning and once at the end of its passage), and furthermore the legs of the cow are obviously considered as the supports of the vault of the cow and each of them is in turn supported by a pair of support gods, the so-called Heh-gods. Anthes further points out that the idea of Nut's carrying the boats of the gods with her when she was raised as the sky, with the gods becoming stars, is attested by the *Pyramid Texts* (Doc. II.1, sect. 785, 909 [and 802, not included in Doc. II.1]), and indeed "in these texts a cow, an ocean, a vulture, and the woman Nut appear among the other concepts of heaven".[10] Regarding this diversity he concludes:[11]

> Nobody in Egypt was supposed to believe in one single concept of the sky, since all the concepts were accepted to be valid by the same theologians. Furthermore, since the Egyptians had as much common sense as we have ourselves, we may conclude with certainty that no one, except perhaps a very unsophisticated mind, took the composite picture of the heavenly cow at its face value. This conclusion is supported by the fact that there exist, in the same royal tombs about 1300 B.C., other pictures of the sky, e.g., in the form of the human figure of Nut and with the sun disk in place of the sun boats. Whoever might have sought for a replica of the actual shape of the sky in these pictures would have become completely confused. Consequently, they were meant to be symbolic of

the sky. The picture under discussion is an artistic combination of symbols, each of which stands for the sky or heaven. We have seen that the composite pictures of heaven in the royal tombs about 1300 B.C. have their counterparts in a drawing of about 2900 B.C. [where the sky is simply depicted as the wings of a vulture or a falcon perhaps supported by two *w's*-staffs; see Fig. II.3] and in the Pyramid Texts. There is no question that at the very beginning of their history, about 3000 B.C., the Egyptians were aware that the concept of the sky could not be understood directly by means of reason and sensual experience. They were conscious of the fact that they were employing symbols to make it understandable in human terms. As no symbol can possibly encompass the whole essence of what it stands for, an increase in the number of symbols might well have appeared enlightening rather than confusing.

A somewhat different method of "saving" Egyptian thought, but which points in the same direction as Frankfort's "multiplicity of approaches", is Hornung's suggested possible use of a "principle of complementarity" to confirm the view that the divine in Egyptian thought is both "one and many":[12]

In the act of worship, whether

it be in prayer, hymn of praise, or ethical attachment and obligation, the Egyptians single out one god, who for them at that moment signifies everything; the limited yet colossal might and greatness of god is concentrated in and focused on the deity who is addressed, beside whom all other gods vanish into insignificance and may even be deliberately devalued .... The god who is addressed is superior to the gods, he is more than they are.... However one describes the emphasizing of the one among the many, the phenomenon itself leads us straight to the problem of logical thought. According to the principles of western logic it would be an impossible contradiction for the divine to appear to the believer as one and almost absolute, and then again as a bewildering multiplicity; we find it surprising that in Egyptian thought these two fundamentally different formulations are evidently not mutually exclusive but complementary [descriptions of reality]. Did the Egyptians think wrongly, imprecisely, or simply in a different way?

This question about Egyptian thought, which we must consider here, has been answered in the most various ways. Egyptian thought has long been said to be "illogical" or at the least "prelogical," and in this way the contradictions that are encountered have been set aside as imperfections in its structure. Édouard Naville's remarks about Egyptian "soul" concepts, published in 1906, are typical of many judgments: "All these doctrines are very vague and ill-defined; here, as with all Egyptian ideas, there is an absolute lack of system and logic."

In sharp contrast, Rudolf Anthes points to the "undeniable role that rational thought and action played in public and private life in Egypt". He wishes to find this "rational" (*vernünftig*) thought, which he relates to timeless "common sense" (*gesunder Menschenverstand*), in Egyptian religion and mythology, and he rejects the assumption that there is a different mode of thought which is "mythopoeic," as Henri Frankfort termed it .... But formal thought in theology, philosophy and science, which is governed by well-defined calculi, is quite another matter. Here problems cannot be solved by "common sense" [as were the problems of social mores and conduct in Egyptian wisdom literature reported by Anthes], and this is just as true of ancient Egypt. The highly systematic theology of the New Kingdom is a formal conceptual structure, which must be studied according to strict, formal criteria that cannot be derived from a loose concept of "reason" or "common sense".

Any application of a two-valued logic, which is based on *a* / not-*a* distinctions and on the law of the excluded middle, to Egyptian philosophical and theological thought leads at once to insoluble contradictions .... Either we equate truly logical thought with two-valued logic, in which case Egyptian thought is undeniably "illogical" or "prelogical"; or we admit the possibility of a different type of logic which is not self-contradictory, which can only be a many-valued logic....

So long as the intellectual basis of a many-valued logic remains uncertain, we can indicate only possibilities, not definite solutions. If the basis is not established, Egyptian thought and all "pre-Greek" thought will continue to be open to charges of arbitrariness or confusion. If it is found, we shall be able to comprehend the one and the many [both with respect to the divine and to existence and non-existence in the worldl as complementary propositions, whose truth values within a many-valued logic are not mutually exclusive, but contribute together to the whole truth: god is a unity in worship and revelation, and multiple in nature and manifestation.

Hornung had already said that the Egyptologist is not competent to decide the validity of the concept of many-valued logic, but he also had said that the Egyptians did not move carelessly in their thought and that the Egyptologist can "sense that their system of thought has a coherence of its own which can often convince the emotions, even though it cannot be analyzed without contradiction according to western criteria, or defined in formal terms".[13] In spite of Hornung's criticism of previous efforts to isolate the characteristics of Egyptian thought as being too concerned with the "general 'cast of mind' of the Egyptians and not enough with the formal side of their thought", his own feeling or sense of the coherence of Egyptian thought, which I have just quoted, is not so far removed from the opinions expressed by those he criticizes. Furthermore one can surely doubt that the Egyptians themselves intentionally used or realized they were using anything so subtle as the modern Principle of Complementarity used in physics, as seems to follow from Hornung's analysis.

One last feature common to the various cosmogonic systems may be alluded to, namely the use of the doctrine of the creative word, that is, the view that to name something and then to pronounce the name can bring it into existence. This was the central view of creation in the *Memphite Theology* (Doc. II.9), but it played at least some role in other systems, as we shall see. Indeed this doctrine accounts for the extraordinary use of punning in almost all Egyptian theological accounts, as is evident for example in the statements by which Re creates the sky, the Field of Offering, the Field of Rushes, and the evershining stars in the *Book of the Divine Cow* (see Doc. II.6). Sauneron describes this doctrine of the creative word and its relationship to punning in a very lucid statement:[14]

> The Creation of the world has been imagined by the Egyptians in many fashions. Each was wont to conceive it on the basis of his own ideas but, as is logical, left the principal part in it to his own local god. One creative "technique", however, seems to have brought together a unanimity among the theologians: namely, that the agent is the *word*. The initial god, in order to create, had only *to speak* and the beings or things evoked were born at the sound of his voice. Speech in the Egyptian mind is not in effect a simple social tool facilitating human relations but is the audible expression of the intimate essence of things. It remains what it was at the origin of the world, the divine act which brought the matter forth. In the articulation of the syllables resides the secret of the existence of

the things evoked. To pronounce a word, a name, is not only a technique allowing an image to be born in the mind of the listener which reflects that of the speaker, it is a procedure of acting on the thing or the being which was mentioned in order to reproduce the initial act of the creator....

The Egyptians never considered language--namely that which corresponds to the hieroglyphs--as a social tool. It always remained for them a sonorous echo of the essential energy which gave rise to the universe, in fact a *cosmic force*. Also the study of this language allowed them an "explication" of the world. This explication was the "play on words" which furnished it to them. From the moment that one considers words as intimately tied to the essence of the beings or things they define, the resemblances of words could not be simply fortuitous but rather convey a relationship of nature, a subtle rapport which the science of the priests will have to define. Names of places, names of divinities, terms designating sacred objects, all ought to be explicable by means of a phonetic etymology--and the door is open to the most extravagant fantasies.

Now we may turn to the various cosmogonic systems.

The Heliopolitan System

While the Heliopolitan system of creation is the first one for which we have extensive literary evidence, we can suspect the existence of a preliminary scheme that featured Horus as the creative god. That the king was the god Horus incarnate ("the Good God, Lord of the Two Lands")[15] seems probable in the writing of the Horus name of the king in a *serekh* (palace facade) having a Horus-falcon above (see Chapter One above), a practice that goes back to the time of the Scorpion king and Narmer just before the beginning of the first dynasty.[16] The cosmic role of Horus seems to be neatly indicated on a comb of the Horus Wadji found at Abydos (see Fig. II.3). We see, at the top of the figure, Horus in a boat, a boat which rides upon two outstretched wings, and below these wings the figure of Horus appears once more, this time placed above the *serekh*-name of Wadji. It seems certain that the wings represent the sky (and since a boat is riding upon them, the wings become the heavenly waters, an image later applied to the solar boat riding on the heavenly underbelly of the Divine Cow, as was already noted in our quotation from Anthes's account of the various treatments of the sky). The wings could be those of a vulture representing Nekhbet, the goddess of Nekheb across the river from Hieraconpolis, or they could be the wings of the Horus falcon.[17] The heavenly wings seem to be supported by two *wꜣs*- or *dꜥm*-staffs[18] shown on the sides of the *serekh*. A remark in the *Pyramid Texts* (Doc. II.1, Sect. 1156) similarly represents the king after death and ascension to the sky as supporting the sky with a *wʾs*-staff. The same passage also speaks of his holding up the sky with a life- or ankh(*ꜥnḫ*)-sign (and we see in our Fig. II.3 that an ankh-sign is also beneath the wings of the sky and to the right of the *serekh*). Another passage (*ibid.*, Sect. 1456) mentions the imperishable stars "who traverse the Land of Libya, who lean on their *dꜥm*-staffs. This Pepi leans with you on a *wʾs*-staff and on a *dꜥm*-staff". The *wʾs*-staffs on Wadji's comb not only seem to serve as supports for the sky, but they also help to frame the *serekh*-name. Thus it appears that the whole picture offers good evidence that the God Horus, incarnate in the king, travels the sky in the same manner that the deceased king is pictured later as traveling in a solar boat.

It could be that the boat pits found near tombs of the first dynasty (as for example, that found north of tomb 3357 at Saqqara, which dates from the time of Horus Aha at the beginning of the first dynasty) contained boats that were to be used by the deceased king when he traversed the sky as Horus. Thus it could be that these early boat graves contained Horus-boats rather than solar boats. But surely when the solar cosmology which developed at Heliopolis absorbed that of Horus, these boats became solar boats.

Possible evidence bearing on the identification of the sky-god Horus and the king occurs in the entry for the year X + 8 of the second-dynasty king Ninetjer found in the early royal *Annals* translated above as Document I.1 (see and consult note 60 to that document): "The Year of the First Occurrence [in

Ninetjer's reign] of the Festival of the Worship of Horus of Heaven." It certainly seems probable that the Horus-king would want to celebrate the festival of his heavenly counterpart. Horus is again and again called Horus of the Horizon in the *Pyramid Texts* (e.g., see the many passages presented in the work of T. G. Allen cited in note 15).

Incidentally, it should be remarked that there is plentiful evidence in the *Pyramid Texts* that the king as Horus became at death an imperishable star "which illumines the sky...[for] the sky will not be devoid of this Pepi and this earth will not be devoid of this Pepi forever" (Sects. 1454-55). In another passage where the king propitiates the primeval gods he asks them to tell their father that King "Wenis knows him and knows his name: 'Eternity' is his name, 'The Eternal One, Lord of the Years' is his name. The Armed Fighter Horus, who is over the gods of the sky, is he who vivifies Re every day" (Sect. 448-49). Some other passages referring to Horus as a star are to be found in the following sections of Document II.1: 330, 1301, 1508-9, 1636 (and again see Allen's work referred to in note 15).

It is also thought that the symbol of the Eye, which plays such an important part as the Eye of Re and also as the Eye of Horus in the later mythology arising at Heliopolis, had as its predecessor a falcon's eye which was conceived when Horus was the supreme god. Thus Clark remarks:[19]

> At the beginning of history the High God of the Egyptians was a falcon which was shown either as perching on a building or emerging from the Primeval Waters. Its right eye was the sun and its left one the

> moon...Certainly, whenever the
> Egyptians pictured the eye of their
> god they depicted a falcon's eye, not
> a human one.

One last point is worth making concerning the possibility that Horus was conceived as the creator god at the time of the union of the two lands. As early as the second dynasty we note that the Horus Sekhemib apparently added to his name Perenmaat, i.e. "the House of Maat", indicating perhaps that Horus established and preserved maat, i.e. the "world order". It must be remembered that, in the later Heliopolitan system, maat was personalized as the daughter of Re and as such was one of the principal creations of the solar deity. Hence it would be surprising if maat had not been considered in the pre-Heliopolitan days to be Horus's creation. Significant evidence that this was so may well be the Horus name of Sneferu, the first king of the fourth dynasty; it was Nebmaat ("Lord of Order"). Also, one of the epithets of the solar god under his name of Reharakhti (Re-Horus of the Horizon) was Nebmaat at the solar temple built in Abu Ghurab by the fifth-dynasty King Niuserre.[20]

The earlier views concerning Horus and the role they may have played in the later Heliopolitan solar scheme is well (but somewhat speculatively) summarized by Anthes:[21]

> In the course of this summary
> of ancient Egyptian mythology the
> reader will have become aware that,
> in Egypt, we have the unique
> opportunity of determining the time
> and the circumstances in which the
> most substantial sector of her

mythology, the myths centering on Horus, originated. The time was the beginning and the middle of the third millennium B.C., starting with the earliest documentation of history, and the circumstances were prompted by the establishment of the kingship in Egypt. The myth of Horus encompassed the concepts of the lineage of Horus [starting from Atum and proceeding through the Ennead to Horus] which then became the Heliopolitan cosmogony of Horus and Seth; of Osiris and Isis, and of the Eye of Horus; it became the prototype of the concept of Re, the sun who was the king of heaven. This myth was rooted in the first known conception of the highest god, the ruler of All, who appeared in the trinity of the Horus falcon, the Horus king of Egypt, and the heavenly Horus. It came into existence through speculations that were conducted in a clearly logical manner, based upon the faith in the universal and eternal character of the king of Egypt, and enriched by cosmogonic ideas that had been transmitted from prehistoric times; and it was made effective by the amalgamation with the rites that were performed in the service for the divine king and, particularly, for his ascension to the throne and his interment.... However...in Egypt, mythology arose out of the creation of a new form of society whose structure was expressed in theological terms. It is true that a few mythological concepts of the sky and the sun, the earth and vegetation had been carried over from the prehistoric period into the myth of Horus and, later, that of Re. Other ideas of the cosmos, however, originated as replicas of ideas of the kingship on earth. One of these more recent cosmic concepts was that of the heavenly king Horus who was incorporated in the sun and a star.

It seems probable that the old Horus of whom we have just spoken merged with the Horus who became the heir of Atum and of the company of nine gods, the Ennead, and who played an essential role in the Heliopolitan system we are about to describe.

There is some evidence of the importance of a solar deity to the united kingship as early as the second dynasty. The reader will recall that one of the kings of that dynasty was named Re-neb ("Re is the Lord") and his stela has been given as Fig. I.14a. Such a title of course tells us little if anything about any cosmogonic role the sun god Re might have been thought to have played at such an early date. But surely it was significant for the king on earth to declare by his Horus name that Re was the lord in heaven, if that is what the title signifies. However, it was later, in the third dynasty, that specific evidence of the importance of

Heliopolis to the kingship undeniably appeared, for the great vizier and architect of Djoser, the famous Imhotep, bore the title of "The Great Seer", the title of the High Priest of the solar city Heliopolis.[22]

Evidence of the interest in Re on the part of the kings was also present in the fourth dynasty, particularly in the names of some of the kings after Cheops: Djedefre ("Re is enduring"), Chephren (i.e. Khafre, "Re rises [or appears in glory]"), Mycerinus (i.e. Menkaure, "Established are the kas of Re"). All three of these names can be regarded as reflecting the Heliopolitan doctrines of Re.

But it was in the fifth dynasty that the greatest influence of Re seems to have been exercised on the monarchy, evident not only in the names of the monarchs: Sahure ("Re is [his or my] protection"), Neferirkare ("Beautiful is the action of the ka of Re"), Niuserre ("Re has power"), and Djedkare ("Enduring is the ka of Re"), but in the fact that several of the kings built solar temples as well as pyramids, the temples apparently being in imitation of the temple at Heliopolis, and they endowed solar shrines and temples throughout Egypt (see my discussion of Doc. I.1 in the preceding section). Furthermore, the use of the designation "Son of Re" for the king, found in the fourth dynasty and more frequently in the fifth and sixth, no doubt has great significance for indicating the spread of the Heliopolitan doctrines.[23] It was also at the end of the fifth dynasty that the *Pyramid Texts* were first inscribed on the walls of the chambers in the pyramid of Wenis, and it is these texts that recount some details of the Heliopolitan doctrines. It now behooves me to describe some of these doctrines.

The creator God Atum (identified in the

ANCIENT EGYPTIAN SCIENCE

Heliopolitan system with Re, Khepri, and the syncretic god Re-Atum) lay inert and formless in the Abyss or Deep of Nun (see Fig. II.4), the primitive waters, which were limitless and dark, and this was before the "first time" when nothing yet existed.

Then in some mysterious manner Atum decided to take form, and the king (identified with him) was born *(msiw)* in the Abyss "before the sky came into being, before the earth came into being, before any established thing came into being" (see Doc. II.1, Sect. 1040). Atum in the name of Khepri or Khoprer created himself *(ibid.,* Sect. 1587). He raised himself on a mound or high hill as a *benben*-stone (shaped like a small pyramid and perhaps thought of as a cluster of solar rays joining the emergent sun and its source) in the Mansion of the Phoenix in On (i.e. Heliopolis) *(ibid.,* Sect. 1652). In the *Coffin Texts* (Doc. II.2, Spell 335) and the *Book of the Dead* (Doc. II.3, Spell 17) the Great Phoenix of Heliopolis identifies himself as the examiner of what exists.

After (or in some accounts before) creating the mound, Atum then created a son Shu and a daughter Tefenet by spitting *(ibid.,* Doc. II.1, Sects. 1652 and 1871) or by masturbation *(ibid.,* Sect. 1248). Both are usually considered as physical acts, though Clark, in a rather extravagant interpretation, believed that "the spitting motif expresses creation through the Divine Word or the entry of the breath of life.... The masturbation motif stresses the reproductive aspect of life, but behind lies the mystery of life itself, the breath of the Divine Soul. Hence the generation of Shu and Tefnut [or Tefenet] has to be described in terms of both the masturbation and spitting myths--they are, in fact, complementary, not alternative."[24] For authority he quotes the *Coffin*

Texts (Doc. II.2, Spell 245). A later account (Doc. II.8, 26(21) *et seq.*) is couched in more abstract theological terms. In that account when the creator god came into being "Being' came into being" and indeed all creatures came into being after he came into being.

> Manifold were the beings which came forth from my mouth ere the sky had come into being, ere the earth came into being, ere the ground and reptiles had been created in this place. I created [some of] them in the Abyss as Inert Ones when I could as yet find no place where I could stand....I alone made every shape ere I had spat out Shu, ere I had expectorated Tefenet, ere there had come into being any other who could act with me. I planned with my own heart and there came into being a multitude of forms of living creatures, namely the forms of children, and the forms of their children. I indeed made excitation with my fist, I copulated with my hand, I spat with my own mouth; I spat out Shu, I expectorated Tefenet, and my father Nun brought them up.... After I had come into being as sole god, there were three gods in addition to myself. I came into being in this land and Shu and Tefenet rejoiced in the Abyss, in which they were.

While Shu and Tefenet were still in the Abyss,

they became lost and Re sent his Eye to find them (see Doc. II.8). Meanwhile Re completed his form with all his members and therefore produced a new Eye. The original Eye, on its successful return with Shu and Tefenet, was angry at his displacement. To pacify him Re put the old Eye on his brow in the form of a rearing cobra, the Uraeus, to serve as Re's protector from inimical activities of the serpent Apep (Apophis). The Uraeus was accordingly adopted by all kings of Egypt to act in the same way as their protector. Thus was explained the role that the Eye of Re played in later ritual. It became thoroughly entangled with the Eye of Horus, whose role in the Osiris legend I shall allude to below.

It is apparent that Shu and Tefenet in the earliest accounts were personifications of elements of the environment, Shu being air or space (and then light) and Tefenet being the female counterpart to allow for the future generation of their children, Geb (the earth) and Nut (the sky). By the time of the *Coffin Texts* (Doc. II.2, Spells 79-80) Shu became the mediating creator between Atum and the subsequent creations in a very elaborate and often confusing doctrine. Tefenet becomes identified with another daughter of Atum, Maat, the world order (*ibid.*, Spell 80; see also Fig. II.5):

> Thus said Atum: Tefenet is my living daughter, and she shall be with her brother Shu, whose name is 'The Living One'. Her name is Maat (Order or Righteousness),... Nu said to Atum: Kiss your daughter Maat, put her at your nose that your heart may live .... Maat is your daughter and your son is Shu.... Eat of your

daughter Maat (i.e. thrive on her); it
is your son Shu who will raise you
up [in the sky].

After the creation of Geb and Nut, their father
Shu separated the two, thus raising Nut as the sky. A
later account of this is found in the *Book of the Divine
Cow*, which I mentioned earlier (and see Doc. II.6 for
the modifications of the cosmos described therein).
There Re rules on earth but is tired of it, particularly
so as the result of the rebellious action of man. Hence
he climbs on the back of Nut and commands that she be
raised to the sky, raising him along with her. Geb and
Nut give birth to Osiris, Isis, Seth, and Nephthys,
completing (along with Atum, Shu, Tefenet, Geb, and
Nut) the Ennead (or Company of Nine Gods). The
relations of Nut to her husband Geb, to her mother
Tefenet, and to her son Osiris are briefly mentioned in
a series of short spells in the *Pyramid Texts* which I
have not included among my extracts in Document II.1.

At this point we should note that the great
Osiris- and Horus-myths (originally separate as shown
by the fact that in the original Horus-myth, Horus was
apparently the brother of Seth, while in the Osiris-myth
he was the nephew of Seth), which were incorporated
into the funerary rites of the kings of Egypt, can be
detected, at least in broad outline, in the *Pyramid
Texts*: the killing of Osiris by his brother Seth, the
dismemberment of Osiris's body, the search for his body
by his sister-wife Isis and his sister Nephthys, the
reassembling of Osiris' body, the birth of Horus and the
protection of him, the struggle of Horus with Seth and
the tearing out of Horus's eye by Seth, the final victory
of Horus (with the recovery of his Eye) and recognition
of him as the king of Egypt, the gift of the Eye to

Osiris, and the establishment of the resurrected Osiris as the king of the underworld, a resurrection mirrored by that of the dead king who is addressed as Osiris "So-and-so" and receives from his son Horus funerary offerings symbolically designated as the "Eye of Horus".

I have not included all of these passages in Document II.1 below, but I can here mention some sections that may be consulted in Sethe's edition and Faulkner's translation (see the titles for these works in the Introduction to Doc. II.1 below): 1339, 972, 163, 173, 175, 1007, 584, 1630, 318, 746, 828, 830, 632, 1636, 1463, 1242, 609, 639, 643, 317, 957-58. Other later accounts of the Osiris-legend fill in some of the details (e.g. the Great Hymn to Osiris given as Document II.7a below), but one has to go to the work of Plutarch *On Isis and Osiris* to find a truly coherent account.[25] Important as the Osiris-legend was to the concept of life after death, initially for the king and then later for all deceased, it probably did not play a great part in the invention of the Heliopolitan cosmogony.[26]

One of the most interesting appendages to the Heliopolitan doctrine involves the concept of Atum as a primitive serpent present in Nun before the "First Time". Pepi is identified with such a serpent and is then said to be the scribe of god's book, "who says what will be and what will not be created" (see Doc. II.1, Sect. 1146), as if the ideas go together to present us with a creator-snake.

This is perhaps a reference to the doctrine of creation by the word. Though this may be the allusion, still all we can assert is that in this passage the king is first identifying himself with a serpent and then separately with the scribe of god's book (i.e. with Atum). We should note incidentally that the phrase "what will be and what will not be created" is no doubt equivalent to the phrase "that which is and that which is not", a favorite one of the Egyptians to describe the totality of things: the nonexistent perhaps being that which was temporally prior to the created cosmos or that which was spatially beyond or limiting the cosmos, and the existent being that which was created (that is, the cosmos), posterior to the nonexistent and spatially limited thereby.

The snake-form of the creator is considered as an earlier form of him, a form he shed and constantly combatted when it was considered to be the primitive chaos. It was the snake Imy-Wehaf in Heliopolis or more widely, in later accounts, the serpent Apep. The earlier snake-form of Atum appears to be identical with the serpent who is named the "Provider of Kas" *(Nḥb-kʾw)* (see Fig. II.6a), and indeed in the reference to the snake identified with Pepi a variant reading for the word for snake is Nḥb-kʾw. Presumably, as Clark indicates, this image of the snake-creator was superseded as creation progressed.[27]

It is of interest to our understanding of the cosmological views of the Egyptians that in Document II.3 (Spell 175) it is remarked that at the end of time Atum will revert to the primitive form of a serpent and the world will return to the undifferentiated chaos. Atum speaks to Osiris in order to soothe the unease of Osiris at having to live in the silent necropolis without water, air, and light, and without sexual pleasures. Atum tells him that only he (Osiris) and Atum will remain when at the end of time he destroys creation and returns to his snake forms:

I will destroy all that I have made. This land shall return into the

> Deep, into the flood, as it was
> aforetime. (Only) I shall survive
> together with Osiris, after I have
> assumed my forms of other (snakes)
> which men know not and gods see
> not.

As Clark has observed, this doctrine of the return
to the serpent form at the end of time provided a figure
of speech to the rulers of Asyut when they claimed to
be as important as "that great surviving serpent, when
all mankind has returned to the slime".[28]

To this point I have mentioned mainly the
beginning of creation and especially the creation and
activity of the Ennead, and incidentally the birth of
Horus as the heir of Osiris. I observe that the Horus so
conceived as the beneficiary of the Heliopolitan Ennead
is essentially distinct from the old sky god with whom
the king was first identified, though of course the two
Horuses became thoroughly intertwined. The king
retained many of the characteristics of the sky god as
he was elevated to the sky on his death, which
elevation is described in countless spells of the *Pyramid
Texts.* He was at the same time the devoted son of
Osiris and responsible for his burial so that when he
(Horus) died he would in turn assume the role of Osiris
to assure his afterlife.

Of course, Atum not only created the rest of the
Ennead,[29] the land on which he stood, the light in
which the land was bathed, and the general cosmic
order, but he created man and the animals as well.
Probably, the earliest doctrine concerning man's
creation was that he was produced from the tears of
Atum (see Doc. II.2, Spell 1130), a doctrine appearing in
more than one form. This doctrine seems to have had

its origin in the similarity between the word for human beings *(rmwt)* and that for tears *(rmwt)*, another example of the role of punning in the creative process. In Doc. II.8 (27, 2-3) we find man produced from the tears of the Eye of Re. Still another popular view had man fashioned by Khnum on his potter's wheel (see Doc. II.7e, Text 319, 16 and Fig. II.7).

Not necessarily at variance with these physical descriptions of the genesis of man, which are mythological in character, are the statements we have already quoted from Doc. II.8 that "manifold were the beings which came forth from my mouth" and that the self-begotten creator in his form of Khepri planned in his heart "a multitude of forms of living creatures, namely the forms of children and the forms of their children". Farther along in the same passage (Doc. II.8, 27, 4-5) the Lord of All declares:

> I came forth from the roots, I created all reptiles and all that exists among them. Shu and Tefenet begot Geb and Nut, and Geb and Nut begot Osiris, Horus Mekhantenirti, Seth, Isis, and Nephthys from the womb, one after another, and they begot their multitudes in this land.

The references to the manifold beings which came forth from his mouth and to planning the forms in his heart (essentially the mind in Egyptian conceits) seem like veiled references to the concept of the creative word, as does the passage in the *Pyramid Texts* (Doc. II.1, Sect. 1100) in which king Pepi asserts that the "lips of Pepi are the Two Enneads; Pepi is the Great Word" and the passage in the *Book of the Dead* (Doc. II.3, Spell 17, S 2) describing the great god who came into being by himself, created his names, i.e. the names of his members, and consequently created the gods in his Train or Following. But toward the end of the passage from Doc. II.8 quoted above the author speaks of the next generation begetting the succeeding generations "from the womb", which implies ordinary sexual reproduction rather than creation by the word. Incidentally, we should note that in this part of the document Geb and Nut have five children, a form of Horus being added to the usual four (see note 29 above). In placing the creation of reptiles before the generation of the gods, the account seems to reflect the above-mentioned doctrine that the early form of the creator was that of snakes, a form which he shed in creating living reptiles who were hostile to him. Such creation had importance in this document, for its main purpose was the ritualistic destruction of the snake Apep by rites and spells replete with magical jargon and acts, that snake being the great enemy of Re and thus of his son the king. I have not included these rites and spells in the extract which constitutes Document II.8. But I have given a passage in that document (see 26, 2-7) where careful instructions are given as to how to depict and write down the names of the foes of Re, those of their fathers and mothers, what color ink to use, how to write the names on the breasts of wax effigies, and so on.

The further creative activity of Re or Atum is considerably elaborated in the *Book of the Dead*, where we read (Doc. II.3, Spell 15A3):

> Hail to thee, Re at his rising, Atum
> at his setting. Thou risest, thou
> risest, thou shinest, thou shinest,
> having dawned as king of the gods.

Thou art lord of the sky and earth,
who made the stars above and
mankind below, sole God who came
into being at the beginning of time,
who made the lands and created
common people, who made the deep
and created the inundation, who
made the water and gave life to
what is in it, who fashioned the
mountains and brought into being
man and beast... (As for) him...who
came forth from the deep, Re
triumphant, divine youth, heir of
eternity who begot himself and bore
himself, sole one, great in number of
forms, King of the Two Lands, Ruler
of Heliopolis, lord of eternity,
familiar with everlastingness.... No
tongue could understand its fellow
except for thee alone [i.e. thou art
the creator of language] .... (Spell 85
[not included in my extracts below]:)
I am a soul; I am Re who came forth
from the deep. The god is my soul.
(It was I) who created Authority.
(Cf. Doc. II.2, Spell 307).

Unusual is the spell in the *Coffin Texts* (Doc.
II.2, Spell 1130) in which the four good creative acts of
the Lord of All are specified: (1) he made the four
winds so that everybody might breathe, (2) he made
the great flood so that the poor as well as the rich
might be strong, (3) he made every man equal to his
fellows, and "I commanded them not to do wrong, but
their hearts disobeyed what I had said", and (4) he made

the hearts of men not forget the West (i.e. the deceased) by making offerings to the gods of the nomes (i.e. the local gods). This is unusual since it emphasizes Atum's evenhandedness toward man at the same time that Atum is laying on him the responsibility for his acts, thus proclaiming a nascent doctrine of free will.

Indeed the Heliopolitan scheme became the starting point in the eighteenth and nineteenth dynasties for the detailed accounts that made up the Theban cosmogonies of Amon-Re, expressed in the hymns that were a part of the temple rituals, as we shall see (Doc. II.7b). It was also influential upon the cosmogonic hymns of Akhenaten in Amarna toward the end of the eighteenth dynasty (Doc. II.7c).

We should also note finally that this scheme even invaded the wisdom literature, i.e. the instruction-books, taught in the scribal schools. For example, in the Instruction-book for Merikare we read of the creative activity and power of the god (who is surely Re):[30]

Well tended is mankind--god's cattle,
He made sky and earth for their
    sake,
He subdued the water monster
    (the primitive snake?),
He made breath for their noses
    to live.
They are his images, who came
    from his body,
He shines in the sky for their sake;
He made for them plants and cattle,
Fowl and fish to feed them.
He slew his foes, reduced his
    children,

When they thought of making
      rebellion.
He makes daylight for their sake,
He sails by [i.e. across the sky]
      to see them.
He has built his shrine around them.
When they weep he hears.
He made for them rulers in the egg
      (i.e. destined some to be rulers
      from birth),
Leaders to raise the back of
      the weak.
He made for them magic as weapons
To ward off the blows of events,
Guarding them by day and by night.
He has slain the traitors among them,
As a man beats his son for his
      brother's sake,
For god knows every name.

The Cosmogony of Hermopolis

It is not easy, as I have said, to single out from
the sprawling Heliopolitan doctrines what we can surely
identify as belonging to the priests of Hermopolis. But
it is usually agreed (from the constant association of
the name Hermopolis with them in the documents) that
three key concepts were probably of Hermopolitan
origin: a creator in the form of eight primeval gods in
the Deep, their production of a mysterious primeval
egg, hatched in some accounts from a Nile goose called
the Great Cackler, from which egg the sun and
ultimately all creatures arose, or alternately the creation
by the Eight of a lotus on the initial land from which

the sun burst forth to assume its crucial but supplementary creative role.

The eight ancestor-gods, or Ogdoad, as they are ordinarily called, consisted of four pairs of gods, each pair representing an aspect of the primitive waters, and each pair consisting of a male and a female, so that the single, self-created creator Atum in the Heliopolitan system is replaced by a sexually coupled set of Chaos gods. Hence the passive role of the Deep in the Heliopolitan system was discarded in favor of this actively creating company of gods. From this set of eight gods, which we shall describe in more detail shortly, the town of Hermopolis took the name that it had as early as the Old Kingdom and that it still possesses, namely Eighttown (Khmun [i.e. *Ḥmnw*] in ancient Egyptian, Shmoun in Coptic, and el-Ashmunein in Arabic), although it apparently had an older name (identical with that of its nome), Haretown (*Wenu*).[31]

The earliest trace of the Ogdoad in the literature is perhaps a passage in the *Pyramid Texts* (Sects. 446-47), where we find grouped together: Niu and Nenet, Amun and Amaunet, Atum and Ruti (the Twin-Lions, a dual form embracing Shu and Tefenet), and Shu and Tefenet, where only the first four may represent traces of the Hermopolitan doctrine. The next mention of the concepts that go to make up the four male gods dates from the Heracleopolitan period (in the *Coffin Texts*, Doc. II.2, Spell 76). This same spell and indeed Spells 79-80 also speak of the eight chaos-gods who surely are related in some fashion to the Ogdoad even though they are brought in line with the Heliopolitan scheme when they are described (Doc. II.2, Spell 76) as created from the efflux of the flesh of Shu "whose names Atum made when *nw* (Abyss) was

created, on the day when Atum spoke in it with Nu in
ḥḥ (darkness) and in *tnmw* (gloom)".[32]   Similarly these
chaos-gods seem to be identical with the eight
support-gods that are described in Document II.6 as
holding up the heaven (see also Fig. II.2, where two of
them are shown seated by each leg of the Divine Cow).

 After a fluctuation of names in the Middle and
New Kingdoms, we find the eight Hermopolitan gods
possessing the following names: Nun [or Nu] and Naunet
(the formless waters), Hehu and Hehet (indefinitely
extended space), Kek and Keket (darkness), Amun and
Amaunet [or Amen and Amenet] (invisibility).[33]   They
are depicted in various fashions [see once more note
33].   In the most interesting representations, the male
gods are depicted as having the heads of frogs, the
female those of serpents, both animal forms being
appropriate to inhabitants of the mud and slime with
which chaos is often associated (cf. Fig. II.8).[34]

 In addition to the doctrine of the Ogdoad,
tradition also associated with Hermopolis a company of
Five.[35]   In Doc. II.11 (F, 1-5), we read:

  Hail to you, you Five Great Gods,
  Who came out of the City of Eight,
  You who were not yet in heaven,
  You who were not yet upon
   the Earth,
  You who were not yet illumined
   by the sun.

This company was perhaps comprised of gods
representing the four basic concepts of the Ogdoad plus
the god Thoth (represented as an ibis [Fig. II.9], an
ibis-headed man [Fig. II.9], or a baboon).   Thoth was
held to be the patron of writing and the inventor of
hieroglyphics, the one who planned temples and rituals,

the one who was expert in all kinds of knowledge (and especially magic), and, as a moon-god, the reckoner of time. He was the scribe of maat for the Great Ennead and for the Lord of Eternity himself, and indeed the scribe of God's book. He (with his helper Seshat) was the recorder of the king's name on the sacred Ished tree at Heliopolis (Fig. II.10) as well as the recorder at the weighing of the heart of the deceased before the latter was justified (i.e. "acclaimed right"). In addition, he retrieved the Eye of Horus and was the assistant of Isis and Horus in the resurrection of Osiris. Finally he was the patron god of Hermopolis and hence his possible appearance as part of the Five. Or perhaps the Five were related in some fashion to the five gods of the epagomenal days. The High Priest of Hermopolis was known as the "Great One of the Five in the House of Thoth", and indeed this office was held by several royal princes of the fourth dynasty.[36]

Returning to the Ogdoad and their roles in creation, we should observe that the qualities that they personalized represented the negative qualities of the Deep, which in some mysterious way were thought to have become creatively active. In one view they were believed to have come together in the primitive waters and thereby to have produced the primeval egg "in the darkness of my Father Nun".[37] There is perhaps a reference to this primeval egg in the *Coffin Texts* (Doc. II.2, Spell 307): "I am the soul who created the Watery Abyss and made a place in god's land; my nest will not be seen nor my egg broken, for I am the lord of those who are on high, and I have made a nest in the limits of the sky." Hymns to the Sun-god tell us:[38]

You have lifted yourself up there,
(leaving) from the mysterious egg as

the infant of the Ogdoad....Your place
since the First Time was on the Hill
of Hermopolis; you touched the land
[that arose] in the Island (*or* Lake) of
the Two Knives, and you lifted
yourself up from the primitive
waters out of a hidden egg.

In the Magical Papyrus of Harris (Doc. II.11, K
5-10) there is what is surely a reference to this egg:

Oh you Egg of the water, Seed of
the earth
... (Essence?) of the Ogdoad of
Hermopolis,
Great One in the Heaven and Great
One in the Netherworld,
Residing in the nest in front of the
Lake of the Two Knives,
I have emerged with you from
the water,
I have left with you from your nest.

In another passage of the *Coffin Texts* (Spell
223) the primeval egg is seen as being hatched by a
primitive bird, the Great Cackler or Great Honker: "Oh
Atum, give to N. (the deceased) this sweet air which is
in your nostrils, for I am the egg which is in the Great
Cackler." This is elaborated in the *Book of the Dead*
(Doc. II.3, Spell 54). The egg of the Great Cackler is
referred to Hermopolis in the same work (Spells 56 and
59, of which I have included 56 in Doc. II.3). It was
the Great Cackler who disturbed the silence which
prevailed before creation:[39]

The Ennead was still joined with
your members..., all gods were still
joined to your body.... He cackled,

> being the Great Cackler, in the place
> where he was created, he alone. He
> began to speak in the midst of
> silence. He opened all eyes and made
> them see. He commenced to cry
> when the earth was inert. His cry
> spread about when there was no one
> else in existence but him. He
> brought forth all things which exist.
> He caused them to live. He made all
> men understand the way to go and
> their hearts came alive when they
> saw him.

The Great Cackler's honking in the primitive night is also mentioned in Doc. II.11 (M,9).

Apparently, then, from that primitive egg the Ogdoad produced the sun and the sun then assumed its creative role. The shell (*lit.* the half *or* halves) of this primitive egg was said by Petosiris of Hermopolis (ca. 330 B.C.) to have been buried in Hermopolis.[40]

In a second scheme the Eight produced the primitive land in Hermopolis and it was called the "Isle of Fire".[41] It was so named because of the belief that the sun came forth from a lotus flower on it. In the *Book of the Dead* (Doc. II.3, Spell 15B1) we read about the Great God who lived on the Isle of Flame, "the youth (born) of gold who came forth from the lotus" and again (Spell 15A4) about the child "rising from the lotus, goodly youth who has ascended from the horizon and illumines [the Two Lands with] his light" (see Figs. II.8 and II.11, showing the young sun coming forth from the lotus blossom). Earlier in the *Pyramid Texts* (Doc. II.1, Sects. 265-66) the king identified himself with the lotus blossom (Nefertem) from which Re emerged on the Isle of Fire.

The complexity of the sources concerning the Ogdoad is well illustrated by the absorption of the doctrine into the Theban cosmogony where the Ogdoad are subordinated to Amon-Re as another form of his creation (Doc. II.7b[3] [III, 23]; Doc. II.11 [H and K]; Boylan, *Thoth*, p. 113, n. 1).

Though the God Thoth seems not at first to have been connected with the events recounted concerning the creative power of the Ogdoad, there is evidence of late cosmogonic conceptions in which Thoth appears as the creator god and even as the father of the Ogdoad. From them we learn that "at his command (i.e. assertion) heaven and earth were established".[42] And in the *Book of the Dead*, Spell 130, S 6 (not in Doc. II.3) identifies him as Thoth,

> the great one dwelling in his eye,
> sitting or kneeling in the great bark
> of Khepri. Osiris N comes into being;
> and what he (Thoth) says comes into
> being.

Hence the power of the creative word seems also to have been associated with Thoth, as it was on occasion with Re and systematically with Ptah in the *Memphite Theology*. Incidentally, in connection with the *Memphite Theology*, note that Ini-sw, who is probably Thoth, declares that "Hu (authoritative command or creative word) is in my mouth and Sia (understanding) is in my heart",[43] a passage that may imply that the heart and the tongue were as important in Thoth's creative activity as they were in Ptah's. Gardiner, in an early article, suggested the possible origin of the Memphite doctrine of the creative word in the Heliopolitan doctrine of Hu and Sia (see Fig.

II.12).[44] Boylan, by abundant citation, has made clear that Thoth's epithet "Lord of the Divine Word" implies not merely his invention of hieroglyphics but also his authorship and control of cult-ritual and magical formulae, in short, of creative words.[45] A series of Thoth's epithets points to his special connection with time and its reckoning: He who made Everlastingness (*Dt*), Lord of Eternity (*Nḥḥ*), King of Eternity (*Nḥḥ*), Ruler of Everlastingness (*Dt*), He who increases Lifetimes and multiplies Years, the Reckoner of Time, the Reckoner of Time for gods and men, the Reckoner of Years, the One who reckons All Things, He who knows Reckoning, He who gives Length of Life to him who is in his favor, Determiner of Life-time, Lord of Lifetime, Lord of Old Age, the Orderer of Fate, the One who announces the Morning, the One who distinguishes the Seasons, the Months, and the Years, and the One who gives Life to men.[46]

Like the other great gods who became widely worshiped and venerated, Thoth was credited on occasion with being the creator of all. In the temple of Philae, Thoth is described as he who came into being when nothing that was later to come into being had yet been created, and as the one who created everything.[47] There is also a Ptolemaic inscription in the Bab el-Abd at Karnak in which the Ogdoad is subordinated to Thoth:[48] the Eight created

> light in the Height [of Hermopolis]
> and took their place in Hermopolis
> with their father, the Venerable One
> [=Thoth].

CREATION AND COSMOLOGY

The Memphite Cosmogony

Though references to cosmogony in Memphis, and even to Ptah, are virtually nonexistent in the early dynasties and rare even in the Old Kingdom, a bowl from Tarkhan (60 miles south of Cairo), presumably from the first dynasty, bears a representation of Ptah (Fig. II.13). He stands in an open chapel, beardless and with smooth head, and carries a *w's*-staff. Later he was ordinarily represented with a beard and standing or sitting in a closed chapel (Figs. II.14-15). When sculpted in the round he wears a close-fitting garment, a skulltight headgear, and a beard, and he grasps his staff with both hands (See BM No. 25261). The early *Annals* refer to a sojourn of King Den in a temple, which may be a temple of Ptah's (Doc. I.1, note 39). References to Ptah increase in the Old and Middle Kingdoms, but they tell us little about him as a creator god.[49]

We do know that in the Old Kingdom there were two officials at Memphis, no doubt high priests of Ptah, who held the title of Chief-controller of the Crafts (see Doc. I.3, n. 4). Such a title appears to have had its origin in the quite early assumption of Ptah himself as a master craftsman. And indeed, as we shall see, he was described as having "crafted" the earth. Furthermore in the tomb of the prince Keti at Asyut in Middle Egypt, dating from the ninth or tenth dynasty, the prince tells us that he has restored a temple of Wepwawet, "which Ptah built with his fingers",[50] again emphasizing the craft activities of the god.

From the time of the New Kingdom on, Ptah was widely mentioned as a creator god. It seems clear that unlike Horus and Re, who were sky-gods, Ptah was an earth-god. His chthonic nature is emphasized by his early association with Tatenen (the deified Land-which-rises, as the name means), who no doubt symbolized the first land which emerged from the formless waters of Nun. The most extensive of the New-Kingdom documents which refer to the creative activity of Ptah is a Berlin Hymn to Ptah, extracts from which we have included as Document II.7d, and it often gives us the compound name Ptah-Tatenen (e.g. see Doc. II.7d, A.1, B.1, C.1). Straightway in that document (A.1-2) he is called the father of the gods and the eldest of the primordial gods. He was the one who produced himself before any other creation had taken place (B.19). He crafted *(ḥm)* the earth according to the plans of his heart *(sḥrw n ib-f)*. This seems to unite the idea of the craftsman-creator with the doctrine of creative will (cf. D.20 and E.8), which we shall see elaborated in the *Memphite Theology*. In brief, Ptah gave birth to everything that exists, having begotten and fashioned all the things that are (B.22). A reference to Ptah as the "father of the fathers of all the gods" (C.4) may reflect the Hermopolitan Ogdoad who were known as the "fathers of the gods" or the "ancestor gods", but on the other hand it may reflect the various forms of Ptah that are outlined in the *Memphite Theology*, as we shall see below. Later in the same passage Ptah is hailed as one who carries Nut [to the sky] and lifts Geb (C.9). He is called Khnum (i.e. the Fashioner, masculine) and the Mother who gave birth to the gods (C.12), and so he is thought of as androgynous, as indeed were most of the creator gods. He it was who begot all men and created their sustenance of life (C.13). Furthermore as Lord of Maat (so-called in a fragment from the time of Amenemhet I of the Middle Kingdom and in several sources of the New Kingdom)[51] he no doubt was considered as the creator of cosmic order. At the end of the twentieth dynasty Amenope introduces his list of created things by saying that he will include "what Ptah has created...". (Doc. I.9, Intro.).

Similarly in Pap. Harris I, Ptah-Tatenen is named the father of the gods, the great god from the primeval time, who fashioned mankind and made the gods.[52] His activity in the primordial time before general creation is outlined in some detail in the Berlin Hymn to Ptah (Doc. II.7d, D.3-13). He is described there as existing before those early gods which he created. His was a body that made its own body before the creation of heaven. He put his own flesh together and counted his members. He was at that first time a being who was alone and who created his own place. He had no father, no mother, but formed himself. The details of such an autogenesis, followed by the creation of all things, are described in a novel fashion in the *Memphite Theology*, which we may now examine (see Doc. II.9).

As I have often said in the preceding paragraphs the so-called *Memphite Theology* is of interest to our cosmogonic investigations primarily because it emphasizes that Ptah created everything by thought in the heart and utterance on the tongue, thus extending the doctrine of the creative word to the whole of the cosmos. We have seen that Gardiner proposed that the Memphite doctrine may have been related to the Heliopolitan view of Sia and Hu as goddesses symbolizing the creative Understanding and Utterance of Re. The *Memphite Theology* was written on a block of stone during the reign of King Shabaka (ca. 700 B.C.) and the introduction claims that it was copied from a worm-eaten original by his majesty (Doc. II.9, lin. 1-2). The document goes on to suggest that the copying produced a work "better than it had been before".

Until recently most scholars believed that the original document from which it was copied was composed in the Old Kingdom or even earlier and that the copy was substantially like the original. But recently it has been argued that the document was composed rather at the time that it was written down on the stone in the twenty-fifth dynasty, or, as another author believes, in the nineteenth dynasty (see the Introduction to Doc. II.9, notes 2 and 3). Even if the document was not composed until the time of Shabaka, it is still likely that it was fashioned out of scattered but firmly held older views concerning the creative power of words, views which we have alluded to in the earlier parts of this chapter and which have been admirably summarized by Zandee.[53] Not only does the doctrine of Hu and Sia appear to have been a possible source of the Memphite scheme but ideas expressed about Ptah himself in the Berlin Hymn (Doc. II.7d, B.19-20) seem to suggest that Ptah had already been looked upon as creator of the earth according to "the plans of his heart" (cf. D.20, where we read "your mouth has engendered and your hands have fashioned"). The latter quotation seems to mix both forms of creation that are associated with Ptah: by will and by handicraft. Earlier, on an amulet of the nineteenth dynasty, we read that Ptah provides for the Living Ka that is Memphis and that there is in his mouth "food-creating utterance".[54]

The cosmogonic section of the *Memphite Theology* begins immediately with a list of the forms of Ptah (Doc. II.9, lin. 48-52): "Ptah-upon-the-Great-Throne

...; Ptah-Nun, the father who made (?) Atum;
Ptah-Naunet, the mother who gave birth to Atum;
Ptah-the-Great, who is the heart and the tongue of the
Ennead [i.e. as creator of the Ennead]; [Ptah]...who bore
the gods; [Ptah]...; [Ptah]...; [Ptah]....Nefertem at the nose
of Re every day."[55] It is evident that only the first
four of these forms of Ptah are clear enough to help us
explicate the basic ideas of the Memphite cosmogonic
scheme when we join them to what follows. The first
form, Ptah-upon-the-Great-Throne, singles him out as
the great king, no doubt the creator who was alone
before assuming the other forms. By the next two
forms he is identified with the primeval Abyss, but here
conceived as two different sexual forms, male and
female, and so resembling the first two forms of the
Hermopolitan Ogdoad. This identification of Ptah with
male and female forms of the Abyss allows Ptah to be
called the father and mother of Atum, and hence implies
that, as Ptah, the Abyss has a more active creative role
than it had in the earlier Heliopolitan scheme.[56] I
should remark, however, that as early as a gloss in the
Coffin Texts (Doc. II.2, Spell 335, IV, 188) the Great
One, the self-created, is identified as the *nu*, the Watery
Abyss, and in the same work (*ibid.*, Spell 714, VI, 343-4)
Nu[n] is said to be the

> Sole One who has no equal and who
> was born on the great occasion of
> my flood.... I am the one who
> originated in the Abyss... I brought
> my body into being with my own
> might. I am the one who made
> myself. I fashioned myself at my will
> according to my desire.

In the later solar scheme Nun was also described

as the father of the gods. This is evident in Doc. II.6, at the beginning: when Re assembles his Eye, the next four members of his Ennead, the ancestor gods and Nun, the last one is called "the father of the eldest ones (gods)" and is addressed by Re as "Oh Eldest God in whom I came into being". Nun replies "My son Re, a god greater than he who made him." Nun's parental role is also evident in the Hymns to Amun, as we shall see when discussing the Theban cosmogonies.

Ptah's fourth form is, of course, the crucial one for the creation doctrine: Ptah-the-Great, who is the heart and the tongue of the Ennead. The significance of this form is evident in the passages that follow the list of forms (Doc. II.9, lines 53 *et seq.*). We first see that the form of Atum took shape in the heart and tongue of Ptah, for, as the account says, "Ptah is the very great one who transmitted [life] to all the gods and to their kas (i.e. souls) by means of the heart in which Horus has taken shape [as an agent or form of Ptah] and by means of the tongue in which Thoth has taken shape as [an agent or form of] Ptah." The document goes on to say (line 54):

> [Thus] it happened that heart and tongue gained mastery over [every] member [of the body] according to the teaching that he (Ptah) is in every body [as heart] and in every mouth [as tongue]: [i.e. in the bodies and mouths] of all gods, all men, all cattle, all creeping things, and of everything which lives. Accordingly [as heart] he thinks out and [as tongue] he commands what he wishes [to exist].

J. A. Wilson interprets this to mean that "the same principles of creation which were valid in the primeval water to bring forth Atum are still valid and operative. Wherever there is thought and command, there Ptah still creates."[57]

The next part of the account contrasts (but apparently without abandoning the former) the Heliopolitan creation of Atum's Ennead by masturbation and consequent ejaculation of semen with the verbal creation of Ptah's Ennead, which is before him as teeth and lips. For Ptah's Ennead are the teeth and lips of Ptah's mouth "which pronounced the name of everything, from which Shu and Tefnut came forth, and which gave birth to the Ennead". The faculties of seeing, hearing, and smelling report to the heart, which causes "every understanding (i.e. completed concept) to come forth, and it is the tongue which repeats what the heart has thought out (i.e. devised). Thus all gods were born and his Ennead was completed." Thus the divine order ("every word of the god") was created "by means of what the heart thought out and the tongue commanded". Thereafter the full range of creation is mentioned as being "thought out by the heart" and as coming forth "on the tongue" and so creating "the performance of everything": food, provisions, justice to the good, punishment for the bad, life to the peaceful, death to the criminal, labor, crafts, actions of the hands and feet, and indeed all physiological processes. In short, Ptah has created everything, and hence Ptah, identified with Tatenen, "is the mightiest of the gods". The cosmogonic section of the document is then completed with a poetic passage that celebrates Ptah's creation of gods, towns, and nomes, his putting gods into shrines with offerings to them, and his letting their bodies enter the statues made of various materials. Confusing though this document may be in places, it remains one of the most remarkable compositions from ancient Egypt and justifies Wilson's evaluation of it:[58]

> The creation texts which we have discussed earlier have been more strictly in physical terms: the god separating earth from sky or giving birth to air and moisture. This new text turns as far as the Egyptians could turn toward a creation in philosophical terms: the thought which came into the heart of a god and the commanding utterance which brought that thought into reality. This creation by thought conception and speech delivery has its experiential background in human life: the authority of a ruler to create by command. But only the use of physical terms such as "heart" for thought and "tongue" for command relate the Memphite Theology to the more earthy texts we have been considering. Here, as Professor Breasted has pointed out, we come close to the background of the Logos doctrine of the New Testament: "In the beginning was the Word, and the Word was with God, and the Word was God."

CREATION AND COSMOLOGY

Theban Cosmogonies of Amon-Re

The most prolific and extravagant development of cosmogonic ideas was evident in the hymns and prayers associated with Amon-Re, the most powerful god in Egypt in the New Kingdom when Egypt's military and political influence in the Near East was highest. Amun, whose name means the Hidden One, emerged as a significant god in Thebes at the end of the eleventh dynasty, at least in court circles. Indeed a vizier of the last king of the eleventh dynasty was named Amenemhet ("Amun is at the fore"), and he was elevated to the kingship as the first king of the twelfth dynasty.[59] The god and his wife Amaunet could have been introduced into Thebes from Hermopolis, for we later find their names among the four pairs of gods of the Ogdoad of that city, or he may have been a special form of the early fertility god Min of nearby Coptos.[60]

The cosmogonic ideas which swirl about Amun, particularly in his syncretic form of Amon-Re, are a pastiche of borrowings from the three systems already described: Heliopolitan, Hermopolitan, and Memphite. For example, we find Amun identified with Tatenen, i.e. Ptah-Tatenen, in Doc. II.7b[3] (III,23-25 and IV, 12-15), where he (Amun) is also described as having created the Ogdoad, the primordial gods, and Re himself. But his attributes are not limited to those of the three systems, for we find that he has borrowed from Min (if indeed he was not in fact Min) the latter's ithyphallic form (see Fig. II. 16 and also note 60).

Before considering four great hymns to Amon-Re, it will be useful to list some of the many epithets and brief statements that bear on the cosmogonic aspects of Amon-Re's activities and which are found in tombs from the early part of the eighteenth dynasty:[61]

The Primordial God of the earth from the beginning, the Illustrious Power who created himself, the Lord of All in each of his places, the One who is complete in form, the Primordial God who created what is and brought forth being, the Lord of millions having no equal, the One who determined the origin of gods and men, the One from whose mouth the earth came forth, He who arose from the primitive waters to the heaven, the Hawk who flies around day and night, the One who gives birth to himself every hour and comes forth from his mother every day and goes to rest in her at his time, the King of unending time or eternity *(nḥḥ)*, the Lord of everlastingness *(ḏt)*, the One who speaks and what should come into being comes into being *(ḏd-f ḫpr ḫprt)*, the Bull (i.e. Progenitor) of the Corporation [of the great Ennead], the One who fashions man and joins together the gods (i.e. produces their forms and joins man's limbs together), the One who illuminates the Two Lands with his right eye, the One who has driven out the ambient darkness when he arose in Nun, the Great Phoenix on his shore when he shines forth as Re (?), the

> Great Goose [of creation], the Eldest
> of the heaven and earth who created
> himself to appear in the primitive
> waters, and so on.

The four hymns I have selected to present as Document II.7b add substance to the various short epithets I have just given. In Document II.7b[1], we first see Amun described in column I as the "eldest of heaven" and the "first-born of earth", the "Lord of what is, [who is] enduring in all things" (repeated). He is called "Lord of order $(m^{jc}t)$". It is he "who made mankind and created beasts" as well as fruit trees and herbage, giving life to cattle. He is the "Divine Power whom Ptah made". Furthermore in column II Amun is called "the chief one who made the entire earth". He is celebrated in the two primitive shrines: the Per-wer of Upper Egypt and the Per-nezer of Lower Egypt (see above, Doc. I.1, n. 35). It is his fragrance which the gods love when he descends from Punt. In column III he is identified as "Min-Amon", and called "Lord of eternity, who made everlastingness". In IV he is identified as Re and again said to be Lord of order $(m^{jc}t)$, "whose shrine is hidden" (thus playing on the fundamental meaning of Amun as the one who is hidden), and he is further identified as Khepri "in the midst of his bark", the one who gave commands and the gods accordingly came into being (a clear indication of the presence of the doctrine of the creative word in cosmogonic accounts long before the *Memphite Theology* was inscribed on the Shabaka stone). He is "Atum, who made the people $(rhyt)$, distinguished their natures, made their life, and separated [their] colors, one from another". In column V he is said to be "great in appearances in the Benben-house", i.e. in Heliopolis, the "lord of the New Moon Feast" and celebrated one of the festival of the Sixth Day and that of the last day of the third Quarter-Month. As is the case with all creator gods he is called the "sole one" and the one who made everything that is (Column VI). Mankind came forth from his eyes and the gods from his mouth, both doctrines familiar to us from the Heliopolitan scheme. He creates herbage for cattle, fruit trees for man, nourishment for fish and birds, for that which is in the egg, for serpents, insects, rodents, and so on. In column VII there is possible reference to his creation of the Ogdoad since he is called there "father of the fathers of all the gods". He is the "sole one, without his peer" (column VIII). He lives on maat every day, i.e. he thrives on the cosmos which he has established.

In the next of our hymns to Amon-Re (Doc. II.7b[2]) Amun is addressed as Re, who is described as having a radiance that exceeds that of electrum (d^cm), and he is a Ptah who fashioned or modeled (nbi) his body. With use of the verb msi in its meaning to fashion or shape statues he is said to be a "shaper who is not [himself] shaped". Once more we learn that he is unique. In one day he races millions and hundreds of thousands of leagues. Each day is but an instant to Re. He is the creator and vivifier of all.

The third document illustrating the cosmogonic prowess of the great syncretic god of Thebes includes sections or stanzas from Leiden Papyrus I 350 (Doc. II.7b[3]). In Sect. II,26 of this document Amun is described as crafting (hm) himself, reminding us of the form of creation ordinarily associated with Ptah. No one knows his shapes or forms. He made his own images as he created himself. In II,27 we are told that he joined his seed to his body to create his egg within his secret self. He became a form and image of birth. I have already mentioned the passage (III, 23) in which it is stated that the Ogdoad was Amun's first form and that he made his further transformation as Tatenen in order to give birth to the Ennead or to the primitive gods (III,24), suggesting once more the doctrines later found in the *Memphite Theology*. This was followed in III,25 by the removal of Amun to heaven and his establishment there as Re. It is emphasized in III,26 that Amun came into being when there were no other beings and no land existed outside of him. All gods came into being after him. When the Ennead was created it was part of Amun and joined to his body (IV,1), i.e., all the gods of the Ennead were united in his body. Then Amun separated himself to produce the beginning of things and he was Tatenen, shaping himself as Ptah (IV,2). The fingers or nails of his members or body were the gods of the Ogdoad, the account once more reflecting the Hermopolitan scheme. Note that I have already quoted much of this passage to throw light on the Hermopolitan doctrine of the Great Cackler who at creation broke the silence with his cackle. In IV,9 Amun's solitude at creation is stressed. Being the first to come into being, Amun's secret image was unknown:

> No god came into being before him;
> there was no other god with him [to
> whom] he might tell his form. He
> had no mother to produce his name
> and had no father who begot him
> [and thus] could say "This is I" (*or*
> "He is mine").

It was Amun who shaped or modeled his own egg (IV,11), this "mysterious force of births, who created his

[own] beauty". In the following passage (IV,12-21) many of the cosmogonic ideas already recounted are repeated or elaborated: the mysteriousness of his forms, his shining appearance, the multiplicity of his forms, Re's union with his body, his primacy in Heliopolis, his identification with Tatenen (i.e. Ptah), his coming forth from Nun, his initial form as the Ogdoad, his procreation of the primeval gods who brought Re into being and the completion of himself as Atum, his indefinable mystery, his unapproachable and unknowable power, and his "hidden" name.

In IV,21 we find the remarkable statement that "all gods are three: Amun, Re, and Ptah, and there is none like them" (i.e., they have no equal). Their cities on earth are Thebes, Heliopolis, and Memphis (IV,22), and these will last forever. This is not a trinitarian grouping implying monotheism. Rather it is a statement that reflects the widespread holdings and influence of these three gods and their priesthoods at the end of the fourteenth century B.C. when this passage was written. Finally we should notice the remarks of V,16-17 where Sia (creative understanding or knowledge) is described as Amun's heart and Hu (authoritative command) as his lips, a doctrine probably originating in the Heliopolitan scheme and taken over in the *Memphite Theology*, as I have often noted above.

Our last hymn to Amon-Re (Doc. II.7b[4]) is the most recent since it is probably contemporary with decrees benefitting the High Priest of Amun Pinudjem II (990-969 B.C.) and his wife. It expresses many of the ideas found in the other hymns. It demonstrates Amun's position as the first primeval god from whom all others came into being. Being the Unique One at the beginning, he created what existed in the first time. He was "mysterious of births, numerous of appearances, whose secret image is not known". All forms came from his form. He gave light to the earth at the first time with the solar disk. He sails the sky. He is the Ancient One who early arises as a youth and "who brings forth the limits of eternity". He passes through the Netherworld to give light to it. He is the "Divine God who formed himself, who made heaven and earth in his heart". "Eternity comes under his might, he who reaches the end of everlastingness." He is identified with the Watery Abyss. He makes himself manifest at his time in order to vivify the clay of man that comes off the potter's wheel. This surely is what he has done in the birth scenes of Hatshepsut and Amenhotep III in Deir el-Bahri and Luxor. His eyes are identified as the sun disk and the moon. Man is born from the creator's eyes and the gods from his mouth. He makes food, creates nourishment, and shapes everything that is. He is the everlasting one who wanders through the years without an end to his existence. He traverses eternity, and when old he makes himself young. With his daily reappearance as the circling sun over so many years, he guides the millions by giving light. He created earth by his plans and receives obeisance from the gods and goddesses. "His form is that of every god." He is decisive and unwavering, his word firm, his decree perfect and unfailing. "He grants existence, he doubles the years of him who is in his favor, he is a good protector of him whom he has placed in his heart."

The various doctrines of creation expressed in these hymns are found widely circulated in many other hymns. For example, notice how the doctrine of the creative word is expressed in a Ramesside hymn to Amon-Re-Atum-Harakhti,[62]

> who spoke with his mouth and there
> came into being men, gods, cattle and
> all goats in their totality [and
> further] all that flies and alights.

The hymn goes on to assert that he created all the regions including the Ḥa-nebu (the Mediterranean Islands), "the fertile meads made pregnant by the Nun (i.e. the Nile) and later giving birth; and good things without limit of their number to be sustenance for the living".

Temple rituals reflected in reliefs and in papyri repeat many of the cosmogonic ideas of the hymns. There is a ritual called the "opening (or uncovering) of the face of Re" which is addressed to Amon-Re. It is accomplished "in the House of Benben" in Heliopolis.[63] The ritual includes a hymn to Amun in which he is described as having constructed (ḳd) himself, and Moret notes a parallel passage to the effect that he (Amun) has "constructed his own members".[64] In another chapter of the ritual book we are discussing,[65] Amon-Re is again addressed in the course of speaking of the divine image

> that came into being in the first time
> (sp tp) when no god was [yet]
> created, when the name of nothing
> was known, when you (Amon-Re)
> opened your eyes to see with them
> and everybody became illuminated
> by means of the glances of your
> eyes, when the day had not yet
> come into being. You opened your
> mouth and spoke with it to establish
> the heaven in the West with your
> two arms in your name of Amun, the

image of the ka of all the gods, the
image of Amun is that of Atum and
that of Khepri..., you are the one
who bore all living things, you are
more alive and more powerful than
all the gods, you have seized for
yourself the whole Ennead..., you are
a god who has created with your
fingers and a god who has created
with your nails what has been
created [i.e. you are the one who
created things by touch], [you are]
the Lord of all, Atum, created in the
first time.

We can also mention the hymn appearing in the
Harris Magical papyrus (Doc. II.11, G), which reappeared
in expanded form in the Temple of Amun at the oasis
of el-Kharga in Persian times.[66]   We read in the
magical papyrus:

Adoration of Amon-Re-Harakhti, who
came into being by himself,
Who founded the land when
he began
[to create],
Made by the Ogdoad of Hermopolis
(or Who made the Ogdoad
of Hermopolis) in the first
primordial time...
Amun the primordial god of the Two
Lands
When he arose from Nun and
Naunet.
What was said [came into being] on
the water and on the land.

> Hail to you, the One who
>   made himself
>   into millions,
> Whose length and breadth are
>   without limit,
> Ready Power who bore himself,
> The primordial serpent who is
>   powerful of flame,
> The One rich in magic with secret
>   forms.

This is a hymn which ties Amun to the Ogdoad of Hermopolis and it may even say that he was made by the Ogdoad, though most often in later inscriptions Amun was said to have created the Ogdoad, usually first creating Ptah (see n. 66). It also seems to say that what he asserted came into being, as in the case of the *Memphite Theology*. It stresses the limitlessness of his creation: one into millions and one who is without limit of length and breadth. It also puts forth the doctrine of the primitive snake-form of the creator god, and it stresses the creator's magical powers and secret forms. In short, this hymn is rich in all kinds of cosmogonic ideas and is a fitting document with which to complete this brief summary of the Theban cosmogonies of Amon-Re.

The Cosmogony of Aten

The hymns to Amon-Re which we have just discussed indicate the elaborate adoration that developed for that syncretic god, reflecting most prominently the Heliopolitan scheme but also revealing strains of the other cosmogonies with which we have been concerned. All of these hymns and eulogies to

Amon-Re (and indeed to the various other gods as well) are addressed to a single god either in simple or syncretic form, but never to such a god considered to be the only god. That is, all the documents we have considered still reflect the polytheistic views of the ancient Egyptians. In our next document, however, the Great Hymn to the god Aten (Doc. II.7c), we find that the hymn is addressed to Aten considered as the only god. It is then a truly monotheistic document. We say this even though it is clear that already, before Akhenaten, the solar disk (i.e. the Aten) had begun to be addressed as an individual personalized god in his own right.[67] It should also be realized that though Aten is considered as the one god, the form and content of the account of his creation are prosaic and everywhere reflect the cosmogonies we have already described, with the obvious exception that he is not credited with creating other gods, he being the only god. We note that in the first part of the hymn it is said that all nature and beings awaken and respond to the Aten on his rise each day, an idea that was conventional by the time this hymn was composed. The hymn (Doc. II.7c,6) treats Aten in largely customary fashion as the one

> Who makes seeds grow in women,
> Who creates people from sperm;
> Who feeds the son in his mother's
> womb....
> To nourish all that he made.

The list of Aten's creations is commonplace: the earth when he was alone and acting on his own desire, people, herds, all that walk upon the earth on legs, all that fly on wings, the various lands and places of the world where men were set, their sustenance, the reckoning of their life-spans, their different languages and colors, the river in the underworld, the rain from the sky, and the mountain rivers, and indeed the millions of forms that spring from him alone.

Cosmogonies at the Temple of Esna

We have already described the cosmogony assigned to Ptah and reflected in the Berlin Hymn to Ptah (Doc. II.7d), and so we may turn immediately to the late cosmogonic material found at the Temple of Khnum at Esna (Doc. II.7e), all of these extracts dating from the time of the Roman Emperor Trajan (98-117 A.D.). I give these extracts as representative of the cosmogonies described on temples built in the Ptolemaic and Roman times. Similar accounts could also be constructed by examining inscriptions in the temples at Philae, Edfu, Kom Ombo, and Dendera.

It is perfectly clear that the god Khnum-Re is credited with the same creative acts as the other creator gods. Indeed he is identified with Ptah-Tatenen as the creator of the primordial gods (Doc. II.7e [Text 394,23]). He is called the "great god who came into being at the very beginning" and the "magnificent ram, at the first time". We hear of his lifting the earth and supporting the sky, and of his shining forth with the form of luminous brightness. "He installed the soul of the spirits in the midst of the waters (?)." "He acted the god when he began to come into being" (Text 394,25). He is "mysterious of aspect" and is called the "modeler of the modelers" (Text 394,26), an obvious reference to his creation of living things on the potter's wheel (a function sometimes also ascribed to Ptah).[68] He is called the "eldest of the primordial gods". He is also the "father of the fathers" and the "mother of the mothers". He made both superior and inferior beings, cities and countries, and the Two Lands (Egypt). He made firm the mountains (Text 394,27). He brought to life those he had modeled on his wheel and he provides continued sustenance for them. "He comes forth at the right time without cease." His most frequent identification is as the lord or god of the potter's wheel (see Texts 319, 378 and 394). Like the other creator gods he is unequaled and he "made that which is and that which is not", i.e. everything (Text 378,9).[69] He is the omnipotent one (Text 319,16):

> You have modeled men on the
> potter's wheel,
> You have made the gods,
> You have modeled large and small
> cattle,
> You have formed everything upon
> your wheel, each day,
> [In] your name of Khnum the Potter.

Also like other creator gods he is described as the "mysterious one whose form no one knows" (Text 378,10). Khnum came forth from the Abyss and appeared with the form of the [solar] flame. Not only does the Nile arise from two caverns under his feet [at Elephantine] but he likewise produces the north breeze "for the nostrils of gods and men". His right eye is the sun and his left the moon. Again we see him identified with Ptah-Tatenen (Text 378,13). He is also described (in Text 378,14) as a Heh god, i.e. a support god (no doubt symbolizing all of the eight Heh gods holding up heaven--gods whom I have mentioned earlier). Finally note that he is also identified with the eldest son of

Atum, i.e. with Shu.

A great many other details concerning Khnum's activities can be milked from these extracts, but I set them aside and pass on to a rather detailed account of creation by the goddess Neith, an account also appearing in our extracts from the Temple of Esna (Doc. II.7e [Text 206]).

Like other Egyptian gods cast in the role of the demiurge, Neith is called "father of the fathers" and "mother of the mothers". Similarly, like other such gods she is described as having come into being from herself at the beginning of time in Nun. This was the time when the land was still in the shadows of the Abyss, i.e. when the land had not yet emerged. In the beginning Neith took more than one form. First she gave herself the appearance of a cow in order to hide her divine form. "Then she changed herself into a lates-fish $(^{c}\dot{h}^{j})$." Then she went forth and gave illumination with her eyes to what she saw (i.e. to say, when she looked at something she illuminated it), as is said of other creator gods and particularly of solar gods. "Then she said, let this place (where I am) become for me a platform of land in the midst of the Abyss in order that I might stand on it." This is one of the many such commands in this account which show the device of the creative word in action. The first land to be created was Esna, which is equated here in some way with the early home of Neith in Sais.[70] Thereafter Neith created thirty gods, again by using the technique of the creative word, i.e. by pronouncing their names. She ordered her children to stand on the primordial land, which was called "Highland" (k^{j}). Then they asked of her what else was to be created. She answered by describing the creative process of conception and verbal command: (1) the enumeration of the four [creative] spirits (*3ḥw*), (2) the giving of form to that which is in the stomach (perhaps the magical forms and concepts in the heart?), (3) the pronouncing (*šd*) of what is on the lips, (4) the recognition or knowing of the resultant beings that will arise that very day. Thereupon they did everything which she described. Then Neith considers what she will produce next. She declares that a god will come into being who will produce light by opening his eyes and darkness by closing them. Men will be produced from the tears of this god and gods will be created from his saliva. Neith will fortify this god by means of her power, making him effective through her own efficacious spirit. She predicts that men will rebel against this god, thus recalling the account of such a rebellion in *The Book of the Divine Cow*, and further that this rebellion will be defeated. The name of this powerful god she will create will be Khepri in the morning and Atum in the evening, and he will be a god who shines forth every day forever in his name of Re. To this the gods reply "we are ignorant" (*ḥm-n*). From this reply arose their name of Eight (*ḥmnw*), from which Hermopolis takes its name. So it is clear that here we have Hermopolitan influence.

Further evidence of Hermopolitan influence is seen in the details of the sun-god's birth as described by Neith. He was born from excretions of hers which she had placed in the [primordial] egg, which egg we have already mentioned in connection with Hermopolitan doctrines. Accompanying the birth of the sun came light and the first day of the year (i.e. the solar year came into being). At this point are described the details of the creation of man from the god's tears brought on by his not seeing his mother and the creation of the gods from his saliva produced when he salivated on seeing her again. Finally we notice a passage which states that, from seven commands which Neith pronounced, the Seven Goddesses of Methyer (a cow goddess) were created, one more reference to the Memphite type of creation by spoken command.

Of the various hymns which we have examined in the extracts designated as Document II.7, only one is not primarily cosmogonic. This is the Great Hymn to Osiris (Doc. II.7a), which I have included as an example of a quite extensive account of the Osiris-Horus legend. But in this account we see that even the chief funerary god Osiris has attracted to himself some of the epithets of a creator god. Thus he is described as lord of eternity, king of gods, one whom the sky and stars obey, the ruler of the imperishable stars, the one who sets maat throughout the Two Shores, and, as heir of Geb, the ruler of the land, the water, the wind, the plants, the cattle, all flying creatures, reptiles, desert game, "the leader of all the gods" and the one "effective in the Word of Command".

Cosmology

To this point our concern has been almost entirely with the various views held by the ancient Egyptians concerning the way the world came into being. In the part that remains we shall try to see what kind of world resulted, what sort of visible and invisible beings populated this world, and what was the nature of the forces which were believed to keep the world and its parts functioning harmoniously and of those which were dangerous and threatened the desired harmony of the cosmos. In this brief examination of cosmology I shall deliberately exclude any discussion of Egyptian astronomy, since that topic will be reserved for the next volume.

The world after creation was inhabited by gods (*netjeru*), spirits (*akhu*), man, and all kinds of animals. However present the gods might have been on the visible water and land during the first creation, it is evident that by the time of the first dynasty, when the Two Lands came under pharaonic rule, the gods retreated to the heavens (and soon also to the netherworld beneath the visible world of man). That the gods formed and continued to control the natural world is evident from all of the cosmogonic accounts we have described. But how were they manifest to man? What was their essential nature?

It is sometimes thought that the prehistoric Egyptian saw gods first in fetishes, then in animal forms, then in anthropomorphic forms, and finally in mixed animal and human forms. But surely we have no hard evidence of the course of belief in the early stages of civilized man.[71] True, early Egyptians throughout the Gerzean or Naqada II Period (ca. 3500-3100 B.C.) fashioned palettes in animal forms that may have signified deities. For example see the palette of Fig. II.17, which is in the form of a fish, and remember that one of the forms that Neith took in the account of creation which I discussed at the end of the last section was that of the lates-fish.

Similarly we see numerous vases decorated with ships that have standards which seem to bear emblems symbolic of gods (see Fig. II.18a). Standards of this kind, which probably represented the divine patrons of towns or regions, are later found on a mace-head of the Scorpion king (Fig. I.3). There we see two emblems that may consist of the animal form (a wild dog?) that represents the god Seth, one that is a jackal representing some divinity not surely known to us (compare also Fig. II.18b) and one that is the notched symbol representing the god Min. Before the king a bearer carries a standard with a curious oval-like symbol that sometimes is said to represent the royal placenta and later appears to be called *khons* (but is not to be identified with the god Khons).[72]

On the mace-head of King Narmer and also on his celebrated palette (see Figs. I.8 and I.9) we find the procession of four identical standards: with two emblems consisting of falcons, which no doubt represent the god Horus; one that is a jackal, and one the unidentified oval emblem mentioned above. These fetish and animal forms do not of course tell us how the gods were truly conceived. It is obvious that from the first dynasty (and no doubt sometime before) statues were made to represent gods and celebrations followed in some kind of religious services, as is perfectly plain from the *Annals* we have translated and discussed as Document I.1 (see particularly note 6). The statue was kept in a sanctuary that was the earthly "House of the God". The purpose of the cult-ceremony performed was no doubt to make the statue suitable for its occupation by the god. This was certainly the case later when we have detailed written descriptions of cult ceremonies.[73]

Incidentally, I remind the reader that we have already given a crudely drawn representation of the god Ptah enclosed in an open naos or shrine on a bowl from Tarkhan which can be dated approximately to the first dynasty (see Fig. II.13). Even the depiction of gods by mixed animal and human forms goes back to the first dynasty or before, as is evident from the divinity shown in relief at the top of both sides of the Narmer palette, who has a human face with cow horns and ears and so is probably a figure of the goddess Hathor. See also the figure which has what may be Seth's animal head and a human body, the figure appearing on a sealing that dates from the reign of Peribsen in the second dynasty (Fig. I.19a, middle figure). Examine other anthropomorphic divinities depicted on early-dynastic objects in Fig. II.20, where it is particularly noteworthy that they often carry signs of life and power, the two signs attesting to important attributes of gods throughout Egyptian history. Hence all the main ways of depicting the gods seem to go back at least to the beginning of dynastic history if not before.

I might add to this discussion that the gods depicted in human form often carried on their heads hieroglyphs indicating their names or other symbols by which they could be identified.

Though the gods are shown in forms that are well represented in nature, perhaps exhibiting characteristics that can be related to powerful, dangerous, or clever animals, to men, to plants, to extraordinary natural events like the flooding of the Nile, and even to inanimate objects, most of the cosmogonic accounts we have discussed mention that at least the creator gods have a mysterious divine form or name that nobody knows (or presumably can know). This is illustrated by the well-known tale that centers on Re's secret name and the efforts of Isis to extract that secret name from the mighty god.[74] She succeeds by a stratagem worthy of her magical prowess, causing in him a painful snake bite which she will not cure until the name is revealed to her.

Not being able to know the god's true divine form, Egyptian theologians tended to multiply the god's describable attributes or forms. Thus we notice a work entitled the *Litany of Re* (Doc. II.5), which commences with seventy-five invocations to Re. They allude to manifestations, activities, or properties of Re-Osiris (as "The One Who is Joined Together in the West"), i.e. in connection with his journey in the Netherworld, and his renewal of bodies in the earth, and the consequences thereof for the kings in whose tombs (in the New Kingdom) this litany was inscribed. To these invocations there is added in a rambling way much that tends to identify Re with Osiris. This added material has been omitted in my extracts in Document II.5.

As the attributes and manifestations of the gods proliferate so do the ways of depicting them, that is, iconographic plurality tends to accompany the developing complexity of theological attributes and forms. Hornung comments on this succinctly:[75]

> There is an astonishingly rich variety of possibilities; only to a very limited extent can one speak of a canonically fixed iconography of a god. The goddess Hathor is a good example (Fig. II.21). Her normal iconography...shows a slim woman who wears a wig covering her head and on top of it a pair of cow horns with a sun disk between them. There are, in addition, three more ways of imagining Hathor. In direct contrast to her completely human form, the goddess may have a pure

animal form, as, for example, in the Hathor shrines of Deir el-Bahri, where she is depicted as a cow from whose udder the king drinks or as a cow stepping forth from the western mountains of Thebes and taking the deceased into her protection. In between there is the unusual form of the capital of a Hathor column or pillar [Fig. II.22]...: a cow with human face, whose ears are animal and whose eyes, nose, and mouth are human. Finally the form of a human body with animal head is not lacking....

There are other ways in which the Egyptian theologians attempted to express the manifold manifestations and complex nature of the gods. Gods were said to have multiple kas and bas, i.e. vital forces and souls, which, in their singular forms found in man, I have already mentioned often in these pages (see particularly Doc. I.6, notes 16 and 17). Thus Re is ordinarily said to have fourteen kas and seven bas.[76] In his analysis of multiple kas, Vandier distinguishes them from an individual ka possessed by man. He labels them as *génies*[77]

who personify a certain number of qualities. These *génies*, when they are mentioned, are generally seven in number and have seven companions who are called *hemsou*. The fourteen qualities that they personify are, according to the most current list: force, power, honor (i.e.

venerableness), prosperity, nourishment, duration of life, radiance, brilliance, glory, magic (=*hike*), creative will (=*hu*), sight, hearing, and understanding (=*sia*). These qualities are properly divine and are those which yield the eternal spiritual force. It is without a doubt that Re possesses them so eminently that he is judged to have fourteen *kas*.

According to Bonnet,[78] these multiple kas are, so to speak, divisions or separate aspects of an individual ka. For our purpose they reveal the characteristics and powers that were thought to be divine by the Egyptians. We notice that the emphasis is placed on power and force, on radiant and brilliant manifestation, on faculties of perception and will (supplemented by hearing and sight) that play a part in the creative word and in magic. Though they are the most important divine qualities when possessed in the degree in which they are present in Re, they are also obviously the most admired qualities when possessed by man even in a lesser degree.

It is clear that from the very beginning of the pharaonic period Egyptians believed that one of the most precious possessions of the gods was the ability to give the breath of life to man. Recall from Figures I.19c and II.20 that among the earliest depictions of gods in anthropomorphic and mixed animal and human forms, the gods were wont to carry the sign of life, the well-known *ankh*-sign, in one of their hands. This persisted throughout Egyptian history, so that when Akhenaten's artists represented the solar disk, the Aten, his rays ended in hands that held signs of life to the nostrils of Akhenaten and the members of his family (see Fig. II.23). This is essentially what transpired when one of the creator gods whom we have already discussed brought to life the clay model of a person which had been turned out on Khnum's wheel.

Associated with the divine power of vivifying is magical power, which gods possessed in varying degrees, but all certainly in some degree. As Zandee has shown, magical power is essentially the same as the power of the creative word.[79] Magical power was possessed by the king when at death he joined the stars, for "his magic has equipped him" (Doc. II.1, Sect. 250). Indeed when the king has become the Bull of the sky, he lives on the being and entrails of every god, "even of those who come with their bodies full of magic" (*ibid.*, Sect. 397). King Pepi identifies himself as a magician possessed of magic (Sect. 924). In still another spell of the *Pyramid Texts* (Sect. 1324), the deceased king asserts that it is magic and not the king himself that speaks to the gods, the magic being in his belly (Sect. 1318). Presumably the deceased king in these passages, if not a full-fledged god, is at least an *akh* whose ba or ka possesses magical powers in the manner of all gods. Boylan's remarks concerning hike [heka] or magic (and especially the magic of Thoth) are worth quoting:[80]

> Looking at the use of the word *hike* generally, we find that it seems to include the whole field of what might be called magical. It is a sort of mysterious power which can produce effects beyond the sphere of man's achievements:...it is something

before which demons of sickness, and of poison--evil demons in general, must give way. Hence the uraeus serpents, the deities who guard the king, are "great in *hike*...". Isis and Nephthis, as protectors of Osiris, are also "Great in *hike*". Thus *hike* seems to include the whole sphere of magic, and to extend beyond it .... Thoth then, when he is called "Great in *hike*" or *Hike* [himself], appears as a god of magic.

Though magical power is in the particular province of gods, by its equivalence to the creative powers of words it can be effectively employed by man (particularly if he calls on the gods for help). It is assumed, of course, that the spells that dominate the documents I have included with this chapter originated with Thoth or some other god; but if a man speaks them correctly after he has become deceased (or even before he is deceased in some cases), then magical effects are accomplished. The effectiveness of spells is declared again and again. For example Spell 83 of Document II.2 tells us:

As for anyone who knows this spell,
he will not die again, his foes will
not have power over him, no magic
will restrain him on earth for ever.

In Spell 146 [not given in our extracts below], we are told at the end that it is "a spell a million times right", a testimonial that should have been impressive to the man who was overseeing the preparation of his coffin. Similarly in Spell 228 [not given in our extracts] we are told that the spell is beneficial for any one who knows it, for he "will complete 110 years of life," this being the desired ripe old age for a man. Spell 335 concludes with the statement that if the deceased goes to the west (i.e. dies) ignorant of this spell, "he shall not [be able to] go in or out [from his tomb], being ignorant". [This conclusion is not included in our extracts below.] There are many similar testimonials and declarations of effectiveness of the spells in the *Book of the Dead*, but we may omit them at this point. I should remark, however, that in the *Coffin Texts* there are a number of spells designed not to let a man's (i.e. a deceased person's) magic be taken from him (e.g., Spells 349, 350, not included in Doc. II.2 below).

As I have suggested, the spell may be effective even if the person is not in the realm of the gods or the dead. For example we may quote a spell against the evil eye:[81]

Sakhmet's arrow is in you, the magic (*ḥkʾ*) of Thoth is in your body, Isis curses you, Nephthys punishes you, the lance of Horus is in your head. They treat you again and again, you are in the furnace of Horus in Shenwet, the great god who sojourns in the House of Life. He blinds your eyes, oh all you people (*rmṯ*), all nobles (*pʿ.t*), all common people (*rḫy.t*), all the sun-folk (*ḫnmm.t*) and so on, who will cast an evil eye (*ir.t bin.t*) against Pediamunnebnesuttowi born of Mehtemwesekhet, in any bad or ominous (*dšr*) manner (*ḳd*). You will be slain like Apap, you will die and not live for ever.

Often such spells for the living call upon the gods to perform some action or make some cure. For example,[82] we find a spell for dispelling the *akhu*-demonic spirits from the belly:

> Come to me, mother Isis and sister
> *(sn.t)* Nephthys. See I am suffering
> inside my body....

Or it may be that the recitation of the god's name and his relatives, coupled with the simplest commands, suffices as in the case of conjurations for any evil swellings in any limb of a man or a woman:[83]

> Hail to you Re, in [that] name of his,
> 'He who has given birth to his
> children': the children of Re in
> heaven, the children of Re on the
> earth, the children of Re in the
> western desert, the [children] of Re
> in the eastern desert, the children of
> Re in the south, the children of Re
> in the north.... The Great Ennead are
> children of Re, the Little Ennead are
> children of Re, (and) Thoth, the
> great one, sojourning in heaven, the
> scribe...of Re (and) of the Ennead,
> the first-born child of Re, who
> ensures an infinite period for all the
> gods... He has made the spell applying
> to you, you evil swelling! He has
> taken away (the effects of) [your]
> utterance *(ts)* ! Oh do stand still, do
> stand still at the throne and the
> great Ennead, which is next to it.
> Oh do stand still, fluid of the evil
> swelling....

To this point I have examined magical power primarily as an attribute of a divinity, though I have briefly mentioned that a person may use a spell to invoke gods on his behalf to cure a disease. A few general remarks on the relationship of heka to religion are in order, since it is obvious that, for the Egyptians, magic was one of the key forces by which the world came into being and continued to run. In a brilliant article[84], Gardiner has made the perfectly valid point that

> So far as Egypt is concerned, there cannot be the slightest doubt that *hike'* was part and parcel of the same *Weltanschauung* as created the religion which it deeply interpenetrated.

He adds:

> It would have been strange if the practice of *hike'* had been restricted to the narrow circle of the living, when the living shared with the gods and the dead all their other modes of intercourse. In point of fact, it was *hike'* more than anything else that welded together the seen and the unseen worlds. The self-protective rites of the living...are full of trafficking with the gods and the dead.... Nothing could better prove the wide range of *hike'* than to observe its transference from secular to funerary or divine employments and *vice versa*. In the *Pyramid Texts* and the *Book of the Dead*,

compilations intended to ensure the well-being of the departed, one may often come across spells that must originally have been composed for earthly use--spells against the bites of snakes... or of crocodiles...; for example even erotic charms may be found inscribed on coffins .... It may therefore be taken as proved that *ḥike'* was intimately associated with the presumed existence of the gods and the dead as it was with the real existence of the living. But further than this, a greater or less element of *ḥike'* may have been inherent in all the dealings between men on the one side and the gods and the dead on the other. The two last classes of being were, after all, creatures of a world apart, elusive in their nature and hard to reach by ordinary, matter-of-fact means. The very idea of their existence puts a strain upon the imagination, and for this reason set forms of words, indicative of an effort to break down mystical barriers, had to accompany even such simple deeds of homage as the presentation of food offerings. In other terms, the gods and the dead could hardly be approached save by the medium of what is known as 'ritual', and the attribute which distinguishes ritual from ordinary performance may have been just that attribute which the Egyptians called *ḥike'* .... From the Egyptian point of view we may say that there was no such thing as 'religion'; there was only *ḥike'*, the nearest English equivalent of which is 'magical power.' The universe being populated by three homogeneous groups of beings--the gods, the dead, and human living persons--their actions, whether within a single group or as between one group and another, were either ordinary or uncanny *(ḥike')*. But the gods and the dead were somewhat uncanny themselves, so that all dealings with them or performed by them were more or less *ḥike'*.

I shall make no effort here to examine in any detail the practices of individual magicians as they attempted to defend someone or even themselves, from hostile gods or forces, or as they attempted to practice black magic against some enemy. However, the reader might find interesting the conclusion of Sauneron to the effect that the magical practices of those defending themselves against enemies employed a kind of internal logic beyond the mere sterile repetition of the formulas.[85] First they attempted to mobilize the power of Amun or some other great god through the image of his bas, for it was believed that the gods could project their bas a great distance to serve as a destructive force. So it was the forces of the gods rather than the gods themselves that were being so invoked. The second part of the operation consisted in convincing the adversary that any hostile act toward the magician would be in fact an act against the great god. Then finally the magician had to convince his adversary that the result would be a terrifying response from the great god.

Sauneron also remarked that "each great divinity has the power to act, even at a great distance".[86] This is evident by the supposed fact that the god, who might well reside in heaven, can be persuaded to occupy a statue, at least if that statue has been properly prepared. In the course of the cult activities that prepare that statue, it is purified, washed, fed, and clothed, and hence one sees that the god is in certain respects being treated like an important and respected human being. Needless to say, the priest could not have failed to see that the offerings of food presented to the gods in their temples (or indeed to the Blessed Dead in their tombs) were not in fact consumed. But the fiction was maintained that the god took his fill and the remainder was divided among the attending priests. All of the many cult rituals and their magical overtones have often been described[87] and are not of further importance for our brief cosmological remarks. In connection with the gods' occupation of cult statues it is worth noticing that when the appearance of gods is mentioned it is often in terms that would be applied to a jeweled statue or one composed of precious metals. Thus in the beginning of Doc. II.6 the aging Re is described as having bones like silver, flesh like gold, and hair like real lapis lazuli.

In addition to the sundry manifestations and powers which I have alluded to briefly, we can mention the oft-noted fragrance associated with the gods.

Earlier we spoke of the fragrance of Amon-Re when he descended from Punt, the land of myrrh and incense (see Doc. II.7b[1], II,4). Similarly Ahmose, the mother of Queen Hatshepsut, when Amun came to lie with her in order that the great queen might be sired by the god to assure her right to the throne, first noticed Amun's presence by the fact that the "palace is flooded with divine aroma," which awakened her.[88] In the early dynastic period an unguent had the name "aroma of Horus".[89] Of course the fragrance of the gods is tied to the fact that in their temples they are constantly censed. Indeed it may not be unimportant in this regard that the gods were said to have come into being from the exhalations of Atum while men came from his tears.

A god's presence might also be signaled by some great natural uproar or event. It is well known that when the shipwrecked sailor of an early tale was cast upon a desert island, the presence of a great snake-god was made known to him by "a thundering noise".[90]

Other attributes of the gods seem remarkably human. Even the greatest gods were not omniscient: witness the above-noted story of Isis's tricking Re's secret name from him. Furthermore it is clear that Egyptian gods could grow old and even die. It was, of course, the initial death of Osiris that set out the great Egyptian doctrine of resurrection in the other world. Furthermore, other gods were conceived of as dying. Such was the case of the form of Amun and the gods of the Ogdoad who went to Djamet at Medina Habu where they were worshiped as dead gods.[91] In the story of the shipwrecked sailor, the snake-god recounts a pitiful story of how his family was wiped out by a fire produced by a falling star. He ends the story: "I was not with them in the fire... I could have died for their sake when I found them as one heap of corpses."[92] Finally, recall the already mentioned passage in Spell 175 of the *Book of the Dead* (Doc. II.3) which held that all of the gods would die at the end of time and only Atum and Osiris would survive.

Not only were gods thought of as capable of dying but of course they were capable of some kind of sexual reproduction. We have mentioned the various generations between Atum and Horus, and we have also mentioned that Osiris was bored with the Underworld in part because he was deprived there of sexual pleasure.

We have seen that the world (heaven and earth) was full of gods. It also was filled with the Blessed Dead, that is, those who had at one time lived on the earth. When the funerary literature in the First Intermediate Period began to offer deceased human beings the same benefits as those enjoyed by the kings in the Old Kingdom, namely their resurrection or elevation to the heavens, the deceased acquired a kind of divinity as *akhu*. They assumed all the same goals as the kings had earlier, namely to join the sun-god in his solar boat, to have their kas and their bas soar to heaven or back to earth for them. They were given the same kind of spells in the *Coffin Texts* to make their passage to and life in the other world safe and assured.

But there was an intermediary being between the gods, whose powers and manifestations we have briefly described, and human beings, who successfully made the transition from their mortal lives here on earth to their future lives--lives to be lived partly in the Netherworld (where their bodies resided nearby in their tombs in the necropolis) and partly in the heavens (in the forms of their bas and kas). This intermediary being was the king himself. Many scholars have stressed the divinity of the king, for example Moret and Frankfort.[93] Others, like Posener and Hornung, have suggested that he was far from being considered a god himself, though the kingship in some fashion represented divinity.[94] In regard to the latter view, Posener has collected a fair number of statements which reveal that the king was hardly considered a god by those people close to him. Still the king and the priests themselves made much of his relationship to the gods. Indeed, it would seem to me to be significant that when the king's artists fashioned statues and reliefs of the gods they often substituted the face of the king for that of the god, thus underlining the divine nature of the king, though it might better be argued that the objective of this substitution was not to divinize the king but to establish his dominant position by creating an idealized physiognomy of the king that might be applied "not only for the representation of other members of the royal household but also for those of deities as well", as my friend Robert Bianchi writes in an unpublished communication. A spectacular case of this is illustrated by a piece in the Cairo museum where Tutankhamen is shown as the divine son of Amun standing next to Amun (on his left) and the young king's face appears on both statues (see Fig. II.24).

In addition to substituting features of the king's person in representations of the gods (a practice that was no doubt at least as early as the probable use of King Chephren's head on the Great Sphinx in the fourth dynasty and the near replication of the king's face in the statue of the goddess Hathor as she appears in a dyad now in Boston of King Mycerinus of the same dynasty), the king and his priests and artists certainly strove to propagate the idea that the king was of divine birth and that the gods (especially Re) gave to him power, enduring existence, and life like Re's forever by means of his (the god's) magical protection *(s')*.[95] Everywhere on the reliefs in the temples the king is represented as talking directly to the gods, who grant him his regalia. Even if the king is only a subordinate of the gods as the recipient of their help in protecting and extending the kingdom, he is clearly shown as their intimate. His figure is of the same size as theirs on the monuments, unlike the figures of other mortals. They communicate with him directly in dreams by manifesting themselves and usually in terms that are clearly understandable rather than in the innuendos one finds in the dreams of other mortals. Thus we find on the so-called Sphinx stela, a red granite tablet between the paws of the Sphinx, an account of the appearance of the god of the sphinx (Harmachis) to the future Tuthmosis IV in a dream as he slept near the Sphinx:[96]

> One of those days it came to pass that the king's-son, Tuthmosis, came, coursing at the time of midday, and he rested in the shadow of this great god [Harmachis]. A [vision] of sleep seized him at the hour (when) the sun was in the zenith, and he found the majesty of this revered god speaking with his own mouth, as a father speaks with his son, saying: 'Behold thou me! See thou me! my son Tuthmosis. I am thy father, Harmachis-Khepri-Re-Atum, who will give to thee my kingdom on earth at the head of the living. Thou shalt

wear the white crown and the red
crown upon the head of Geb, the
hereditary prince. The land shall be
thine in its length and breadth, that
which the eye of the All-lord shines
upon. The food of the Two Lands
shall be thine, the great tribute of all
countries, the duration of a long
period of years. My face is thine,
my desire is toward thee. [I do this
in order that] thou shalt be to me a
protector, (for) my manner is as if I
were ailing in all my limbs....The sand
of this desert upon which I am has
reached me; turn to me, to have that
done which I have desired, knowing
that thou art my son, my protector...

In short, the god promises Tuthmosis the kingdom if he
will keep the sand from engulfing him!

As I suggested above, the gods also
communicated with other human beings through dreams,
but in a more indirect way. It was well established
that the gods were the source of dreams.[97] The
meanings of dreams were collected together in
dream-books, in which each dream was delineated, its
prognostication as good or bad in consequence given,
and its significance or meaning revealed. I have
included extracts of this kind as Doc. II.10.

Notice first of all that similarity of the sounds of
key words in the dream with those of key words in the
meaning given to the dreams plays an important part in
the dreams recorded in Doc. II.10. For example (2,23)
the dream of being "up a growing tree (nht)...means his
loss (nhy) of...[something]" and (3,4) a dream of "white

(ḥḏ) bread being given to him...means [that] something [will happen] at which his face will brighten up (ḥḏ) and (9,10) a dream of "uncovering his own backside (pḥwy)... [means that] he will be an orphan later (ḥr pḥwy). The reader will discover that many of the dreams and their interpreted meanings concern prosaic matters. Thus one dream means his good fortune, another his elation at something, a third his finding something he has lost, and a fourth his sitting among his [distant] townsfolk. Others may be interpreted with more serious omens: the killing of his enemies, the absolution of all his ills, the pardoning of him by his god, the passing of a judgment against him, and so on. Notice that the list of dreams and their interpretations is followed by an incantation to Isis to protect the dreamer after he awakens.

There is another class of dreams in which the dreams are brought about by sleeping in a holy place with the hope that the god of that place will come to the sleeper in a dream and give him the answer to some question or help him in some fashion.[98] There are also many examples of the statues of gods being addressed for oracles as they passed in procession, and there is one well-known case recounted in a speech of Tuthmosis III, inscribed in the Karnak temple of Amun in one of the king's additions. In this case the god Amun sought out Tuthmosis III, when he was a young man serving in the temple and thus long before his elevation to the throne. When the young Tuthmosis was found, Amun, i.e. his statue, made him stand in the "Station of the King," i.e. the place usually occupied by the king.[99] The purpose of this story was obviously to demonstrate the selection of the king by divine oracle.

I have already mentioned that ordinary people

sought divine help by magical means when sick.
Needless to say, they also called upon the gods by
magical means for other purposes. Late examples of
this occur in the Leiden and London magical papyrus,
and I quote only one to illustrate the use of a lamp and
a small boy as a medium:[100]

> An inquiry of the lamp. You go to a
> clean dark cell without light and you
> dig a new hole in an east wall and
> you take a white lamp in which no
> minium or gum water has been put,
> its wick being clean, and you fill it
> with genuine Oasis oil, and you
> recite the spells of praising Ra [Re]
> at dawn in his rising and you bring
> the lamp when lighted opposite the
> sun and recite to it the spells as
> below four times, and you take it
> into the cell, you being pure, and the
> boy also, and you pronounce the
> spells to the boy, he not looking at
> the lamp, his eyes being closed,
> seven times. You put pure
> frankincense on the brazier. You put
> your finger on the boy's head, his
> eyes being closed. When you have
> finished you make him open his eyes
> towards the lamp; then he sees the
> shadow of the god about the lamp,
> and he inquires for you concerning
> that which you desire. You must do
> it at midday in a place without light
> .... The spells which you recite <to
> the lamp> to the wick previously

before you recite to the boy [are the following]... 'Art thou the unique great wick of the linen of Thoth? Art thou the byssus robe of Osiris, the divine Drowned [one], woven by the hand of Isis, spun by the hand of Nephthys? Art thou the original band[age] that was made for Osiris Khentamente? Art thou the great bandage with which Anubis [the embalmer] put forth his hand to the body of Osiris the mighty god? ....The spells of the great Sorcerer are those which I recite to thee. Do thou bring me the god in whose hand is the command today and let him give me answer as to everything about which I enquire here today truly without falsehood.'

To this point I have emphasized the magical power of the gods and its consequences for relations between the gods and the Blessed Dead on the one hand and human beings on the other, and I have discussed various attributes of the gods, focusing on those mentioned among the fourteen kas of Re. But we have not said much about the first two attributes of Re in their general expressions as force and power. Černý has some interesting general remarks about the attribute of power:[101]

The most conspicuous attribute of a god was his power; thus gods are sometimes referred to as *sekhem* "powers" and are then represented materialized in a sceptre of special form also called *sekhem* (lit. "powerful one"). Osiris is a *sekhem* or *sekhem o*, "power" or "great power", and at the same time also a "sekhem-sceptre", or "great sekhem-sceptre", since despite his entirely human nature he was given the material form of a sceptre in his temple at Abydos. A large *sekhem*-sceptre was kept there in a shrine and carried through the town in religious processions. A golden cap with a human face was placed over the top of the sceptre to recall the original human form of the god. But surely to the ancient Egyptian the god's power was known primarily by its effectiveness, just as a human ruler's power was judged. Cosmologically speaking, the god's power was most importantly tested in his ability to ensure the continuation of the cosmic order, that is its maat. And presumably maat was present as long as the flooding of the Nile was the right height for an effective yield of crops and as long as social conditions were tranquil. Again and again we hear, in times of crisis such as in the first or second intermediate periods or when the Nile was too low and there was drought, that the gods had deserted Egypt.[102] The rituals to be followed by the king and his priests to preserve the life of the king and hence the maintenance of the cosmos were constantly studied and practiced both at the Houses of Life and the various temples.[103] A corollary of a great god's cosmic power was his power to assist the king in his political and military ventures. The power of such assistance is expressed in

the famous poetical stela of Tuthmosis III, whose
prologue I now partially quote:[104]

> Speech of Amon-Re, Lord of the
> Thrones-of-the-Two Lands;
> You come to me in joy at seeing my
> beauty,
> My son, my champion, Menkheperre,
> everlasting!
> I shine for love of you, my heart
> exults
> At your good coming to my temple.
> My hands have endowed your body
> with safety and life,
> How pleasant to my breast is
> your grace!
> I placed you in my temple and
> did signs for you,
> I gave you valor and victory over
> all lands.
> I set your might, your fear (i.e. fear
> of you) in every country,
> The dread of you as far as heaven's
> four supports.
> I magnified your awe (i.e. awe of
> you) in every body,
> I make your person's fame traverse
> the Nine Bows.
> The princes of all lands are gathered
> in your grasp,
> I stretched my own hands out
> and bound them for you.
> I fettered Nubia's Bowmen by
> ten thousand thousands,
> The northerners a hundred thousand

captives.
I made your enemies succumb
    beneath your soles,
So that you crushed the rebels and
    the traitors.
For I bestow on you the earth, its
    length and breadth,
Westerners and easterners are under
    your command.

We may conclude our treatment of the gods and their relations to the other classes of beings by reiterating that Egyptians, at least by the time that religious literature was composed in the Old Kingdom, were compelled to treat the gods, no matter how exalted, in human terms (for they had no others). Again we can quote the observations of Černý:[105]

> In the Old Kingdom a god is said to "abide" or to "appear (like the sun)". Gods "live" (or "belong to life", are "lords of life"), they are "great", "powerful" and "strong", "good" ("beautiful"), "merciful", "noble", "high" and "just". Like men they have a *ka* or several *kas* and these are also powerful, good, pure, great noble and abiding. The *bais* (external manifestations) "appear (like the sun)" and are "great" and "good". They "make" or "bring" a child, "love it", "bring it up", "protect" it while they stand "behind" it, "keep it alive" or "nourish" it, "clothe" it, and "make it happy", "favour" it, "come" to it, "lift it up", in short "its life is in the hand

of the god, for man is his *"servant"*
and *"adores"* the god.... The Middle
Kingdom names have little to add to
this picture. The gods are also
"sweet" and *"pleasant"*, people are
their *"sons"* and *"daughters"* and the
gods *"make them good"*. For the first
time we hear that gods are *"in
festival"* and though they are said to
be in the *"columned hall"* or in the
"courtyard" of the temple, they also
appear in public *"on the lake"* or
"navigating" on the Nile.... The
prayers--or "hymns", as they are
u s u a l l y   c a l l e d   b y
Egyptologists--down to the end of
the Middle Kingdom are singularly
void of any reference to the relation
between the worshipper and his
god.... It is only in the New Kingdom
that we meet with prayers referring
to the personal feelings of
individuals towards divinities.

In the previous discussion we have learned a
certain amount about the role of human beings in
Egyptian cosmic schemes: their divine origin in the tears
of a creator god or in the plans of his heart, their
consequential designation as "god's cattle" or the "cattle
of Re", the formation of their overall form and shape on
the potter's wheel, their animation by the god's gift of
the breath of life, their possession of spiritual elements
or separate personalities called kas and bas (no doubt
reflecting godlike attributes), their death upon the earth
followed by their reappearance as Blessed Dead in the

Beyond if they received the judgment "acclaimed right" *(m¹ᶜ-ḫrw)*. Indeed a few notable human beings achieved a status approaching that of a divinity. The wise-man Kagemeni was worshiped at the end of the Old Kingdom (though never unequivocally called a god).[106] On the other hand, the great architect Imhotep achieved the status of a god, a god of healing like the Greek Asclepius, by the time of the Saite dynasty.[107]

A word must now be said concerning the geography and character of the regions occupied by the three kinds of beings we have described, the gods, the Blessed Dead, and live human beings. We have seen that the gods had their chief habitations in the stars (witness our remarks earlier concerning Horus), though they of course inhabited properly prepared statues in their temples in the real world. In this connection it is evident that they often had their special regions in this world: Re in Heliopolis, Ptah in Memphis, Thoth in Hermopolis, and so on. Now it also is obvious upon reading the *Pyramid Texts* that the gods had special regions in the sky. In one spell (Doc. II.1, Sect. 480) the god Geb says to the king: "The mounds of my mound (i.e. realm) are the Mounds of Horus and the Mounds of Seth...."

The deceased king is described again and again in the *Pyramid Texts* as becoming a star. In Sect. 263 the king is called a "star brilliant and far-traveling, who brings distant products to Re daily". In Sects. 934-36 the king says that he goes up "on this eastern side of the sky where the gods were born...Sothis (i.e. Sirius) is the sister of Pepi, and the Morning Star is Pepi's child." The king is also often said to join the solar bark. For example, in Sects. 710-11, the king declares: "Teti will assume his pure seat which is at the bow of the Bark of

Re." The sailors who row the bark will "convey Teti round about the horizon." In Sects. 906-07, the king says that he takes his oar and he "occupies his seat, this Pepi sits in the bow of the ship of the Two Enneads. Pepi rows Re to the West." During this journey "The doors of *B¹-k¹*, which is in the firmament, are opened to this Pepi, the doors of iron which are in the starry sky are thrown open for this Pepi, and he goes through them." So evidently the king is considered to be a star and traveling companion of Re.

In the course of the king's celestial travels he visits many ill-defined regions, such as the High Mounds, the Mounds of Seth, and above all the Field of Rushes and the Field of Offerings. In Sect. 936, he says "This Pepi will come with you and wander with you in the Field of Rushes; he will serve as herdsman with you in the Field of Turquoise." In Sects. 1086-1087 the Field of Rushes is distinguished from the Field of Offerings:

> He (Pepi) takes for himself his throne which is in the Field of Rushes and he descends to the southern region of the Field of Offerings.

In another spell (Sect. 918) the king reports that "This Pepi has bathed in the Field of Rushes, this Pepi is clothed in the Field of Khoprer, and Pepi finds Re there." At times the king reaches the mysterious Field of Rushes by means of a ferry (Sect. 1188 [not given below in Doc. II.1]: "O you who ferry over the righteous boatless as the ferryman of the Field of Rushes."); the same is true for the Field of Offerings (Sect. 1193), which may be identical or at least connected in some way with the Field of Rushes, perhaps the prototype of the Greek Elysian Fields.[108] In Sect. 1196 the king identifies himself as "the potter (*or* plumb-line) of the Two Enneads by means of which the Field of Offerings was founded". A further detail of celestial geography is given in Sects. 1199-1200, when the king asks Osiris to commend him "to the supervisors of the Causeway of Happiness north of the Field of Offerings.... Let Pepi eat of the fields and drink of the pools within the Field of Offerings."

Though these fields may on occasion have been shifted to the Netherworld by the time of the composition of the *Book of the Dead*, that work retains and expands on the theme of their agricultural plenty. In Doc. II.3, Spell 99, we read:

> If one recites this spell, he goes forth (by) day from the Field (of Rushes). Given him are a cake, a jar, a *pz(n)*-loaf, a chunk of meat, and (fields of) barley and ⌈Upper Egyptian⌉ wheat 7 cubits (high). It is the Followers of Horus who reap them for him.

In Spell 110, the two fields are linked together in its title:

> Beginning of the spells for the Field of Offerings..., attaining the Field of Rushes, existing in the Field of Offerings, the great settlement..., gaining control there, becoming a blessed one there, plowing there, reaping (there), eating there, drinking there, copulating there, doing everything that is done upon earth.

Hence it is obvious that by the time of the *Book of the Dead*, when the funerary practices and benefits for the dead were extended widely to other than the royal family and their select courtiers, the ideal life of the Beyond was that which they left behind but without trials and tribulations. This latter conclusion is emphasized by the invention in the Middle Kingdom of the so-called shawabti, washabti, or shabti figures, figurines of wood or wax or stone, to be buried with the deceased in order to be available to perform hard labor in the afterlife when called upon by the deceased.[109] The manner in which the life in the Fields of Offerings and Rushes was to be passed by the deceased is illustrated by vignettes that are included with Spell 110 in various copies of the *Book of the Dead* (see Figs. II.25-26) or by paintings on the tomb walls (see Fig. II.27).

Another topographical feature of the celestial regions which was mentioned often in the *Pyramid Texts* was the so-called Winding Waterway. This was conceived as meandering through or around the sky in some fashion and giving access to the Field of Rushes. Thus in Sect. 340 we read: "The Fields of Rushes are filled [with water], and Wenis ferries across on the Winding Waterway." For other mention of the Winding Waterway, see the following Sections of the *Pyramid Texts* which I have not given in the extracts of Doc. II.1 below: 594-97, 599-600, and 1084. Incidentally, in one further reference to the Winding Waterway in Doc. II.1, Sects. 2172-73, the king is said to board the bark "like Re on the banks of the Winding Waterway" and the king "shall be rowed by the Unwearying Stars and shall give commands to the Imperishable Stars". This is fine confirmation of the view that the gods in the stars acted as oarsmen in the Bark of Re, with the Imperishable Stars being the circumpolar stars that do not set and the Unwearying Stars being the planets and the decans of stars (as we shall see in Chapter Three in the next volume). In the hymns of the New Kingdom the day-journey of the solar bark is said to be accomplished by the Imperishable Stars and the night-journey by the Unwearying Stars.[110]

Still another place frequently mentioned in the *Pyramid Texts* is the Jackal-Lake in which the king is purified. It is mentioned once in Sects. 1164-65 together with the Lake of the Dat when the king's son says: "...wash yourself in the Jackal-Lake, be cleansed in the Lake of the Dat, be purified on top of your lotus-flower in the Field of Rushes. Cross the sky, make your abode in the Field of Offerings among the gods who have gone to their kas." This last phrase is often used as a euphemism for dying, and if that is what it means here, we have one more instance of gods being said to die. As we move from the Old Kingdom and its references to the places located in the other world, we find an increasing specification of the locales and inhabitants of the Netherworld. There was a rather vaguely expressed belief in the Old Kingdom that the Dat or Duat (the land where the deceased lived in company with the gods) had an upper part in the skies and a lower counter-sky in the Netherworld, though surely the term Dat originally must have been principally used in connection with the celestial regions, since it was written with a star enclosed by a circle. Be that as it may, the doctrine of the daily coursing of the solar bark through the sky in the daytime and an inferior world at night was clearly present. Indeed the passing of the sun into the western horizon to begin its night course and its subsequent passage until rising again in the morning not only was thought of as counting the hours and delineating the days and nights but was probably looked upon as evidence of rebirth in nature and thus of human resurrection.[111]   At any rate, the doctrine of the death and birth of the sun each day exerted enormous influence on funerary practice in the Middle and New Kingdoms. Indeed it was the nighttime journey which stimulated the imagination of the priestly authors of the New Kingdom, who recounted that journey in word and picture in the tombs of the kings in the Valley of the Kings at Thebes. Among these accounts are those entitled the *Book of Amduat* (so called, but perhaps better *Amdat* or *Amdet*), the *Book of Gates*, and the *Book of Caverns*; the first of these works will serve as a model for this type of work.

In fact these works of the New Kingdom were to some extent anticipated by a composition found among the *Coffin Texts* which is usually called the *Book of Two Ways*.[112]   This work consists of an obscure journey of the solar bark. The goals of the deceased in undertaking this journey, aside from the purpose of his resurrection, are by no means clear, but an ultimate sojourn with Osiris appears to be involved.[113]   The deceased seems to embark on the boat at Heliopolis (Doc. II.2, Spell 1030):

> See, its starry sky (or collection of stellar deities?) is in On (i.e. Heliopolis), the sun-folk are in Kheraha.... I will go with them aboard the lotus-bark at the dockyard of the gods.... I will ascend in her to the sky. I will sail in her in the company of Re.... I will act as pilot

in her to the polar region of the sky,
to the stairway of the bark of
Mercury.

The journey is beset with dangers and the hostility of demons. Thus in Spell no. 1034, the deceased recites an incantation against hostile snakes, and he declares that Re's protection is his protection. In the very next spell the traveler says that he has passed along the ways of Rostau, i.e. the paths from the gate of the other world (perhaps originally thought to be at Saqqara). We are told that these are the paths or ways of Osiris and they form the border of the sky. The rubric of Spell 1131 indicates that it is a guide to the double doors of the horizon and it emphasizes the importance of the written expression of their keepers' (collective) name. Similarly knowledge of the names of doorkeepers and their gates and doors was important for the deceased when he entered the judgment hall in Spell 125 of the *Book of the Dead* (Doc. II.3). There is a reference to Hu, the personalization of the creative word, in Spell 1136 of the *Coffin Texts*, a reference that may refer to the original creation in darkness: "Hu who speaks in darkness belongs to me". That is, by Creative Utterance, or magical words, the way can be opened for the traveler. Spell 1146 speaks of an image of a mansion, presumably one of the goals of the voyage. It is the "place of a spirit, the place of excellent magic". A rubric for Spell 1162 relates the Field of Offerings to Osiris and Thoth: "Spell for being in the Field of Offerings among the Following of Osiris and among the Following of Thoth every day."

The most interesting feature of the *Book of Two Ways* is the map of the two paths around which the spells are grouped. It varies from coffin to coffin (see

Fig. II.28). I shall make no effort to treat this bizarre
map of the two ways but will confine myself to a brief
comment from Lesko's work:[114]

> The map is divided horizontally into
> two compartments each having a
> zigzag path .... The blue path or way
> in the upper compartment meets or
> almost meets the black lower way at
> the far end of the two compartments
> .... The upper way of the map has
> Osiris, who is in the sky, as its
> central figure.

Continuing our geographical survey of the
nightly other world in which resided gods, spirits of
various forms, and the dead, we can now turn to the
three works of the New Kingdom which we mentioned
above. The first of these works chronologically was
the *Book of Amduat*, i.e. the *Book of What Is in the
Netherworld*. I shall give a short summary of this
work, referring to the extracts of Doc. II.4, which are
accompanied by Figs. II.29-II.40.

The contents of the *Book of Amduat* are
epitomized in the introductory text, where we are told
that the work will inform us of the locations of the
various beings of the Netherworld (souls, gods, shades,
and spirits), their power, their activities, their rituals
for Re, their mysterious powers, their conversations
with Re as he travels through their regions, the
movement of the hours and their gods (i.e. the gods
who guide the Great God through their hours), who are
the powerful ones and who are the annihilated. We can
note that the names of the gates of entrance for the
hours are usually given but they are not emphasized
and are not depicted in the accompanying illustrations

as they are in the *Book of Gates*.

The bark of Re often changes form in the course of the journey and its crew of gods varies somewhat as to number and names, though it is of some interest that Sia and Hu are usually present, as seems appropriate for the main purpose of the voyage, the rebirth of Re. Some of the same activities occur in each successive cavern. The sun-god communicates with the resident gods, he illuminates them, and in a sense he brings them alive each night, even though he leaves them behind. He also often commends to them various fields, as he does to the gods of his own Following. Sometimes he greets the resident gods simply as corpses and other times as disembodied souls. The sun-god when shown in his bark throughout the journey, until its end, is represented as ram-headed and he is called Flesh, i.e., he too is a kind of corpse, though the use of the ram-form seems to imply that it is the ba (or soul) of Re that completes the night journey since the ram is one of the signs for ba as soul. There are, however, constant indications in the various hours that he will be reborn at the eastern horizon: the beetle is introduced into the text as a portent of the sun-god's rebirth in the beetle form of Khepri (e.g., the third register of the illustration of the very first hour contains a beetle in a bark with two gods adoring him).

The journey starts at the western gateway, and he makes his transformations [to his ram-headed form] after passing that gate. The distance through the gateway to the next cavern is 120 *iteru* (each *iteru* being about 10.5 km.). Re's bark is towed by Two Truths (represented as goddesses). He dispenses fields to the gods in his Following. The name of this field of the first cavern is Water of Re.

ANCIENT EGYPTIAN SCIENCE

The second cavern is called Wernes and is 309 *iteru* in length and 120 *iteru* in breadth. In this cavern Re makes further grants of fields to the gods and talks to the inhabitants. In one of the boats that precedes the sun bark a beetle replaces one of the customary symbols at the prow, again indicating the ultimate purpose of the voyage. The resident gods call out to the sun-god and live through his voice when he replies. Indeed they are said to breathe by means of his voice, when he grants them their duties *(irw)*. They are accordingly able to supply plants, i.e. vegetables, to the gods in Re's Retinue. And so we see here and elsewhere that agricultural activities are as important in the Netherworld as they were in the Field of Rushes when it was located in the sky. In the third register depicting this hour we see gods carrying signs for the year *(rnpt)* and branches that represent the ages, and the text tells us: "These gods...present this Great God with the year-signs... [and] with ages-branches which they hold in their hands."

The cavern in the third hour is called the Field of the Offering-dwellers in the realm of Osiris. It too is 309 *iteru* in length. Above the gods depicted in this cavern, that is, in its first register, we read: "They (i.e. the gods) are like this in the Netherworld in the flesh of their own bodies. Their souls (i.e. bas) speak for them and their shades (or shadows) rest in them." In view of this description it is not surprising that they are shown standing on water. We are further told that these gods must overthrow the enemies of Re by causing them to be drowned in the flood of water from the Abyss and blown about by winds drawn from the earth. Re stops at the city where Osiris resides and gives orders to him and his followers.

A mysterious cavern of the West called Cavern of the Life of Forms is reached in the fourth hour. Its name is not surprising since we find there the creator-snake mentioned earlier in this chapter, the Provider of Forms (Neheb-kau). As I suggested, he perhaps represents an early stage of creation in the Abyss. We must also remember that among the crew of the bark the gods of the creative word, Sia and Hu, are likewise present. Both this cavern and the next are a part of the abode of the dead Sokaris, the ancient god of the dead at Memphis. In the fifth cavern we see this great mound which constitutes his burial ground. But he is depicted sticking his head out of the top of the mound to see Re pass. The third register accompanying this part of the journey shows the inside of the great mound of Sokaris. The diagram of the whole cavern is referred to as the plan of access to Sokaris. It is said to be executed "according to the plan which is drawn in the hidden region of the Netherworld". We note the zigzag descending tunnel that appears on the illustrations of both the fourth and fifth caverns, and we are reminded of the similar zigzag route in the earlier *Book of Two Ways*, though there the zigzag path no doubt represented a route in a horizontal plane and here we seem to have a vertical tunnel being depicted.

In the sixth hour Re arrives at a field close to the body of Osiris. Thus in the third hour we had the living Osiris as the ruler of the Netherworld and here we have his body. Here it is emphasized that Re talks to the resident gods (who are of course also dead) and illuminates their darkness. They can hear his voice and they can accordingly breathe. The creative activity of snakes is emphasized in the second register of the accompanying depiction where a five-headed snake with tail in mouth encloses a god with a beetle on his head, another portent of the sun's rebirth. In the seventh hour the bark is drawn through the Cavern of Osiris, along which it is towed by the magical utterances of Isis and by the magical utterances of the Elder One, the magic being directed against Re's archenemy, the snake Apep. In the middle register of this cavern's depiction the snake is given the name Neha-her, i.e. Evil Face, and is said to be on a sandbank named Sadjau, which is 440 cubits long and 440 cubits wide and which is filled by the snake's coils. It is an interesting feature of these fanciful accounts that often great precision as to size and form is given, though as usual with scribes and artists in all ages there is considerable variety in the reporting of these precise numbers from one copy to the next. Knives are taken to Apep by four goddesses. Horus is cast here as a protector of Re when he orders the help of gods and goddesses who bear stars on their heads.

In the eighth hour the sun-god comes to other banks or mounds of dead gods, and the name of the city here is "Sarcophagi of her Gods". Here Re "calls to their souls [i.e. those of the deceased gods] after he has entered this city of the gods who reside on their sand". Further, "it is his voice that the gods hear after he has called them while the forms of their bodies remain on their corpses which are under the sand". "When this Great God calls them, what is in them comes alive." Here is further evidence of the general belief that gods die and may be reborn like Osiris or indeed like Re on this very trip. The dead gods hide after Re passes. Darkness returns.

In the ninth hour the Great God and his crew

rest from rowing. The bark is shown as preceded by a file of twelve gods with paddles. In both the first and the third registers gods are shown seated on stools that are the hieroglyphs for "clothing", which means they are "in their own flesh". In the tenth hour Khepri the beetle appears in the presence of Re, a sign of what is to come at the end of the journey. We read: "They are like this in the Netherworld in the forms and births of Khepri when he carries his egg toward this city in order to come forth afterwards in the Eastern Horizon of Heaven." In this hour, in the third register, we see depicted Horus addressing the "drowned ones" who are shown stretched out on their backs in the water. They are said to be in the Abyss, i.e. the primordial water that enters the Netherworld. It is clear, from this and preceding references in this book, that always near at hand in the regions of the underworld the Abyss threatens. I have already mentioned early in the chapter that the cosmology of the Egyptians assumes the tension and balance that are maintained between the created, existent cosmos and the threatening nonexistence of Chaos.

The eleventh hour takes the sun-god through another mysterious cavern. Here we see anticipations of the climactic rebirth to take place at the end of the next hour. In this penultimate hour the rebels, who constantly threaten during the journey, are disposed of by The Starry One, the Mistress of the Bark. Portentous is the statement that "Everlastingness swallows up her images before the Seeing One, who is in this city, and she gives them back [again] upon the birth of Khepri on earth." The figure of swallowing images to regurgitate them is also used with respect to a goddess who is mounted on the back of a serpent above the constellation of Shedu. The text tell us that "what she has to do, is to live by the voice of Re every day. She swallows her images in this city [of the eleventh hour]."

Now the travelers are in their last nocturnal hour, the twelfth, when the glorious rebirth is to take place. The sandy borders at the top and the bottom of the depiction of each hour are shown to meet in the middle of the semicircular arc at the end of the picture. Where they meet we see the rising Khepri and the sun's disk. Clearly the rebirth is accomplished. The action of this hour is epitomized in its introductory text:

> Born will be this Great God in his forms of Khepri at this cavern. Nun and Nunet, Heh and Hehet (i.e. four of the primordial Ogdoad of Hermopolis) arise in this cavern at the birth of this Great God that he may come forth from the Netherworld, let himself down in the Daybark, and arise from the thighs of Nut.

The importance of the creator-snake in the final process of rebirth is evident in the middle register of the depiction of the hour where the towers of the bark are shown leading it to the tail of the serpent and we read:

> This god travels...along the backbone of this secret image of the serpent Life of the Gods, when his gods tow him. He enters into its tail and comes out of its mouth, being born in his forms of Khepri...

The completion of the renaissance of the sun-god

is emphasized in the illustration which shows the discarded mummified form of the night traveler cast aside at the bottom of the arc that provides the eastern boundary of the underworld.

Needless to say, I have only skimmed the surface of this curious and difficult text. Further help with its contents may be had by consulting the studies of Piankoff and Hornung mentioned in the introduction to Document II.4 below.

The nightly journey to the Netherworld by the solar bark which we have just described by examining the contents of the *Book of Amduat* brings us to Egyptian concepts of time and eternity. The calendaric achievements that resulted from their attempts to measure time will be discussed in the next volume. Here we wish only to note that at a very early period the Egyptians conceived of time as measurable in terms of repetitive natural events: the day by the continually repeated rising and setting of the sun, the month by the waxing and waning of the moon, and the year by the repeated inundations of the Nile (and only later in pharaonic times related by them to the heliacal rising of the star Sirius). They also felt time to continue in a linear way indefinitely into the future.[115] This is clear from their very early conception of the continuous and successive rule of the gods and men on the earth since the time of creation. This conception is embodied in the lists of kings and in early annals like Document I.1 given in the first chapter. The remarkable point in this view of time was that it tied time to the cosmos and specified that time began with the activity of creation. As we have seen again and again the cosmogonies reported creation as taking place in the "first time". It certainly seems likely that the Abyss, inert as it was,

was timeless before the creator first established his form within it. And no doubt when a creator god was designated as the Lord of Time, it was his invention of time as an aspect of creation that was implied. The connection between existence (which is indeed the cosmos) and time is asserted in a gloss given in Spell 335 of the *Coffin Texts* (see Doc. II.2 below):

> The Examiner of what exists. *Who is he? He is Osiris. As for what exists, it is eternity (nḥḥ) and everlastingness (ḏt). As for eternity, it is day. As for everlastingness, it is night.*

Thus if we consider eternity and everlastingness as time in its full sweep, i.e. all the days and all the nights, then this gloss certainly identifies it with existence (i.e. "that which is"). These components of time were personified and in one figure shown as props of the sky and thus as a major aspect of the cosmos (see Fig. II.52). The logical corollary of considering time as a function, or perhaps better as a limit, of the cosmos that pertains to it, is that if the creator did decide to destroy everything that he had made and return to the Abyss, as Spell 175 of the *Book of the Dead* declares, then time would also be destroyed.[116] But such a possibility must have seemed unlikely to Egyptians who were so convinced or at least so hoped that the deceased king (and perhaps later any dead person who had been justified at Osiris's hall of judgment) would "live like Re forever (or forever and ever: *nḥḥ ḏt*)", as the conventional phrase had it. And indeed the "eternity" *(nḥḥ)* so desired throughout the whole history of pharaonic Egypt is occasionally used in the same context as the word "million" *(ḥḥ)*, which is often written with the palm-branch sign for "year" as a determinative or mounted on the head of the goddess who represents the word and accordingly means a million, or in fact an almost countless number, of years.[117]   Even the most fastidiously logical Egyptian priest would be glad to settle for this as a definition of eternal life.   Indeed we read in the *Book of the Dead* (Doc. II.3, Spell 42) of the deceased expressing his confidence in or perhaps his hope for lasting a million years or even millions of years:

> I am one who has ascended sound,
> whose name [is not known].   I am
> yesterday; my [name] is He Who
> Sees a Million Years, who has gone
> ...along the roads of the Chief
> Examiners.   [I am lord of
> everlastingness]....   I am your
> protector for millions ⌜(of years)⌝.... I
> shall not die again.

A similar hope was implicit in naming Re's bark "The Bark of Millions of Years" and in the designation given to the mortuary temples of Seti I, Ramesses II and others in western Thebes and elsewhere: "The Mansion of Millions of Years".   But in identifying time with existence, the glossator of the *Coffin Texts* surely did not mean to exclude the other aspects of existence which express its great variety.   We have already described the rich complex of divine, quasi-divine, and human beings that constituted the heart of the cosmos. But of course the Egyptian cosmos included far more: the animate and inanimate beings of the visible world, or as Amenope put it in his Onomasticon (Doc. I.9, Introduct.):

> all   things   that   exist: what   Ptah

created, what Thoth copied down,
heaven with its designs, earth and
what is in it, what the mountains
belch forth, what is watered by the
flood, all things upon which Re has
shone, all that is grown on the back
of earth....

Indeed we may conclude this summary of
Egyptian cosmology by urging the reader to study the
wide range of entities listed by Amenope in the
document just quoted and from which I have given
extensive extracts in the previous chapter. He will
find there none of the mythological cosmology that I
have mentioned often in this chapter: the sky as a
cow's belly or the wings of a vulture or a falcon, the
sky's support as the legs of the cow, the morning sun as
the beetle, and so on. Instead he will find a sober
listing of all the categories of existence which the
author thought his scribal readers should know. No
doubt if he had wished to describe the entities in
addition to listing them, he would have resorted to
mythological description, as his remarks concerning Ptah
and Thoth suggest.

Notes to Chapter Two

1. This conclusion is evident enough when one studies the documents that appear at the end of the chapter. P. Derchain, *Le Papyrus Salt 825 (BM 10051), rituel pour la conservation de la vie en Égypte* (Brussels, 1965), Chap. 1: "Physique et théologie", p. 4, rather turns my conclusion around when he declares that religion in Egypt "n'était pas une mystique, mais une physique". But as one studies that physics it is evident that it is unlike any physics for which we now customarily use the term, since it regularly included contradictory symbols to represent natural entities and events, expressed contending forces by conflicting gods, made wholesale use of divine agencies to describe creative acts, and everywhere employed magical terms and pronouncements to bring things into existence and to effect communication between human and divine beings. It is of some interest that Champollion, at the very beginning of modern studies of religious documents in Egypt, made some remarks concerning the inscriptions and figures in the royal tombs which have some pertinence for the whole of Egyptian religious literature: *Lettres écrites d'Égypte et de Nubie en 1828 et 1829* (Paris, nouv. ed., 1868), p. 201, describing the burial chamber of the tomb of Ramesses V: "Les parois de cette vaste salle sont couvertes, du soubassement au plafond, de tableaux sculptés et peints comme dans le reste du tombeau, et chargées de milliers d'hiéroglyphes formant les légendes explicatives; le soleil est encore le sujet de ces bas-reliefs, dont un grand nombre contiennent aussi, sous des formes emblématiques, tout le système cosmogonique et les principes de la physique

générale des Égyptiens."

2. It should be observed, however, that the so-called Heliopolitan cosmogony, reflected so dominantly in Documents II.1-II.3 and indeed in most of the other documents, tended to absorb and alter the doctrines devised at the other religious centers. This was particularly true in the case of the Hermopolitan system, whose paramount beliefs have to be sifted from essentially Heliopolitan accounts. See the remarks of S. Sauneron and J. Yoyotte, "La naissance du monde selon l'Égypte ancienne," *Sources orientales:* Vol. 1, *La naissance du monde* (Paris, 1959), p. 52 (whole article, pp. 19-91).

3. H. Frankfort, *Kingship and the Gods* (Chicago, 1948), pp. 151-54, and particularly p. 152: "In fact, each and every temple was supposed to stand on it [i.e. the Primeval Hill]....The identity of the temples with the Primeval Hill amounts to a sharing of essential quality.... Each temple rose from its entrance through its successive courts and halls to the Holy of Holies [i.e. the sanctuary], which was thus situated at a point noticeably higher than the entrance. There the statue, barge, or fetish of the god was kept, resting upon the Primeval Hill." Primeval Hills are mentioned by Frankfort for Heliopolis ("The High Sand"), Hermopolis ("The Isle of Flames"), and Karnak ("The Venerable Hill of the First Time"). Note in Leiden Papyrus I 350 the description of Thebes as the place of creation: "Waset [i.e Thebes] is the pattern for every city. Both the flood and the earth were in her in the First Time [i.e. the beginning of time], the sands came to delimit her fields, to create *(shpr)* her ground upon the mound when earth came into being *(hpr)*. Mankind came into being *(hpr)* within her....." (See C. F. Nims, *Thebes of*

the Pharaohs, Pattern for Every City [London, 1965], text on the page opposite the title page and translation [slightly modified] on p. 69; cf. J. Wilson's translation of it found in J. B. Pritchard, ed., *Ancient Near Eastern Texts Relating to the Old Testament* [Princeton, 1950], p. 8). For the full text, see Zandee, in the work cited below in the introduction to Doc. II.7b[3] (Pl. II, lines 10-11). Incidentally, in his dedicatory inscription to the forecourt, pylon, and obelisks at the temple of Amun in Luxor, Ramesses II says: "Now His Majesty did research in the office of archives, and he opened the writings of the House-of-Life. He thus knew the secrets of heaven and all the mysteries of earth. He found Thebes, the very Eye of Re, as being the original plot of earth which arose in the beginning." (See K. A. Kitchen, *Pharaoh Triumphant* [Warminster, 1982], p. 48.) The concept of the primeval land or hill in creation was examined at length in the remarkable study of A. de Buck, *De Egyptische Voorstellingen betreffende den Oerheuvel* (Leiden, 1922). Cf. the brief summary by H. R. Hall in a review in *The Journal of Egyptian Archaeology*, Vol. 10 (1924), pp. 185-87. Also, see R. David, *A Guide to Religious Ritual at Abydos* (Warminster, 1981), p. 2: "Thus, the first shrine symbolised creation. Each 'new' temple which was built throughout Egyptian history was regarded as being a representative of the island where life was thought to have started, and each religious locality in Egypt later claimed to be the actual island of creation. The sanctuary of each temple was envisaged as the primaeval mound where the falcon [or whatever form the demiurge was supposed to have had] alighted, a centre of great sanctity and power....[Therefore Egyptians came] to regard every temple as the 'House of the God'. The form which the deity took varied in the different localities, but the temple was the sacred place where the statue of the god could rest, where the deity could be attended to, and where the gods, otherwise formless and inaccessible, could be approached by the king or mediator [i.e. priest]."

4. Sauneron and Yoyotte, "La naissance du monde," p. 27. Notice that I have included *hpr* (or its causative form *shpr*) in parentheses everywhere it occurred in the preceding footnote in the passage concerning Thebes as the place of creation.

5. E. Hornung, *Conceptions of God in Ancient Egypt: The One and the Many* (Ithaca, N. Y., 1971).

6. *Ibid.*, p. 91, quoting H. Bonnet, *Reallexikon der ägyptischen Religionsgeschichte* (Berlin, 1952), p. 239.

7. A. Erman, *A Handbook of Egyptian Religion* (London, 1907), p. 3.

8. H. Frankfort, *Ancient Egyptian Religion: An Interpretation* (New York, 1948), pp. 3-4.

9. See R. Anthes, "Mythology in Ancient Egypt," in S. N. Kramer, ed., *Mythologies of the Ancient World* (New York, 1961), pp. 1-92. Cf. Anthes, "Egyptian Theology in the Third Millennium B.C.," *Journal of Near Eastern Studies,* Vol. 18 (1959), pp. 169-212.

10. Anthes, "Mythology," p. 21.

11. *Ibid.*, pp. 21-22.

12. Hornung, *Conceptions of God*, pp. 236-39, 242. Cf. E. Drioton, "La religion égyptienne," in M. Brillant and R. Aigrain, eds., *Histoire des religions* (Paris, 1955), pp. 37-38, commenting on what he believed to be the apparent contradiction of the simultaneous presence of monotheistic and polytheistic views in the Egyptian instruction-books: "L'attachement aux affirmations traditionnelles--C'est que l'Égyptien jugeait moins une

doctrine sur son fond tiré au clair que sur les circonstances d'ancienneté, de tradition, de présentation qui l'entouraient. La tradition surtout était, pour une société qui plaçait tout son idéal dans le passé, à l'âge d'or des dieux, la touche décisive de l'autorité d'une assertion. Que si deux traditions contradictoires se présentaient avec des titres équivalents, force était bien de les admettre toutes deux comme valables. Cela faisait simplement pour les Égyptiens un mystère de plus, dont le sentiment non seulement ne leur causait aucune inquiétude, mais leur donnait l'impression de toucher aux arcanes de la science cachée aux mortels par la divinité. Et puisque, croyaient-ils, les deux assertions se conciliaient dans l'inconnaissable, la sagesse la plus haute--nous dirions de nos jours la science la plus avertie--permettait de parler de l'une en empruntant des expressions à l'autre. Voilà ce qu'il y a au fond de ce 'conciliatisme' foncier, qui est une des marques de la pensée égyptienne et qu'on lui a trop souvent reproché comme un manque de fermeté dans le jugement, une phobie de rien abandonner, voire un scepticisme inconscient. C'était en soi une position rationnelle, mais viciée par une fausse information. Ainsi s'explique que dans une série d'écrits monothéistes il se trouve partout des infiltrations polythéistes. Le fait prouve que les anciens Égyptiens qui ont composé ces livres de sagesse tenaient les deux croyances contradictoires pour également vraies, parce que toutes deux bénéficiaient d'une longue tradition dans le pays." Hornung and others would deny that the wisdom literature is fundamentally monotheistic but would hardly deny Drioton's basic view that the role of tradition is fundamental in what Drioton calls the doctrine of "conciliationism".

13. Hornung, *Conceptions of God*, p. 239.

14. S. Sauneron, *Les prêtres de l'ancienne Égypte* (Paris, 1957), pp. 123-25.

15. This "expression" was common for the king throughout the whole period of Egyptian history, at least in written accounts from the sixth dynasty on. The identification of the king with the god Horus appears in countless references in our earliest sustained religious literature, i.e. the *Pyramid Texts*. Such references were collected in the useful dissertation of T. G. Allen, *Horus in the Pyramid Texts* (Chicago, 1916). Hornung, *Conceptions of God*, pp. 141-42, throws doubt on the simple identification of Horus as king with Horus as god. In general Hornung reflects the views of G. Posener, *De la divinité du pharaon* (Paris, 1960), who marshals the evidence distinguishing the pharaoh from a divinity. Hornung's summary (*Conceptions*, p. 192) is of interest: "Proponents of this view [of the declining power of the king after the early dynasties] have always assumed that in the early dynastic period the king had been, as 'Horus,' entirely a god, so that in the Fourth and Fifth Dynasties he subordinated himself as the 'son' of the Highest god. But I have shown... that it is quite uncertain that as 'Horus' the king was identical with or an 'incarnation' of Horus. All we can safely say is that from the beginning of Egyptian history the king was termed Horus, just as from the Fifth Dynasty on the deceased king was termed Osiris; for the early dynastic period and Old Kingdom we do not know precisely in what form the relationship between Horus and the king was envisaged. The extensive textual evidence of the Middle and New kingdoms suggests that at his accession Pharaoh took on the role of Horus, and at his death he took on the role of Osiris, adopting the

attributes of these gods without being identical with them. This form of divinity does not relate to him as an individual but to his office."

16. See the various early examples of the use of a *serekh* surmounted by the Horus falcon collected by S.A.B. Mercer, *Horus, Royal God of Egypt* (London, 1942), pp. 9-12, 18-24. Perhaps the earliest *serekh* with a falcon is one on pottery that may be of Amratian date (*ibid.*, pp. 8-9).

17. A. H. Gardiner, "Horus the Beḥdetite," *JEA*, Vol. 30 (1944), pp. 46-52, discusses the winged figure on this comb in relationship to the winged solar disk that was to become identified with Horus the Beḥdetite in Edfu. Full article pp. 23-60.

18. A. H. Gardiner, *Egyptian Grammar*, 3rd ed. (London, 1957), p. 509, sign S 40, notes that the same sign stood for both the w'ȝs-staff and the ḏꜥm-staff. Both phonetic readings are explicitly given in the passages from the *Pyramid Texts* quoted below in the text. A. Moret, *Du caractère religieux de la royauté pharaonique (Annales du Musée Guimet: Bibliothèque d'études*, Vol. 15) (Paris, 1902), p. 293, had long since remarked that in temple reliefs the w'ȝs-staffs were shown as supporting the heavens.

19. R. T. R. Clark, *Myth and Symbol in Ancient Egypt* (London, 1959; paperback ed. 1978), p. 219.

20. H. Kees, *Der Götterglaube in alten Ägypten* (Berlin, 1956), p. 234, n. 1. Cf. F. W. von Bissing, ed., *Das Re-Heiligtum des Königs Ne-Woser-Re*, Vol. 3: H. Kees, *Die grosse Festdarstellung* (Leipzig, 1928), p. 41 and plate 21, Nr. 346.

21. Anthes, "Mythology in Ancient Egypt," pp. 85-86.

22. The pedestal of a statue of King Djoser

includes the name and titles of Imhotep: "The chancellor of the King of Lower Egypt, the first after the King of Upper Egypt, administrator of the great palace, hereditary lord, Great Seer (the High Priest of Heliopolis), Imhotep the builder, the sculptor, the maker of stone vases...." (see J.-P. Lauer, *Saqqara: The Royal Cemetery of Memphis. Excavations and Discoveries since 1850* [London, 1976], p. 92 and plate 89).

23. For the epithet and (later) title "Son of Re", see H. Müller, *Die formale Entwicklung der Titulatur der ägyptischen Könige* (Glückstadt/Hamburg/New York, 1938), pp. 63-73.

24. Clark, *Myth and Symbol*, p. 44.

25. For a penetrating consideration of the original Osiris and Horus myths as reflected in the *Pyramid Texts,* see J. G. Griffiths, *The Origins of Osiris and his Cult* (Leiden, 1980), Chap. 1, where a good many passages from the *Pyramid Texts* are translated and analyzed. See also Plutarch, *De Iside et Osiride*, edited, translated, and commented on by J. G. Griffiths (Cardiff, 1970). A brief account of these myths with some attention to the Middle Kingdom and later sources was given by Anthes, "Mythology in Ancient Egypt," pp. 68-85. See also Clark, *Myth and Symbol*, pp. 97-180. Though it is often said that from the earliest days Osiris was a god of vegetation, the first evidence of an unambiguous nature indicating any connection of Osiris with vegetation or fertility appears in the *Coffin Texts* (Spell 80, II,90; not included in our extracts below) where the deceased identifies himself with Osiris and says that he is one "whom Atum made into Neper (the Grain-god)...when I became Osiris". Similarly in Spell 300 (also omitted from our extracts) entitled "Becoming Neper" the deceased says (IV, 167-69): "I am Osiris....I

have grown fat through you, I flourish through you....I live and grow as Neper....I live and die, for I am emmer and I will not perish." Griffiths, *The Origins of Osiris*, pp. 165-66, throws some doubt on the usefulness of this passage for identifying Osiris and Neper, suggesting that the deceased is identifying himself with a series of gods of which Osiris and Neper are two. But there is no doubt from this passage that the deceased, who is most commonly identified with Osiris from the fifth dynasty, is at the same time identified with the grain itself. This idea of connecting rebirth with vegetation may be very old, but the identification of Osiris with vegetation may have followed the practice of also identifying the reborn deceased with Osiris. For Seth's role in the legends, see the thorough study of H. Te Velde, *Seth, God of Confusion* (Leiden, 1977).

26. This is borne out by the fact that there is no mention of Osiris in the *Annals* for the first four dynasties and part of the fifth which we have presented as Document I.1. The earliest known pictorial representation of Osiris is on a block of the fifth-dynasty king Djedkare (see the photograph opposite the title page in Griffiths, *The Origins of Osiris*). Furthermore Griffiths in the same work summarizes on page 41 what we know about the cult of Osiris in the early period: "The earliest evidence for the cult of Osiris is still confined to the closing phases of the Fifth Dynasty. His name occurs in the Pyramid Texts of Wenis in Saqqara, and Wenis is the last king of the Fifth Dynasty. His name also occurs in private mastabas of this dynasty, especially in the offering formulae.

"While there is every likelihood that the Osirian material in the Pyramid Texts derives in part from a much earlier date, so far it has not proved possible to track down the god or his symbols tangibly to the First or Second Dynasty. An interesting find was made by Zaki Y. Saad in a tomb of the First Dynasty at Helwan--two ivory pieces which, in his opinion, form the most ancient example of the *djed*-pillar, which in turn he describes as 'the symbol of the god Osiris'. The *djed*-pillar, however, is not consistently an Osirian symbol, at least not in its early forms. Only in the New Kingdom does it become unambiguously Osirian. Even so, do these pieces represent *djed*-pillars? They seem to imitate papyrus-columns, with four papyrus flowers superimposed one over the other. It may equally be doubted whether the ivory toilet spoon [another of Saad's pieces] is at this time a symbol of the goddess Isis, or again the knot represented on the adjacent box lid [a knot later associated with Isis]." In a footnote to this passage Griffiths acknowledges that some scholars, such as Drioton, Vandier, Westendorf, Emery, and Clark, accept Saad's claims concerning these pieces, though Helck does not.

27. Clark, *Myth and Symbol*, pp. 51-52. Hornung, in his *Conceptions of God*, p. 178, stresses the connection of snake-forms with the Abyss: "These limitless depths also house the enemies of the gods, in particular Apopis, who daily stretches out his snake's head to attack the sun god, and must forever be driven into nonexistence. For the Egyptians the form of a snake... has a particular affinity with nonexistence. The primeval god who outlives the existent world changes into a snake at the end of time, and after the Amarna period the Egyptians devised the image of the snake coiled back on itself, called 'tail-in-mouth'; in the Roman period this image was called Ouroboros, 'the

tail-swallower' [see Fig. II.6b]. The complete circle of the snake's body illustrates--so far as it is possible to depict it--the nonexistent which encompasses the world continually on all sides.... There is a similar image in the New Kingdom underworld books; a snake, out of which one hour is mysteriously 'born,' and then 'swallowed' again when the hour is passed--which thus embodies the limitless, vertiginous aspect of time." See pp. 172-85 for Hornung's brilliant excursus on the nonexistent identified temporally and spatially with chaos as opposed to the existent identified with the cosmos. See also his "Chaotische Bereiche in der geordneten Welt," *ZÄS*, Vol. 81 (1956), pp. 28-32. Incidentally the snake with his tail in his mouth is also depicted in the *Book of Amduat* (see Fig. II.34) and at the beginning of the *Book of Gates* (Fig. II.41). For the Kematef-snake in Medina Habu and the Earth-Creator snake, see the end of note 66 below. See also K. Sethe, *Amun und die acht Urgötter von Hermopolis: Eine Untersuchung über Ursprung und Wesen des ägyptischen Götterkönigs* (Berlin, 1929), pp. 88-89.

28. Clark, *Myth and Symbol*, p. 52.

29. I may have given the impression above that the four generations of gods embraced by the Ennead were neatly and canonically described in the *Pyramid Texts*, including just the nine gods we have mentioned: Atum, Shu, Tefenet, Geb, Nut, Osiris, Seth, Isis, and Nephthys. This is far from the truth. I have already mentioned that at times Horus appears as the brother of Seth rather than as his nephew, thus increasing the number of gods of the basic four generations to ten instead of nine. Furthermore, this augmentation persists in the tradition that the children of Nut, whose births were celebrated on the epagomenal days, included Horus

the Elder along with Osiris, Seth, Isis, and Nephthys (see Doc. I.1, n. 112). I have also indicated above that Maat was often described as another daughter of Re, though she was at times identified with Tefenet. For other divergencies from the constitution of the Ennead described in the text, see Anthes, "Egyptian Theology," pp. 194-202. As he notes (p. 196): "The deities who belong to the pedigree [of Horus], with the exception of Horus, are called 'the great Ennead' as early as the Pyramid Texts (Pyr. 1655 [see Doc. II.1, Sect. 1655]). For the later periods, H. Brugsch, *Thesaurus inscriptionum aegyptiacarum* (Leipzig, 1883-84 [reprint, Graz, 1968]), pp. 724-30, quotes thirty-nine enumerations of the members of the Ennead. One of them, No. 11, is the duplicate of Pyr. 1655..., but none of the other thirty-eight lists contains the same nine gods completely and exclusively." Whatever diversity there might have been in the composition of the Ennead, in the Heliopolitan system it was surely considered as having issued from Atum as the creator god either directly or indirectly and the gods making it up were considered as a single corporation or company. G. Jéquier, *Considérations sur les religions égyptiennes* (Neuchâtel, 1946), pp. 114-16, stresses that the nine signs for god *(nṭr)* used for "Ennead" were arranged in three columns of three signs and thus constituted a kind of super plural (three signs being an accepted form of the plural in hieroglyphic writing). As such, "Ennead" merely stands for a company of gods without specification of the number of gods in the company. This suggestion had been made in a preliminary form long since by C. R. Lepsius, *Über die Götter der vier Elemente bei den Ägyptern* (Berlin, 1856), p. 227. Hornung's remarks on the Ennead (*Conceptions of God,* pp. 221-23) also stress its variable number and composition. Cf. Kees, *Götterglaube*, pp. 150-51, and K. Sethe, *Dramatische Texte*, p. 49 (see full title in Introduction to Doc. II.9 below). As E. Otto notes in his *Egyptian Art and the Cults of Osiris and Amon* (London, 1968), pp. 46-47, the Great Ennead of Abydos in the time of Tuthmosis I consisted of Khnum of Herewer, Khnum of Kebehu, Thoth of Heseret, Horus of Letopolis, Horus the defender of his father, Wepwawet of Upper and Lower Egypt (the dualized person of the Upper Egyptian royal god).

30. M. Lichtheim, *Ancient Egyptian Literature*, Vol. 1 (Berkeley/Los Angeles/London, 1975), p. 106. For a discussion of this work of Merikare's, editions of the text, other translations, and studies, see Miss Lichtheim's introduction to this document, pp. 97-98. The references to "shines in the sky" and to "sails by [across the sky]" confirm that the god understood here is Re. Cf. Drioton, "La religion égyptienne," pp. 36-37.

31. For a discussion of the Egyptian names for Hermopolis and the doctrines of the Ogdoad, see above all Sethe, *Amun und die acht Urgötter von Hermopolis*, pp. 36-38. Also consult G. Roeder, "Die Kosmogonie von Hermopolis," *Egyptian Religion*, Vol. 1 (1933), pp. 1-37 (passim); Roeder, "Zwei hieroglyphische Inschriften aus Hermopolis (Ober-Ägypten)," *Annales du Service des Antiquités de l'Égypte*, Vol. 52 (1954), pp. 315-74, and particularly pp. 371-72. For evidence of the town name Khmun in the Old Kingdom, see L. Borchardt, *Das Grabdenkmal des Königs Ṣa'ḥu-Reᶜ*, Vol. 2 (Leipzig, 1913), pp. 99 and 101 and Pl. 21, and C. R. Lepsius, *Denkmäler aus Ägypten und Äthiopen*, Abt. II (reprint, Geneva, 1972), 112d (column on extreme left). Cf. Jéquier, *Considérations sur les religions égyptiennes*, p.

162, fig. 58.

32. Cf. Sauneron and Yoyotte, "La naissance du monde," pp. 52-54. J. Assmann, *Zeit und Ewigkeit im alten Ägypten* (Heidelberg, 1975), p. 21, interprets the negative characteristics of pre-existence as "lightlessness" *(kkw)*, "placelessness" *(tnmw)*, and "endlessness" *(ḥḥw)*.

33. Sauneron and Yoyotte, "La naissance du monde," p. 53. But see the classic treatment of the Ogdoad in Sethe's *Amun und die acht Urgötter*, where the variation in the names of these primitive gods over the whole period of pharaonic history is given in Tafel I appended to the volume. An interesting collection of illustrations of the Ogdoad was added by Lepsius, *Über die Götter der vier Elemente*, Taf. I-V. It shows the Ogdoad represented in a number of different forms of eight gods: (1) all sitting--four frog-headed male human forms and four serpent-headed female human figures, (2) all standing--four short-skirted male figures and four long-skirted female figures, (3) all sitting--four bearded males and four beardless females, (4) eight baboons, no distinction between male and female, (5) all standing--grouped as pairs, each pair consisting of a frog-headed male and a serpent-headed female, the first couple followed by Ptah, the second by Thoth, and the last two by no other god; (6) all sitting--four serpent-headed male gods and four lioness-headed female gods; and a number of other representations that are variations on the five listed above. As an example, see my Fig. II.8a.

34. See Sauneron and Yoyotte, "La naissance du monde," p. 53; Kees, *Götterglaube*, p. 307; and Clark, *Myth and Symbol*, p. 55.

35. Sethe, *Amun und die acht Urgötter*, pp. 39-40,

doubts Maspero's theory that the Five consists of four gods personalizing the basic concepts of the Ogdoad plus Thoth. For an extended treatment of Thoth, see P. Boylan, *Thoth: the Hermes of Egypt* (Oxford, 1922). Still another view of the company of Five is given by Jéquier in *Considérations sur les religions égyptiennes*, pp. 53-59. There we find an interesting account of the development of the Ogdoad. He begins with the study of four cynocephalic baboons, which he sees as the earliest representatives of the four qualities of the Primordial Abyss. Then next, he believes, followed the creation of the Five by the addition of a female goddess, namely, the nome and city goddess, The Hare. From this then developed a more satisfactory arrangement where instead of one goddess each of the principal male gods was matched with a female counterpart, and so the Ogdoad was born. He believes this to have taken place at the end of the prehistoric period but he has no evidence earlier than the evidence I have already mentioned that shows the name of Hermopolis as The Eighttown appearing in the Old Kingdom. Of course, he may be correct in assuming that the name goes back to a time just prior to the first dynasty but there is no way to test the assumption.

36. See H. Kees, "Die Feuerinsel in Sargtexten und im Totenbuch," *ZÄS*, Vol. 78 (1942), p. 43 (full article pp. 41-53).

37. So Clark, *Myth and Symbol*, p. 55, interprets this spell. But Faulkner, *Coffin Texts*, Vol. 3, 167 (Spell 1129) [full bibliographical reference in Doc. II.2 below] translates it very ambiguously: "This portal has closed on me as on the body of the egg from which I emerged. It is the darkness of my father Nu which does it, being far from this portal." Faulkner in note 7 says: "This

passage is unintelligible to me." But even if Clark is wrong in so interpreting this spell, the passage above the next note clearly conveys the basic ideas expressed by Clark here, especially as regards the Ogdoad's parentage of the Primeval Egg. For this parentage also see Sethe, *Amun und die acht Urgötter*, pp. 62 and 66 n. 2.

38. Sauneron and Yoyotte, "La naissance du monde," p. 61.

39. See Doc. II.7b[3] (IV,1-IV,8). For its association with the Hermopolitan doctrine of the Great Cackler, see Sauneron and Yoyotte, "La naissance du monde," p. 61.

40. Sauneron and Yoyotte, "La Naissance du monde," p. 61; Roeder, "Die Kosmogonie," p. 10, and G. Lefebvre, *Le tombeau de Petosiris* (Cairo, 1923-24), I, p. 140 and II, pp. 56-57. See also Lefebvre, "L'oeuf divin d'Hermopolis," *Annales du Service des Antiquités de l'Égypte*, Vol. 23 (1923), pp. 65-67.

41. Kees, "Die Feuerinsel," pp. 41-53, and his *Götterglaube*, p. 309.

42. Roeder, "Kosmogonie," p. 11. In the hymns to the gods found in Haremhab's tomb at Saqqara we see that Thoth, who is entitled "Lord of Hermopolis", has some characteristics of a creator god, for he is said to be "a self-created one who was not born". But in the same verse he is mentioned as "he who gives prescriptions to the Westerners who are in the Following of the sun god", which seems like a purely scriptorial function. See J. Assmann, *Ägyptische Hymnen und Gebete* (Zurich and Munich, 1975), p. 168.

43. Boylan, *Thoth*, p. 73; H. Brugsch, *Religion und Mythologie der alten Ägypter* (Leipzig, 1891), p. 49, gives a passage in which Thoth was designated not only

as the "Heart of Re" (as he was often called in other Ptolemaic sources) but also as the "Tongue of Atum". Cf. A. H. Gardiner, "Some Personifications. II. *HU*, 'Authoritative Utterance.' '*SIA* ', 'Understanding,'" *PSBA*, Vol. 38 (1916), pp. 50-51 (full article pp. 43-54, 83-94), where he gives examples in which courtiers assert to the king (whose powers on earth match those of his solar father) that he has Hu in his mouth and Sia in his heart.

44. Gardiner, *ibid.*, p. 52: "Wherever, in fact, Hu and Sia' appear as attributes of kingship, it is by virtue of the old legend which tells how Re^c, on emerging from the Nun, invented 'Utterance' and 'Understanding' to aid him in his creation and governance of the world. The myth...whereby Re^c fashioned Hike', 'Magic,' for the identical purpose, now appears as a variant version of the same idea, Hike' [or Heka] implying both the power of knowledge and the power of speech, but adding to these conceptions the notion of mysterious efficacy." P. 53: "It is not unjustifiable, therefore, to suppose that the legend of Sia' and Hu provided the model for that curious analysis [in the *Memphite Theology*] of Ptah into 'heart'..., the seat of the intelligence, and into 'tongue'..., the organ of speech, the interaction of which two sources of purposeful will is said to have resulted in the creation of all things." See also Gardiner's postscript in Vol. 39 (1917), p. 139: "I have tried to prove that the so-called philosophical text from Memphis (*Brit. Mus.* No. 135), in explaining the creation of the universe as due to Ptah, who took shape in Thoth as 'heart' or 'intellect' and in Horus as 'tongue,' has merely adapted to its special purpose the Heliopolitan legend according to which Atum divided into Hu 'Creative Utterance' and 'Sia' 'Intelligence' (see

pp. 53-54 of my first article)." Note that here Gardiner (following a common misinterpretation of the relevant passage of the *Memphite Theology*) reverses the roles of Thoth and Horus. The proper roles are Horus as heart and Thoth as tongue. For a recent summary of the creative role of Hu and Sia, see the article "Hu" by H. Altenmüller in W. Helck and W. Westendorf, eds., *Lexikon der Ägyptologie*, Vol. 3 (Wiesbaden, 1980), cc. 65-68. We should also note the scene at Dendera where Hu and Sia appear in the following of Thoth and Maat and are accompanied by Seeing and Hearing. They are preceded by the Ogdoad (see Fig. II.8f). Hu and Sia often appear in the solar bark as personalized gods (see Fig. II.12b). On more than one occasion Hu is replaced by Heka, the god of Magic or Magical Word (cf. Jéquier, *Considérations sur les religions égyptiennes*, p. 80). Jéquier correctly associates Sia and Hu with the rebirth of the sun every morning after its night voyage through the Underworld (see also Doc. II.4, n. 2).

45. Boylan, *Thoth*, pp. 92-97.

46. *Ibid.*, pp. 180-200. I have sometimes translated epithets differently from Boylan, but without omitting any important ones.

47. H. Grapow, "Die Welt vor der Schöpfung," *ZÄS*, Vol. 67 (1931), p. 37 (full article pp. 34-38).

48. Boylan, *Thoth*, p. 109, n. 1.

49. In the *Pyramid Texts* there are only two direct mentions of Ptah (Sects. 560 and 1482). M. Sandman Holmberg, *The God Ptah* (Lund and Copenhagen, 1946), pp. 24-25, says that these citations are interesting not because of their content but because they are the only ones in this large collection. After all, Ptah was the principal god in the capital city and hence one might expect more references to him. There

are perhaps two explanations. By the time the *Pyramid Texts* were put upon the walls of the pyramids, Re had become the most important god and references to him far outnumber or tend to mute references to other gods. Furthermore, since Ptah's center was in Memphis, the residence of the king who was the Living Horus, it is not surprising in the early period to find Ptah taking a position inferior to Horus.

50. Sandman Holmberg, *The God Ptah*, p. 27. His craft activities were still being emphasized at Edfu (E. Chassinat, *Le Temple d'Edfou*, Vol. 2, [Cairo, 1918], p. 37): "Words to be said by Ptah 'south of his wall', the Lord of the Life of the Two Lands (=Memphis or a district of it), the great god at Edfu, the father of men and women, the splendid god who came into being alone at the beginning. He has no equal, he who has molded himself in the beginning without having father or mother. He made his body completely alone, he who created the gods, he who created without being created, the one who lifts the heavens by the work of his hands...."

51. Sandman Holmberg, *The God Ptah*, pp. 30, 33-34.

52. *Ibid.*, p. 33.

53. J. Zandee, "Das Schöpferwort im alten Ägypten," *Verbum: Essays on Some Aspects of the Religious Function of Words dedicated to Dr. H. W. Obbink (Rheno-Traiectina*, Vol. 6, 1964), pp. 33-66.

54. Sandman Holmberg, *The God Ptah*, pp. 41, 14* (51).

55. Of all the treatments of the cosmogonic section of the *Memphite Theology*, the most useful is that found in Sethe's *Dramatische Texte*, pp. 46-70 (for full title see bibliography in the introduction to Doc.

II.9 below).

56. Note that in the Hymn to Ptah (Doc. II.7d, K.23) Re is referred to as "his (i.e. Ptah's) son".

57. H. and H. A. Frankfort, J. A. Wilson, T. Jacobsen, and W. A. Irwin, *The Intellectual Adventure of Ancient Man* (Chicago and London, 1946; Phoenix ed., 1977), p. 58.

58. *Ibid.*, p. 56.

59. According to Nims, *Thebes of the Pharaohs*, p. 18: "The name borne by the new king was known as early as the reign of Intef III [eleventh dynasty] and the sheets in which were wrapped the bodies of sixty soldiers who had fallen in the siege of some fortress in a campaign of Montuhotep II [the uniter of the kingdom] have in ten instances names compounded with Amon. One already shows Amon's claims to power, Sankh-Amensekhemtawy, meaning, 'Amon, the (most) powerful (god) of the Two Lands, sustains life'. But while names compounded with Amon were popular in court circles, none occurs in the list of more than eighty persons in the letters of Hekanakht [eleventh dynasty] and another contemporary archive. However, in the latter a blessing is pronounced in the name of Amon-re, the earliest evidence of Amon's identification with the ancient sun god." For the reference in the reign of Intef III, see Sethe, *Amun und die acht Urgötter von Hermopolis*, p. 32. See also pp. 11-12 for references to Amun in the eleventh and twelfth dynasties.

60. Sethe, *Amun und die acht Urgötter von Hermopolis*, p. 79, develops the thesis that Amun was imported into Thebes from Hermopolis. This is vigorously combatted by G. A. Wainwright, "The Origin of Amun," *The Journal of Egyptian Archaeology*, Vol.

49 (1963), pp. 21-23, and earlier in his review of Sethe's book in that journal, Vol. 17 (1931), pp. 151-52. Wainwright argues that Amun was essentially a form of Min. H. Kees, *Götterglaube im alten Ägypten*, p. 345, accepts Sethe's view. No one disputes that Amun took over the ithyphallic form and accompanying fertility aspects of Min (see Sethe, *Amun*, pp. 19-21).

61. Epithets and eulogies are found in German translation in Assmann, *Ägyptische Hymnen und Gebete*, Nos. 72-86, pp. 188-99. The textual sources and studies of them are conveniently given in Assmann's commentaries to Nos. 72-86. Needless to say, I have made no effort to include all the epithets, but I have given the most common and the most distinctive ones referring to Amon-Re's cosmogonic activities. See also H. M. Stewart, "Traditional Egyptian Sun Hymns of the New Kingdom," *Bulletin of the Institute of Archaeology*, University of London, Vol. 6 (1967), pp. 29-74.

62. A. H. Gardiner, *Hieratic Papyri in the British Museum: Third Series, Chester Beatty Gift*, Vol. 1 (London, 1935), p. 32, slightly altered.

63. A. Moret, *Le rituel du cult divin journalier en Égypte* (Paris, 1902), p. 108.

64. *Ibid.*, p. 123, and p. 123 n. 1.

65. *Ibid.*, pp. 129-30. Cf. Assmann, *Ägyptische Hymnen und Gebete*, No. 122, pp. 264-66.

66. Assmann, *Ägyptische Hymnen*, No. 129, pp. 293-99, and N. de Garis Davies, *The Temple of Hibis in El-Khargeh Oasis*, Part III, *The Decoration* (New York, 1953), plate 32. Incidentally, we should note the investigation of the relationship between Amun and the Ogdoad made by Sethe in his *Amun*, where (p. 56) he quotes an inscription to the effect that Amun "created Ptah in order to produce the Ogdoad" and another one

that asserts that he "created Ptah the goldsmith for building the Ogdoad". Still we also find (*ibid.*, p. 49) that Amun is called the "heir of the Ogdoad". As a matter of fact, Sethe concludes (p. 60) that a very complicated relationship between Amun and the Ogdoad developed as the result of the splitting of the primordial form of Amun into four different gods: (1) Amon-Re, the father of the father of the Ogdoad, the one who was the primordial snake Kematef in Medina Habu; (2) Amenapet I, the father of the Ogdoad, equivalent to Ptah-Tenen of Memphis, who every ten days appeared in the Temple of Amun at Luxor, the one who was the Earth-Creator snake; (3) Amun, who was one of the Ogdoad of Hermopolis, and like them was the father of the solar god Re, and who was worshiped as a deceased god in Medina Habu, being the primordial form of the gods; and (4) Amenapet II, the heir of the Ogdoad, the great living god, the chief of the gods, equivalent to Horus the Son of Isis and Osiris. This situation shows how complicated the later developments became.

67. On the history of the personifying of the Aten before Akhenaten, see the brief remarks of A. Piankoff, *The Shrines of Tut-Ankh-Amon* (New York, 1955), pp. 5-6.

68. Sandman Holmberg, *The God Ptah*, p. 48, quotes a statement from the Temple of Horus at Edfu that the king "had created Edfu on a potter's wheel, like 'Him who is south of his wall' (Ptah)".

69. S. Sauneron, *Esna*, Vol. 5: *Les fêtes religieuses d'Esna* (Cairo, 1962), p. 213, translates this phrase as "que fit ce qui est et ce qui n'est pas (encore)". This gives the negative half ("that which is not [yet]") the sense of a potential existence rather than the sense of nonexistence defined as that which is in the realm of

chaos (cf. Hornung's interpretation referred to in note 27 above).

70. Sauneron, *Esna*, Vol. 5, pp. 249-50, 255.

71. For example, Jéquier, *Considérations sur les religions égyptiennes*, pp. 14-25, describes the history of Egyptian religion as passing through three stages: fetishism (=veneration of gods in the form of inanimate objects), zoolatry (=the veneration of gods in animal forms), and anthropomorphism. Hornung mentions this theory in the course of analyzing the hieroglyphs used for "god" in his *Conceptions of God*, pp. 33-42. The three glyphs for "god" discussed by Hornung are shown in Fig. II.19. The wrapped pole on the left may indicate some early veneration of a god by means of an inanimate object, the falcon on a perch veneration by means of an animal form, and the seated figure with the beard veneration by means of a human form. The first two go back at least to the beginning of writing, while the third appears as early as the beginning of the Old Kingdom. But, as Hornung points out, the use of the wrapped pole for *nṯr* (=god) may show us a fetish, but it tells us nothing more about the early Egyptian conception of god (and indeed we might well add that neither do the other two signs). A fourth sign for god, a star, is said by Hornung to be attested only from the beginning of the Ptolemaic period. We can point out, however, that the appearance of a star in the name of a fortified place whose name may be translated "The Might of Horus the God" (F. Petrie, *Royal Tombs*, Vol. 2, pl. 23, 200) and which dates from the second dynasty would (if the translation is correct) indicate the use of the star as a glyph for "god" at a much earlier time. From this we could infer that in this early period the gods were thought of as at least residing in

the stars, as they were somewhat later in the *Pyramid Texts*. Another translation less favorable to my interpretation would be "The Might of Horus the Star", which, however, would still confirm the view that at least Horus was conceived as a star in the sky. We have again and again in this chapter referred to the personification of abstract concepts as gods or goddesses: Maat (Order or Justice) represented as the daughter of Re, Sia (Understanding), Hu (Utterance), Heka (Magic), and so on. Černý has some succinct observations on the phenomenon of personification, *Ancient Egyptian Religion*, pp. 58-59: "Both the deities dwelling or manifested in objects and animals, and the gods who are personifications of natural phenomena and are manifested in the phenomena themselves, underwent at a relatively early date a personification in human form; but besides these two classes there are a number of divinities who are personifications in human form of certain abstract conceptions or activities. Analogy on the original fetishes and animal and nature gods played its part here and underlying human characters were added to the abstract nouns of the language which made it possible to treat them on the same footing as other deities in pictorial art or in myths. In representations they were usually provided with certain marks which made them at once recognizable to an Egyptian. We have already met personifications of Shoy *'Fate'* as a man and of Renenet *'Nursing'* as a woman, and may add here personifications of certain materials, like Napri *'Corn'* as a god, Nub *'Gold'* as a goddess (because for some reason or other *'Gold'* became early an epithet or attribute of the Goddess Hathor), Kebhowet *'Cold Water Libation'* as a goddess, and activities such as *'Weaving'* (the goddess Tayet) and 'Winepressing' (the god Shesmu). Certain conceptions of time are found personified in the goddess Ronpet 'Year', Akhet 'Flood-season' and Proyet 'Spring' and the god Shomu 'Summer', the sex of the deity being decided by the gender of the Egyptian words. There are also geographical conceptions such as the goddess Sokhet 'Cultivable Plain', the god Ha 'Desert' and the goddess Amentet 'West', who bear on their heads hieroglyphic signs for 'field', 'desert' and 'west' respectively. The most important among all the personifications, however, is Ma‘et, personifying both 'Right' and 'Justice'....[her] name appears as early as the IInd Dynasty; very early too she is represented bearing on her head an ostrich feather which for some unknown reason had become her symbol. She is 'daughter of Re', for Re, the sun-god, rules the universe according to the principles of 'right' and 'justice' (both ma‘et) which he has laid down, and Ma‘et is, therefore, regularly seen standing on the prow of the sun-barque accompanying the sun-god on his course across the sky."

72. See Kees, *Der Götterglaube im alten Ägypten*, pp. 101-02; Jéquier, *Considérations*, pp. 175-76, and Černý, *Ancient Egyptian Religion*, p. 18.

73. Moret, *Le rituel du culte divin journalier en Égypte*, passim.

74. For translations of this story of Re and Isis, see that of Wilson in J. B. Pritchard, *Ancient Near Eastern Texts Relating to the Old Testament* (Princeton, 1950), pp. 12-14, and that of A. Piankoff, *The Litany of Re* (New York, 1964), pp. 56-59. The story is in the form of a spell, whose opening lines encapsulate Re's activities as a creator god: "The spell of the divine god, who came into being by himself, who made heaven, earth, water, the breath of life, fire, gods, men, small

and large cattle, creeping things, birds, and fishes, the king of men and gods at one time *(for whom) the limits (go) beyond years*, abounding in names, unknown to that (god) and unknown to this (god)" (Wilson). At one point in the story Isis asks Re his name, "(for) a man lives who is called by his name". This shows once more the creative power of naming.

75. Hornung, *Conceptions of God*, p. 110.

76. Moret, *Du caractère religieux de la royauté pharaonique*, p. 57, n. 2: "Il est bien caractéristique du caractère religieux de la royauté que le Pharaon ait officiellement quatorze 'doubles' [i.e. 'kas'] comme son père Râ. De même, au *décret de Phtah-Totunen* [i.e. *Ptah-Tatenen*] (1.13) on apporte au roi les '14 doubles de Râ'. Le soleil possède en effet sept âmes (*biou* [i.e. *b³w* or bas]) et quatorze doubles (*kaou* [i.e. *k³w* or kas]), ainsi que l'atteste un texte du temple de Philae (Mariette, *Denderah*, texte, p. 219, n. 3). Ces quatorze doubles figurent souvent sur les murs des temples ptolémaïques; ils portent tous un nom d'une faculté ou d'un sens (ouïe, vue, entendement, puissance, etc.) et sont 'comme les émanations de la divinité par lesquelles la divinité vit et qu'elle transmet a l'homme.' (Mariette)." See also J. Dümichen, *Altägyptische Tempelinschriften in den Jahren 1863-1865*, Vol. 1 (Leipzig, 1867), pl. 29, where it is noted that prayers are offered to each ba and ka. In an inscription on a stela at Abu Simbel dating from Year 35 of Ramesses II's reign the fourteen kas of Re are said to have been given to him by Thoth (Lepsius, *Denkmäler*, Abt. III, Bl. 194, line 13; J. F. Champollion, *Monuments de l'Égypte et de la Nubie*, Vol. 1 [Paris, 1835], Pl. 38, line 13). See also Budge's *An Egyptian Hieroglyphic Dictionary*, Vol. 2 (London, 1920) under *k³w*. Gardiner, "Some Personifications...," (1916), pp.

84-85, 94-95, discusses eight lists of the kas of Re, seven of which include the same names. For the full title of Gardiner's article, see note 43 above.

77. J. Vandier, *La religion égyptienne* (Paris, 1949), p. 132.

78. H. Bonnet, *Reallexikon der ägyptischen Religionsgeschichte* (Berlin, 1952), p. 359. For information on the ten bas of Amun, see J.-C. Goyon's treatment in R. A. Parker, J. Leclant, and J.-C. Goyon, *The Edifice of Taharqa by the Sacred Lake of Karnak* (Providence, Rhode Island, and London, 1979), pp. 73-79, where reference is given to the discovery by C. Traunecker of the ten bas of Amun painted in fresco on the walls of the north crypt of the temple of Opet. The identical scene of the procession of the ten bas was apparently present on the south and north walls of room F of the edifice of Taharqa.

79. Zandee, "Das Schöpferwort im alten Ägypten," p. 60: "Das Schöpferwort und das magische Wort sind identisch. Auch eine Beschwörung ist ein Schöpferwort. Es beraubt Dämonen ihrer Kraft. Es errichtet Ordnung statt Chaos. Es beseitigt Gefahren und lässt einen erwünschten Zustand eintreten. Es vertreibt den Tod und bringt das Leben."

80. Boylan, *Thoth: the Hermes of Egypt*, pp. 126-27. Boylan also says (p. 126) that *hike [heka]* "seems...to resemble, somehow, other qualities, or endowments needed for the full perfection of the glorified Dead--something, that is, on the same plane as the so-called Kas of Re. It deifies to some extent. Again it is clear from a number of passages that *hike* implies powers over things demonic and evil.... The power thus employed against evil is often described as if it were something carried in the mouth, and put into

action by utterance."

81. J. F. Borghouts, *Ancient Magical Texts* (Leiden, 1978), p. 2.

82. *Ibid.*, p. 22.

83. *Ibid.*, pp. 33-34.

84. A. H. Gardiner, "Magic (Egyptian)," J. Hastings, *Encyclopedia of Religion and Ethics*, Vol. 8 (Edinburgh, 1915), pp. 262-63 (whole article, pp. 262-69).

85. S. Sauneron, *Le papyrus magique illustré de Brooklyn* (New York, 1970), pp. 16-17.

86. *Ibid.*, p. 13.

87. See the work of Moret cited in note 63, which, despite its relative age, is still the most penetrating examination of Egyptian rituals. For a short, lucid account of temple rituals, see the work of Sauneron mentioned in note 14.

88. Hornung, *Conceptions of God*, p. 133.

89. P. Kaplony, *Die Inschriften der ägyptischen Frühzeit*, Vol. 1 (Wiesbaden, 1963), p. 291.

90. Lichtheim, *Ancient Egyptian Literature*, Vol. 1, p. 212.

91. Sethe, *Amun und die acht Urgötter*, pp. 53-61.

92. Lichtheim, *op. cit.* in n. 90, p. 213.

93. See the work of Moret cited in note 18 and that of Frankfort in note 3. See also G. Steindorff and K. C. Seele, *When Egypt Ruled the East* (revised ed., Chicago and London, 1957), pp. 82-83, who say that from the earliest time the king was considered as an embodiment of the falcon-god or as the son of the sun-god. They mention the depiction of the divine birth of Amenhotep III in the temple of Amun at Luxor. They quote a flowery courtier's speech in which we read (p. 83): "If you sayest to the waters, 'Come upon the mountain,' a flood floweth directly at thy word, for

thou art Re...and Khepri.... Authority is in thy mouth and perception is in thy heart; the activity of thy tongue is the temple of Maat... and God sitteth upon thy lips." These authors concede that not often in the time of Tuthmosis III or his immediate predecessors was the king the object of worship in the temples devoted to him. But they point "to the temple of Soleb in Nubia where Amenhotep III was worshipped 'as the living image of Re on earth,'...".

94. See the references to Posener and Hornung in note 15.

95. Moret, *Du caractère religieux*, pp. 43-48, displays from Lepsius, *Denkmäler*, Abt. III, plate 121a, the figure from the tomb of Haremhab of the solar disk with the signs for power, duration, and life hanging from the bottom of the disk. He translates the inscription somewhat close to the figure as follows: "Re makes magical protection *(s¹)* of your body." He suggests that this magical protection is that involved in the phrase: *s¹ ᶜnḫ ḏd* ("the magical protection of life-power-duration"), which he then interprets as a kind of fluid that the god imparts to the king. I would prefer to translate the inscription *(wnn Rᶜ m s¹w ḥᶜw-k)* "Re will be as magical protection to your body." This would modify the idea that Re transfers some kind of fluid to the king. Still, the basic idea that Re will somehow provide the king with magical protection is certainly intended.

96. J. H. Breasted, *Ancient Records of Egypt*, Vol. 2 (New York, 1906; repr. 1962), pp. 323-24. I have altered the spelling of the king's name to Tuthmosis to conform to my practice in this volume. Furthermore I have changed Breasted's reading of the name of the earth-god from Keb to Geb to follow the now generally

accepted reading. I have added the phrases in brackets.

97. Bonnet, *Reallexikon der ägyptischen Religionsgeschichte*, p. 835.

98. *Ibid.*, p. 837.

99. Breasted, *Ancient Records*, Vol. 2, pp. 60-61.

100. F. Ll. Griffith and H. Thompson, *The Demotic Magical Papyrus of London and Leiden* (London, 1904), pp. 50-53. This papyrus dates from the third century A.D. and shows considerable Greek influence, including passages in the Greek language, though the gods named are mostly Egyptian.

101. Černý, *Ancient Egyptian Religion*, p. 59.

102. For example, see the Prophecies of Neferti, in W. K. Simpson, *The Literature of Ancient Egypt* (rev. ed., New Haven and London, 1973), p. 238, which appears to describe in the literary form of a prophecy the bad conditions preceding the Middle Kingdom: "The land is diminishing, though its controllers are many; he who was rich in servants is despoiled and corn is trifling, even though the corn-measure is great and is measured to over-flowing (i.e. the dispensers of corn are hoarding it). Re separates himself from men; he shines that the hour [of dawn] may be told [lit., exist], but no one knows when noon occurs, for no one can discern his shadow [the sun being too weak]....". This is a poetic way of saying that Re has all but quit caring about the earth and its inhabitants. Similarly in the Admonitions of an Egyptian Sage (*ibid.*, p. 212) we read that "Khnum fashions (men) no more because of the condition of the land", and later after recounting more miseries the sage says "Indeed, the hot-tempered man says: 'If I knew where God is, then I would serve him.'" (*Ibid.*, p. 216). Furthermore, the sage says concerning Ptah (p. 217) "Why do you give to him? There is no reaching him."

103. Derchain in the work cited in footnote 1 of this chapter has a great deal to say about the role of the House of Life and its rituals in the preservation of maat and in the protection of the life of the king. The two are intimately tied together. Posener has expressed this succinctly: "Tout changement sur le trône prend une signification cosmique. Si, à la mort d'un roi, le chaos menace l'ordre de l'univers, l'avènement de Pharaon renouvelle la création originelle, rétablit l'équilibre de la nature. 'Rejouis-toi, pays entier! Les temps heureux sont arrivés. Un Mâitre s'est levé dans toutes les terres,.... l'inondation monte haut, les jours sont longs, la nuit a ses heures exactes, la lune revient avec regularité.' L'harmonie du monde dépend de la condition du roi, il doit garder tous ses moyens pour la bonne marche du cosmos." See G. Posener, et al., *Dictionnaire de la civilisation égyptienne*, 2nd ed. (Paris, 1970), p. 219.

104. Lichtheim, *Ancient Egyptian Literature*, Vol. 2, pp. 35-36.

105. Černý, *Ancient Egyptian Religion*, pp. 54-56.

106. *Ibid.*, p. 49. Cf. Bonnet, *Reallexikon*, p. 858.

107. Bonnet, *Reallexikon*, pp. 323, 858-59.

108. See H. Kees, *Totenglauben und Jenseitsvorstellungen der alten Ägypter* (Leipzig, 1956; 4th printing, 1980), pp. 73-74; Jéquier, *Considérations sur les religions Égyptiennes*, p. 67; Vandier, *La religion égyptienne.* pp. 76-81 (and particularly p. 78); R. Weill, *Le champ des roseaux et le champ des offrandes dans la religion funéraire et la religion générale* (Paris, 1936), and A. Bayoumi, *Autour du champ des souchets et du champ des offrandes* (Cairo, 1941).

109. A. J. Spencer, *Death in Ancient Egypt* (Harmondsworth, England, 1982), pp. 68-70. The

well-known spell from the *Book of the Dead* for putting the shawabti to work (no. 6) runs as follows: "Spell to cause the shawabti to perform work for a man in the Netherworld: O shawabti, if So and So (i.e. the deceased) is called upon or if I am listed to perform any work that is performed in the Netherworld, as a man to carry out his duties, to cultivate the fields, to water the banks [of the river], to transport the sand of the East to the West, 'Present' (i.e. 'I am here'), so shalt thou say." Černý, *Ancient Egyptian Religion*, p. 93.

110. Stewart, "Traditional Egyptian Sun Hymns," p. 38. As Stewart also remarks, "The day journey was thought to begin through the doors of heaven beyond the mythical eastern mountain of Baku, and end at the western mountain of Manu, where he [i.e. the sun-god] was received by the welcoming arms of his mother Nut. According to an early belief, already recorded in the Pyramid Texts (2173), the solar barque was rowed by the stars, who represented the spirits of the dead, and in the hymns of the New Kingdom the day journey is said to be accomplished by the circumpolar stars, 'those which can never set', and the night journey by 'those which can never weary'...."

111. Jéquier, *Considérations sur les religions égyptiennes*, pp. 54-59.

112. L. H. Lesko, *The Ancient Egyptian Book of Two Ways* (Berkeley / Los Angeles / London, 1972), includes an English translation and extensive notes and comments. However, I have for the most part followed Faulkner's later translation (see the introduction to Doc. II.2 below).

113. *Ibid.*, pp. 4, 6-7, Lesko characterizes the *Book of Two Ways* as follows: "Usually the *Book of Two Ways* was copied on the inside bottoms of the nobles'

coffins, probably so that the deceased would have this guide at their feet when walking in the underworld....The *Book of Two Ways* includes a number of different gods and goals....In addition to Re and Osiris...[it] also involves Thoth the god of the moon who is accompanied in the sky at night by the deceased as stars. To some extent he is as important as the other gods providing as he does a goal for the deceased, but since all the coffins having these texts are from his city [Hermopolis] this is not surprising....The last major deity in these texts is Horus....the eldest son of Re, who with wings outstretched represents the whole sky with his eyes as the sun and the moon... Seth...can also turn up unexpectedly as one of the guides on the bark of Re at the end of the book.... For goals of the deceased in the *Book of Two Ways* there is not only the famous 'Field of Offerings'... but also the mansion or palace of Osiris, the mansion of the moon, and the solar bark. Of the demons who guard the flaming red gates many are the same as those known from the two chapters (144 and 147) of the *Book of the Dead* devoted to the subject of gatekeepers."

114. *Ibid.*, pp. 40-41.

115. S. Morenz, *Egyptian Religion* (Ithaca, New York, 1973), pp. 75-80. If we take the end of the cosmos seriously, as does Hornung, *Conceptions of God*, p. 183, then of course we cannot say that time extends indefinitely.

116. Cf. Hornung, *Conceptions of God*, pp. 178-79, who mentions the view that "the primeval god who outlives the existent world changes into a snake at the end of time".

117. Eternity itself is sometimes given as *ḥḥ* (Erman and Grapow, *Wb*, Vol. 2, p. 299; *Belegstellen*,

Vol. 2, p. 435). We find the two expressions joined together as "millions of eternities" (*ibid.*).

Document II.1: Introduction

The *Pyramid Texts*

The earliest collection of religious documents in ancient Egypt is the so-called *Pyramid Texts*, found principally in the royal tombs at the end of the fifth dynasty and throughout the whole of the sixth dynasty.[1] These texts are written on the walls of the chambers of the pyramid of King Wenis in the fifth dynasty and those of the pyramids of Kings Teti, Pepi I, Merenre, and Pepi II in the sixth. In addition they are found in the pyramids of a number of queens of the late Old Kingdom and particularly in that of Queen Nit, the wife of Pepi II. We find them as well in the tombs of some nobles of the Middle Kingdom, the New Kingdom, and the Late Period, even as late as the thirtieth dynasty. In noting their long use, I speak of essentially unaltered texts and not of the funerary spells that developed out of the *Pyramid Texts*, those which we designate as *Coffin Texts* from the Middle Kingdom (see Doc. II.2 below) and those known as the *Book of the Dead* which originated primarily in the New Kingdom but also appeared in many later, altered versions (see Doc. II.3).

The texts consist of collections of individual statements or spells which are ordinarily called Utterances because they are introduced by the expression "Words to be Spoken". In all, the utterances number over 750, though the largest number in any single pyramid is about 712 in the tomb of Pepi II. They have no overall coherence and no single, all-embracing title. They reflect spells used in burial and offering rituals, and their oral character is everywhere evident. Traces of the ritual procedures of purification, offering food and clothing, censing, and anointing the king's body as well as royal and divine statues, can be found in the utterances.

The purpose of the *Pyramid Texts* was to ensure the resurrection of the king, his ascension to the sky as a divine star and as a companion of the sun-god in the latter's bark, and his enjoyment of the life of celestial regions with abundant provisions, regions where the king would mingle with and enjoy the company of the gods. The utterances also reflect the relatively recent doctrine of Osiris's resurrection and his role as king of the Netherworld (though this is not at all stressed as it was to be later).

We also find in these utterances allusions to the Horus- and Osiris-myths, as I have mentioned above in Chapter Two. The dead king had already begun to be identified with Osiris, the live king with Horus as Osiris's son. But despite these harbingers of an Osirian future for the deceased, the overwhelming emphasis of these utterances is on the celestial ascension of the king, an ascension that will result in his complete acceptance as a powerful god.

In selecting the extracts given here, I have limited myself largely to those utterances that bear on the cosmogonic and cosmological ideas discussed in Chapter Two. The most important thing to notice in this respect is that the king was identified with the creator god (sometimes as Re and sometimes as Atum) and his continuing governance of the world. Thus the utterances, though aimed at preserving the celestial future of the king, throw considerable light on the earliest expressed views of the creation of the cosmos and the nature of the gods and spirits that inhabit the Beyond. The identification of the king with the creator is particularly instructive in those utterances that describe the king as created in the Abyss before anything else (see particularly Sects. 1040, 1463, 1466). Among the things said to be not yet created are earth, sky, people, death, punishment, anger, turmoil, and in fact everything else that was later created. The role of magic in the king's ascension and power in the heavens is mentioned (Sects. 1318, 1324).

I stated earlier in my treatment of cosmogony that Heliopolitan doctrines predominated in the earliest documents but that there were also intermingled with these doctrines ideas that probably arose elsewhere than Heliopolis. I have mentioned in the notes the cases where Hermopolitan doctrines peek through the Heliopolitan utterances of the *Pyramid Texts*. This is evident, for example, in Sects. 265-66 and 446. The mention of the king as the Great Word in Sect. 1100 need not be a reference to the Memphite idea of the creative word (which, as we have said, no doubt developed later than the *Pyramid Texts*), since there is evidence for a limited doctrine of this kind of creation in the Heliopolitan system with its acceptance of Sia and Hu as creative conception and pronouncement.

The reader will notice that I have included a number of extracts that describe in a vague way the topography of the sky, including such places as the Field of Rushes, the Field of Offerings, the Field of Turquoise, the Mound of Horus, the Mound of Seth, the Jackal-Lake, the Lake of the Dat or Duat (usually

translated Netherworld, as it later became, though at
this time Dat had both celestial and subterrestrial
components), the Winding Waterway, the Causeway of
Happiness, and so on. Later documents were to fill out
the details of life in these places in the Beyond, but
occasionally we note some down-to-earth narrative
detail that removes us from the mysterious or at least
unknown references to ritual. Such is the case when
King Pepi (Sects. 915-18) declares that he will be passed
from mound to mound until he reaches the "tall
sycamore tree in the east of the sky on which the gods
sit", a naturalistic touch that pictures the gods as birds
perching in a tree. The same kind of homely detail is
evident when the gates of heaven are described as
being of iron (Sect. 907), if indeed that is the correct
translation since, apparently, only meteoritic iron was
likely to have been known this early, and certainly not
enough of that to conceive of great gates being made
from it.

Texts and Studies of the *Pyramid Texts*

The fundamental text is still that of K. Sethe, *Die
altägyptischen Pyramidentexte*, 4 vols. (Leipzig,
1908-22). To these we must add his posthumously
published *Übersetzung und Kommentar zu den
altägyptischen Pyramidentexten*, 6 vols. (Hamburg,
1962), which is the most useful single work of
translation and commentary that we possess, despite the
fact that much has been added both in terms of new
textual material by G. Jéquier, *Le monument funéraire
de Pepi II* (Cairo, 1936-) and *La pyramide d'Oudjebten;
Les pyramides des reines Neit et Apouit; La pyramide
d'Aba* (Cairo, 1928-35) and some significant

improvement in the translation by R. O. Faulkner, *The Ancient Egyptian Pyramid Texts: Translated into English* (Oxford, 1969), with a *Supplement of Hieroglyphic Texts* (Oxford, 1969). Consult also A. Piankoff, *The Pyramid of Unas* (Princeton, 1968). For recent investigations and discoveries, see J. Leclant, "Recherches récentes sur les Textes des Pyramides et les pyramides à textes de Saqqarah," *Académie Royale de Belgiques: Bulletin de la Classe des Lettres et des Sciences Morales et Politiques*, 5e série, Tome LXXI (1985, 10-11), pp. 292-305.

The English Translation

I have found the German translation by Sethe and the English translation by Faulkner the most useful of the various renderings of the *Pyramid Texts*. I have not, however, followed either of them slavishly. I have tended to use the titles of the utterances (or their parts) which Faulkner added to his translation, but occasionally have added my own where I wanted to emphasize the part of the utterance which I have translated. I have not adopted the first-person forms for the utterances, which Faulkner has almost everywhere employed. But instead I have kept to the third-person forms which include the king's name. I do this to maintain the historical sense of the documents, which disappears when the utterances are converted to model, first-person statements. Hence the reader will be able to tell from my translation whether the utterance appeared in the earliest corpus, that of Wenis, or whether it was first used by one of the later kings of the sixth dynasty. I have also abandoned Faulkner's translation of Dat as Netherworld since, as I explained earlier, the Dat in the Old Kingdom (written as it was with a star enclosed by a circle) included both celestial and subterrestrial regions.

I have included Sethe's section numbers within slant lines embedded in my text. The Spell or Utterance Numbers are distinct from the section numbers and are centered in my translation.

Note to the Introduction of Document II.1

1. For an informative but brief summary of the *Pyramid Texts*, see H. Kees, "Toten-Literatur. 20. Pyramidentexte," *Handbuch der Orientalistik,* Abt. 1, Vol. 1, Part 2 (Leiden, 1970), pp. 52-60. See also the clear account of certain aspects of the *Pyramid Texts* given by J. Vandier, *La religion égyptienne* (Paris, 1949), *passim*, but particularly pp. 74-82. The specialized studies mentioned at the end of this introduction will give the reader more detailed information.

Document II.1

The *Pyramid Texts*

Utterance 245
The king ascends to the sky
/250/ This Wenis has come to you, O Nut.[1] This Wenis has come to you, O Nut. He has committed his father to the earth.[2] He has left Horus behind him. His two wings have grown as those of a falcon, his two feathers (i.e. plumes) are those of a sacred falcon.[3] His soul *(b')* has brought him. His magic *(ḥk'w)* has equipped him.
The sky-goddess speaks
/251/ Open up your place in the sky among the stars of the heaven, for you are the Lone Star[4]....

Utterance 248
The king becomes a star
/262/ Wenis is a great one; Wenis has come forth from the thighs of the Ennead .... /263/ a star brilliant and far-traveling, who brings distant products to Re daily ....

Utterance 249
The king at the nose of Re
/265/ ....Wenis has gone to the Island of Fire.[5] Wenis sets Maat *(m'ct)* (the right or order of the cosmos) in the place of Isfet *(isft)* (the wrong or disorder of chaos) .... /266/ Wenis appears as Nefertem, as the lotus at the nose of Re; he will come forth from the horizon every day, and the gods will be cleansed (or purified)

by seeing him.

Utterance 262
The king prays to the gods for recognition
/330/ .... Be not unaware of Wenis, O Har-Sopd. If you know him, he shall know you.[6] ....

Utterance 263
The king ferries across the sky to Re
/337/ The reed-boats[7] of the sky are launched for Re that he may cross on them to the horizon, ... for Harakhti that he may cross on them to Re, ... for Wenis that he may cross on them to the horizon, to Re .... /340/ .... The Fields of Rushes are filled [with water], and Wenis ferries across on the Winding Waterway. /341/ This Wenis is ferried over to the eastern side of the horizon. This Wenis is ferried over to the eastern side of the sky. His sister is the star Sothis,[8] his child is the dawn.

Utterances 273-74
The king hunts and eats all of the gods
.... /397/ The king is the Bull of the sky, who conquers (?) at will, who lives on the being of every god, who eats their entrails, even of those who come with their bodies full of magic from the Island of Fire ....

Utterance 301
An address to the primeval gods
/446/ You have your offering-bread, O Niu (Nun) and Nenet (Naunet), you two embracers (?)[9] of the gods, who embrace the gods in your shadow. You have offering-bread, O Amun and Amenet (Amaunet),[10] you two embracers (?) of the gods, who embrace the gods in

your shadow. /447/ You have your offering-bread, Atum and Ruti, who yourselves created your own bodies. O Shu and Tefenet, who made the gods, who fashioned the gods, who established the gods: /448/ tell your father[11] that Wenis has given your offering-loaves to you... [so that] you will not prevent Wenis from crossing to him at the horizon. /449/ For Wenis knows him and knows his name: 'Eternity' is his name, 'The Eternal One, Lord of the Years' is his name. The Armed Fighter Horus, who is over the gods of the Sky,[12] is he who vivifies Re every day ....

Utterance 306
An ascension text
.... /480/ It is Geb who speaks about it: 'The mounds of my mound (i.e. realm) are the Mounds of Horus and the Mounds of Seth,[13] and the Fields of Rushes worship you in your name of *Dw'w* as Sopd who dwells under his *ksbt*-trees....'

Utterance 325
A lustration text
/525/ The doors of the sky [are opened]. The doors of the watery firmament are thrown open for Horus of the Gods, on the first day. He comes forth into the Field of Rushes. He bathes in the Field of Rushes. /526-29/ [So is the case successively with Harakhti, Horus of the East, Horus of Sezmet, and King Teti][14] ....

Utterance 407
The king takes his place in the Beyond
/710/ Teti purifies himself. He assumes for himself his pure throne in the sky. Teti will endure and the good throne of Teti will endure. Teti will assume his pure

seat which is at the bow of the Bark of Re. /711/ The godly sailors who row Re will row Teti; it is the godly sailors who convey Re round about the horizon who will convey Teti round about the horizon ....

Some adrdesses to Nut
Utterance 427
/777/ O Nut, spread yourself over your son Osiris Pepi that you may conceal him from Seth; protect him, O Nut ....

Utterance 428
/778/ O Nut, fall over your son Osiris Pepi; protect him, O Great Protectress, this great one who is among your children.

Utterance 429
/779/ Thus says Geb: 'O Nut, you have become spiritually mighty. Power *(shm)* was in you [when you were] in the womb of your mother Tefenet at the time you were not yet born, that you might protect Pepi with life and power, for he has not died.'

Utterance 432
/782/ O Great One (i.e. Nut), who came into being in the sky,.... the entire land is yours; you have taken it, for you have enclosed the earth and all things within your arms. You have set this Pepi as an Imperishable Star who is in you (i.e. the sky).

Utterance 434
/784/ Be far from the earth, for to you belongs (?) the head of your father Shu.... /785/ You have taken to yourself every god who possesses his bark, that you may install (?) them in the starry sky, lest they depart from you as stars....

Utterance 469
The king joins the solar bark

/906/ This Pepi is pure. This Pepi takes his oar in hand. He occupies his seat, this Pepi sits in the bow of the ship of the Two Enneads. Pepi rows Re to the west.... /907/ The doors of *Bꜣ-kꜣ*, which is in the firmament, are opened to this Pepi, the doors of iron which are in the starry sky are thrown open for this Pepi, and he goes through them.... /909/ ...I am a star *(nḫḫ)*, companion of a star *(nḫḫ)*. This Pepi becomes a star.

Utterance 470
A collection of spells

/914/ 'O Bull of Offerings, bend down your horn and let this Pepi pass,' so says Pepi. 'Where are you going?' 'This Pepi is going to the sky, for all life and dominion, /915/ that he may see his father, that he may see Re.' 'To the High Mounds [or] to the Mounds of Seth?' /916/ 'The High Mounds will pass him on to the Mounds of Seth, to that tall sycamore tree in the east of the sky on which the gods sit'.... /918/ This Pepi has bathed in the Field of Rushes, this Pepi is clothed in the Field of Khoprer, and Pepi finds Re there....

Utterance 472
A ferryman text

/924/ The sky quivers, the earth quakes before this Pepi, for this Pepi is a magician *(ḥkꜣw)*, Pepi possesses magic. /925/ This Pepi has come that he may glorify Orion and set Osiris at the head, and that he may set the gods upon their thrones. O *Mꜣ-ḥꜣf*, Bull of the gods, bring this (i.e. the ferry) to this Pepi and set him on that [yonder] side.

Utterance 473
The king crosses the Celestial River

... /934/ This Pepi comes forth on this eastern side of the sky where the gods were born, and this Pepi was born as Horus, as Him of that horizon; /935/ this Pepi is acclaimed just (i.e. vindicated) and this Pepi's ka is acclaimed just. So, acclaim Pepi and this Pepi's ka, for Sothis (Sirius) is the sister of Pepi, and the Morning Star is Pepi's child.[15] /936/ This Pepi will come with you and wander with you in the Field of Rushes; he will serve as herdsman with you in the Field of Turquoise ....

Utterance 477
Osiris and the king's services to him

.... /964/ ....This Pepi has come to you, O Osiris. This Pepi will wipe your face, he will clothe you with the clothing of a god.[16] He will do the purification ceremony for you at Djedit. /965/ It is Sothis (Sirius) your beloved daughter who prepares the yearly sustenance for you in this her name of 'Year'(Rnpt)....

Utterance 486
Address to the primitive waters

/1039/ Hail to you, you waters which Shu has brought, which the two sources lifted up, the waters in which Geb bathed his limbs. Hearts were pervaded with fear, hearts were pervaded with terror /1040/ when this Pepi was born in the Abyss before the sky came into being, before the earth came into being, before any established thing came into being, before tumult came into being, before that fear which arose on account of the Eye of Horus came into being ....

Utterance 504
The king's clear way to the sky
.... /1086/ .... He (i.e. the king) takes for himself his throne which is in the Field of Rushes /1087/ and he descends to the southern region of the Field of Offerings.

Utterance 506
The parts of the king are various entities
.... /1100/ The lips of Pepi are the Two Enneads; Pepi is the Great Word....[17]

Utterance 510
A miscellany of spells
.... /1143/.... Pepi takes possession of the sky, its pillars, its stars. /1144/ The gods come to him bowing; the spirits follow (i.e. serve) Pepi because of his power *(b'i)*. They have broken their staffs and smashed their weapons /1145/ because Pepi is a great one and the son of a great one, whom Nut bore .... /1146/ Pepi is the effluent fluid; he has come forth from the creating of the waters; he is a snake of many coils. This Pepi is the scribe of the god's book, who says what will be and what will not be created....

Utterance 511
The king goes to the sky
.... /1156/ .... he (the king) will hold up the sky with life and support the earth with joy; his right hand will support the sky with a *w's*-staff, and his left will support the earth with joy ....

Utterance 512
Speeches by the dead king's son

.... /1164/ Oho, raise yourself, Pepi, receive these four pleasant *nmst*-jars, wash yourself in the Jackal-Lake, be cleansed in the Lake of the Dat, be purified on top of your lotus-flower in the Field of Rushes. /1165/ Cross the sky, make your abode in the Field of Offerings among the gods who have gone to their kas. Sit upon your throne of iron, /1166/ receive your mace and scepter, that you may lead those in the Abyss *(nw)*, that you may give commands to the gods, and establish a spirit in his spirit-state. /1167/ Run your course, row your waterway like Re over the banks of the sky. O Pepi, raise yourself up and go into your spirit-state.

Utterance 518
A ferryman text

/1193/ O *Iw*, ferryman of the Field of Offerings, bring this [ferry] to this Pepi, for Pepi is the one who goes, Pepi is the one who goes.... /1196/ This Pepi has descended with the Two Enneads in the cool waters. Pepi is the potter (*or in other texts*, the plumbline[18]) of the Two Enneads by means of which the Field of Offerings was founded .... /1198/ .... The causeway north of the Field of Offerings is called the Causeway of Happiness. /1199/ Stand up, Osiris. Commend this Pepi to the supervisors of the Causeway of Happiness north of the Field of Offerings just as you commended Horus to Isis on that day you made her pregnant. /1200/ Let Pepi eat of the fields and drink of the pools within the Field of Offerings.

Utterance 527
The Creation of Shu and Tefenet

/1248/ Atum is he who came into being [alone], who masturbated in On (i.e. Heliopolis). He took his penis in his hand that he might ejaculate thereby, and so were born the twins Shu and Tefenet....

Utterance 537
A resurrection text

/1298/ O Pepi, stand up and sit upon the throne of Osiris! Your whole flesh is that of Atum .... /1299/ ....Stand up. You shall not perish, you shall not be destroyed, /1300/ but live, O Pepi! Your mother Nut grasps hold of you that she may embrace you, and Geb takes your hand .... /1301/ May you go out as Horus of the Netherworld who is at the head of the Imperishable Stars, may you live as the living scarabaeus-beetle, as long-lasting as the _dd_-pillar, for ever and ever.

Utterance 539
An ascension text

/1303/ The head of this Pepi is a vulture. He will ascend and rise up to the sky. The sides of the head of this Pepi are the starry sky of the god. He will ascend and rise up to the sky .... /1304/ .... The face of this Pepi is Wepwawet .... /1305/ The eyes of this Pepi are the Great One at the head of the Souls of On....The nose of this Pepi is Thoth....[and so on through the body down to the toes, with each part related to a divinity and with the constant repetition of the ascension of Pepi] .... /1316/ Pepi is the companion of a god, the son of a god; .... /1318/ Pepi was born for Re; he will ascend and rise up to the sky. The magic belonging to him is that which is in the belly of Pepi .... /1324/ It is not Pepi who says this to you, you gods. It is Magic who says this to you, you gods....

Utterance 570
Some miscellanies concerning the king
.... /1453/ This Pepi escapes his day of death just as Seth escaped [his day of death] .... [And successively Pepi escapes his half-months of death, his months of death, his year of death as Seth had escaped his.]

The king becomes a star
/1454/ Do not tread on (i.e. break up) the ground, O you Arms of Pepi which lift up the sky as Shu. [The bones of Pepi are iron] and his limbs are the Imperishable Stars. /1455/ Pepi is a star which illumines the sky. Pepi mounts up to the god that he may be protected. The sky will not be devoid of this Pepi and this earth will not be devoid of this Pepi forever. /1456/ This Pepi lives a life beside you, you gods of the Lower Sky, the Imperishable Stars [who traverse the Land of Libya, who lean on] their d^cm-staffs. This Pepi leans with you on a w^js-staff and on a d^cm-staff .... /1462/.... This one who belongs to the first generation for punishment(?) and for vindication, /1463/ which was born before anger came into being, which was born before noise came into being, which was born before strife came into being, which was born before tumult came into being....

Utterance 571
The king is the son of Atum and is a star
/1466/ The mother of Pepi was [pregnant] with him, he who was in the Lower Sky. This Pepi was fashioned by his father Atum before the sky came into being, before the earth came into being, before people came into being, before the gods were born, before death came into being. /1467/ This Pepi escapes [his] day of death just as Seth escaped his day of death .... /1468/ .... this

Pepi will not die on account of any death, /1469/ for Pepi is an Imperishable Star, son of the Great Sky-goddess who dwells in the Mansion of Selket ....

Utterance 576
A resurrection text
..../1508/ Pepi is conceived for Re, he is born for Re. Pepi is your seed, O Re, in this your name of Horus at the Head of the Spirits, star which crosses /1509/ the sea....

Utterance 587
An address to the sun-god
/1587/ Hail to you, Atum. Hail to you, [Khoprer] who created yourself. May you be high in your name of 'Height'. May you come into being in your name of Khoprer....

Utterance 593
Part of a resurrection text
.... /1636/ Your seed goes into her, she being ready as Sothis.[19] Har-Sopd has come forth from you in your name of Horus Who is in Sothis. You have spirit-state in him, in his name of Spirit Which is in the *Dndrw*-bark ....

Utterance 600
A prayer for the king and his pyramid
/1652/ O Atum-Khoprer, you became high on the Height. You rose up as the *bnbn*-stone in the Mansion of the Phoenix in On (i.e. Heliopolis). You spat out Shu, you expectorated Tefenet[20] /1653/ and set your arms about them like a ka-symbol, that your ka might be in them. O Atum, set your arms about Merenre, about this

construction, about this pyramid like a ka-symbol, that the ka of Merenre might be [in it, enduring forever] .... /1655/ O you Great Ennead, which dwells in On: Atum, Shu, Tefenet, Geb, Nut, Osiris, Isis, Seth, and Nephthys; O you children of Atum, extend his [good] wishes to his child in your name of the "Nine Extensions".[21]

Utterance 660
The king is the son of Atum
/1870/ O Shu, this Osiris Neferkare (i.e. Pepi II) is a son of Atum. You are the eldest son of Atum, his first-born. /1871/ Atum has spat you out from his mouth in your name of Shu....

Utterance 697
A resurrection text
.... /2172/ She bears you, O Neferkare, like Orion. She establishes your abode at the head of the Conclaves. Neferkare shall go aboard this bark like Re on the banks of the Winding Waterway. /2173/ Neferkare shall be rowed by the Unwearying Stars and shall give commands to the Imperishable Stars....

Notes to Document II.1

1.   Faulkner changes third-person forms here to first-person forms, as he does throughout, on the assumption that this was the original practice, though all of the texts we have give the third-person forms in most places. See also Sethe, *Übersetzung und Kommentar*, Vol. 1, p. 237.

2.   That is, he has made the obligatory burial of his father. He goes on to note that his son Horus has succeeded him on earth.

3.   The deceased king takes the form of a falcon so that he can fly up to the sky. In brief, he must have some physical means to get there. The falcon-wings give him that means and in addition stress his basic connection with the sacred falcon of Horus.

4.   R. O. Faulkner, "The King and the Star-Religion in the Pyramid Texts," *JNES*, Vol. 25 (1966), pp. 160-61 (full article, pp. 153-61), suggests that the Lone Star is Venus seen just after sunset.

5.   This spell reflects the Hermopolitan cosmogony of the sun's first appearance on the Island of Fire, which I mentioned in Chapter Two. The king here identifies himself with the lotus-blossom at Re's nose.

6.   I have included this line simply because it is one of the many that establish a relationship between Horus and the star Sirius. Usually Har-Sopd is said to be the offspring of Osiris and Isis as the goddess Sothis. This line comes in the midst of similar addresses to the various gods asking them to take cognizance of the king so that he in turn will take cognizance of them.

7.   Apparently the earliest form of the solar bark

was the simple reed-boat known on the Nile from the very earliest times. These boats are a far cry from the elaborate boats pictured in the *Book of Amduat*, and even more so from the boat discovered at the bottom of the south side of the Great Pyramid, which was presumably the solar boat that Cheops would sail across the sky (see N. Jenkins, *The Boat Beneath the Pyramid: King Cheops' Royal Ship* [New York, 1980]).

8. Here the king's sister is identified as the star-goddess Sothis (Sirius).

9. Sethe, *Übersetzung und Kommentar*, Vol. 2, p. 232, translates this as "Quellen (? Ursprünge)" because of the role played by the primordial creator gods. Faulkner rejects this translation because of the word's root-identity with the verb that follows, so that he translates the noun as "protectors(?)" and the verb as "protect". I prefer "embracers" and "embrace" in order to emphasize that the gods were born within and under the embrace of these primordial gods. See also the next note.

10. We have here the mention of four of the traditional Ogdoad of Hermopolis: Nun, Naunet, Amun, and Amaunet. They are all described as "embracers" of the gods, while the gods Atum and Ruti, which follow, are described as self-creators. Then Shu and Tefenet are said to have fashioned the (succeeding) gods. Here then we have a mixture of Heliopolitan and Hermopolitan doctrines. Incidentally, Ruti is a dual personage embracing Shu and Tefenet, who are, however, individually mentioned immediately thereafter.

11. Who the father is, is not clear.

12. This is probably another reference to Horus and the stars; this time Horus is said to be "over the gods of the sky" (no doubt the stars), i.e. their ruler.

13.   It is made clear in other passages that, though Geb, the god of the earth, is said to be the ruler of these two Mounds of Horus and Seth, the mounds are in fact of celestial location. See particularly Sects. 915-16 below.

14.   Hence it is evident that the king Teti joins (and no doubt is at one with) four forms of Horus: Horus of the Gods, Harakhti (Horus of the Horizon), Horus of the East, and Horus of Sezmet. He meets and bathes with them in the Field of Rushes.

15.   The king is sometimes seen as the Morning Star itself (see Faulkner, "The King and the Star-Religion," p. 161). As Faulkner says, there is little doubt that the Morning Star is Venus as seen at dawn.

16.   Though in heaven, the king is performing the temple ritual that is performed on earth. Parts of that ritual are wiping the face of Osiris, clothing him in the proper dress for a god, and generally performing the purification ritual done at Djedit.

17.   This may reflect the doctrine of the creative word, which, as I have said in Chapter Two, apparently had early roots in the Heliopolitan doctrines.

18.   In Pepi's text, the king is called "potter", but in the texts of Merenre and Pepi II "plumbline" is found.

19.   Here Isis is identified as the goddess in Sothis (Sirius). Notice the play on words involved in the Egyptian name for "Sothis" and the Egyptian word for "ready" since both of them have the root letters *s, p, d*.

20.   A play on words is involved in the doctrine of the creation of Shu and Tefenet by spitting. Similar in sound to Shu is the word *išš*, which means "to spit out". Further, similar to the sound of Tefenet is the word *tf*, which also means "to spit".

21.   The expression "Nine Extensions" refers to

the basic idea that the Ennead represented the nine extensions of the form or being of Atum. Faulkner translates the expression rather as "Nine Bows", the traditional enemies of Egypt, which makes no sense here, though perhaps he thought it to represent some kind of play on words, since the verb "to extend" and the word for "bow" share the same hieroglyph.

Document II.2: Introduction

The *Coffin Texts*

From the end of the Old Kingdom through the establishment and the development of the Middle Kingdom a new form of funerary texts appeared, the so-called *Coffin Texts*. These were written on the sides, the lids, and the bottoms of the coffins of nobles. The coffins from which the modern edition of these texts has been edited come from all parts of Egypt, but the greatest number are from Middle Egypt, especially from the necropolises at Asyut, el-Bersha, Beni Hasan, and Meir, where great nobles exercised much of the power that the kings had held to themselves in the Old Kingdom. Also used were coffins from Lisht, Dashur, and Saqqara. In Upper Egypt a principal source was coffins from Thebes, with some coffins from Aswan, Gebelein, Dendera, and Abydos.[1]

The spells that comprise these texts were of a character not unlike those of the *Pyramid Texts*, which we have already examined in Document II.1. In fact, a significant number of coffins contain spells taken directly from the *Pyramid Texts*; these have not been included in de Buck's great edition of the *Coffin Texts* and will not be discussed here since we have already completed our examination of the *Pyramid Texts*. But the spells, whatever their origins, were composed on the behalf of nobles rather than for kings. This extension from royalty to nobility of the practice of including burial spells with the deceased to help him in the underworld has been rather extravagantly called the democratization of the hereafter, "extravagantly", I say, because the deceased represented in these coffins were not ordinary people but those grandees in obvious control of the nomes or regions of Egypt. For it is clear that with the collapse of the Old Kingdom a significant social change took place that was reflected in the *Coffin Texts*. In view of the exalted position in the Old Kingdom of the king, who was not only the living Horus but also called himself the Son of Re, it is not surprising that there should have developed a special view of resurrection in the hereafter in which the king retained the life and privileges which he enjoyed on earth. But as political power slipped from the king's hand and the nobles absorbed the effective power, they also absorbed the royal benefits, one of which was an elaborate burial accompanied by written spells that would assure to them not only a life in the hereafter and protection from the dangers lurking there (as in the case of Spell 1034), but one in which, like the kings of old, they accompanied the solar god in his bark, talked to the gods, received their magical protection, and even identified themselves with them. Thus the deceased calls himself Shu, the father of the gods (see below, Spell 76, II,5), and even Re (see Spell 1034, where the deceased says in the beginning that he is a "dignitary of Re" and in the end that he is Re himself: "for I am Re, a greater god than you"), and in Spell 335 (IV,186) the deceased declares flatly that he is Re "at his first rising from the horizon".[2]

The ultimate achievement of transforming the deceased into gods is promised in the title at the end of Spell 290 (which I have not included in my extracts):

"The man shall be transformed into any god the man may wish to be transformed into."

More important for us in this chapter, the deceased absorbs the creative functions and might of the gods with whom he identifies himself. So, in the course of assuming the divine person of Shu, he becomes the creator of the Chaos-gods. The association of these gods with a chaos defined in terms of the negative characteristics of indefinite space, the formless Watery Abyss, darkness, and gloom (characteristics embraced by the Egyptian words: ḥḥ, nw, kk, and tnmw) gives us precious information concerning the age of the eight gods of Hermopolis associated with the chaos (Spells 76, 79-80). This Ogdoad of Hermopolis consisted of four pairs that embraced these negative characteristics. We also find evidence in these spells of other Hermopolitan doctrines such as the existence in primordial times of the Great Cackler who bore the primordial egg (see the document below, notes 6, 7, 11). It is perhaps no accident that the coffins that give us the material on the Chaos-gods are almost exclusively from el-Bersha, which was a necropolis associated with Hermopolis.[3] For the most part, however, the *Coffin Texts* contain doctrines that are of Heliopolitan origins, e.g., the autogenesis of the creator as the "Great One who created himself" (Spell 335, IV, 188) and as Khepri in the very meaning of his name and the consequential autogenesis of the deceased, raised on his throne "by means of myself, without a father and a mother contributing to me" (Spell 245). In the former spell the deceased also indicates that he is the one who created his [own] names and that he is "Lord of the Ennead" (IV,190). In another spell he says that he is the "one who originated in the Abyss, and see, the Chaos-gods came out to me....I fashioned myself at my will according to my desire" (Spell 714). We also see expressed the belief that the Lord of All had the form of a snake at the first stages of creation, i.e. when he accomplished his four deeds for mankind (Spell 1130). The description of these deeds is a brief but remarkable account of the creation for man of the very necessities of his life: air, water (for growing), equality of nature accompanied by the creator's injunction to do right (but coupled with the moral freedom of man to disobey that injunction, a freedom which unfortunately many men have exercised), and worship of the gods of the nomes.

In these spells are found both of the Heliopolitan views of the generation of Shu and Tefenet: from Atum's spittle (Spell 76, II,3-4) or as the result of Atum's masturbation (Spell 245). Further, the phoenix of Re is described as the form of Atum that came into being in the chaos, the latter in terms of the negative qualities deified in Hermopolis as the four pairs of the Ogdoad (Spell 76, II,4). We also see the common Heliopolitan doctrine that the gods were created from the sweat of the Lord of All, and mankind from the tears of his eye (Spell 1130, VII,464-65). In Spell 335 the cosmos, i.e. "what exists", is characterized in a gloss as being eternity and everlastingness, i.e. day and night. Thus eternal time is tied directly to the existent and presumably was created as an aspect thereof. Not only do we find the solar voyage as one of the objectives of life in the hereafter described in the spells, but we find a group of spells that constituted a *Book of Two Ways*, which appears to have been the earliest example of books that describe that voyage. It constitutes a kind of guide to the paths to take, with the highlights and dangers of the voyage described (see Spells 1030, 1034,

1035, 1131, 1136, 1146, and 1162). I mentioned this work above in Chapter Two and commented on the fact that it included a map of the two ways (Fig. II.28).

It is in the course of this voyage that we see a number of spells introduced to provide magical protection to the voyager, as for example Spell 1034, which was to help him against the danger of the *Aftet*-snakes. Spell 1146 presents a mysterious mansion, presumably one of the desirable objectives of the voyage, for it was a "place of excellent magic". Magic was blatantly in play in Spell 83, "to be said over the forepart of a lion[-charm] made from carnelian (?) or from the bone of a vulture, to be given to a man for his neck when he descends to the necropolis as a protection from the soul of Shu".

In describing the voyage, the spells mention the Field of Offerings, familiar to us from the *Pyramid Texts*. The mention of this field in Spell 1162 ties it with Osiris and Thoth; that is to say, the deceased will find himself there among the Followings of Osiris and Thoth. This suggests a location of the Field of Offerings in the Netherworld. Incidentally, though the main purpose of the spells was still to ensure for the deceased a celestial future with Re, as was that of the utterances in the *Pyramid Texts*, we find in the *Coffin Texts* increasing reference to Osiris, showing the apparent spread of Osirian ideas and cult. The practice of referring to the dead king as Osiris, which we noted in the *Pyramid Texts*, is extended to the nonroyal deceased in the *Coffin Texts*.

A final remark on the form of the spells is in order: titles are quite often included, as will be evident in some of the extracts I have given below: Spells 76, 335, 1030 (at the end), 1034 (at the end), 1130, 1162. The inclusion of a title was to become the rule in the spells of the *Book of the Dead*, as we shall see (Doc. II.3).

Texts and Studies of the *Coffin Texts*

The earliest edition of the *Coffin Texts* was that of P. Lacau, "Textes religieux égyptiens," *Recueil de travaux relatifs à la philologie et à l'archéologie égyptiennes et assyriennes*, Vols. 26-37 (1904-15), partially reprinted in *Textes religieux égyptiens* (Paris, 1910). This publication was based on coffins from Cairo only.

It was completely superseded by the superb edition of A. de Buck, *The Egyptian Coffin Texts*, 7 vols. (Chicago, 1935-61). Included are 1185 spells. The texts from all the coffins are given in parallel, vertical columns. De Buck's text was translated into English by R. O. Faulkner, *The Ancient Egyptian Coffin Texts*, 3 vols. (Warminster, 1973-78).

Before the publication of Faulkner's work a study with a translation of and a commentary on those spells considered to be part of the *Book of Two Ways* was published by L. H. Lesko, *The Ancient Egyptian Book of Two Ways* (Berkeley/Los Angeles/London, 1972). See the earlier study of H. Schack-Schackenburg, *Das Buch von den Zwei Wegen des seligen Toten* (Leipzig, 1903).

The English Translation

I owe much to the excellent translation of Faulkner mentioned above in the section on tcxts and studies. I do, however, occasionally depart from his work and I have added bracketed phrases when I thought they would clarify the text. The textual references embedded in the text between slant lines are of course to the edition of A. de Buck. It would be difficult to overpraise the clarity and usefulness of that work. It is obvious that I have only skimmed the cosmogonic and cosmological ideas from this extensive collection and the reader will surely find further study of it of great profit if he wishes to gain further knowledge of ancient Egyptian religious thought.

Notes to the Introduction of Document II.2

1. The list of the coffins used by de Buck in his edition is given in the front of each volume.

2. Still of great interest are the comments of J. H. Breasted, *Development of Religion and Thought in Ancient Egypt* (New York, 1912; Harper Torchbook, 1959), pp. 274-75: "The hereafter to which these citizens of the Feudal Age [the Middle Kingdom] looked forward was...still largely celestial and Solar as in the Pyramid Age....There is the same identification with the Sun-god which we found in the Pyramid Texts. There is a chapter of 'Becoming Re-Atum,' and several of 'Becoming a Falcon.' The deceased, now no longer the king, as in the Pyramid Texts, says: 'I am the soul of the god, self-generator....I have become he. I am he before whom the sky is silent,...I have become the limbs of the god, self-generator. He has made me into his heart (understanding), he has fashioned me into his soul. I am one who has breathed (?) the form of him who fashioned me, the august god, self-generator, whose name the gods know not....He has made me into his heart, he has fashioned me into his soul, I was not born with a birth.' This identification of the deceased with

the Sun-god alternates with old pictures of the Solar destiny, involving only association with the Sun-god. There is a chapter of 'Ascending to the Sky to the Place where Re is,' another of 'Embarking in the Ship of Re when he has Gone to his Ka,' and a 'Chapter of Entering into the West among the Followers of Re Every Day.' When once there the dead man finds among his resources a chapter of 'Being the Scribe of Re.'"

3. One merely has to note the coffins used by de Buck in Spells 76, 79, and so on, to see the truth of this statement.

Document II.2

The *Coffin Texts*

Spell 76

/II,1/ Going forth to the sky, going aboard the bark of Re, and becoming a living god. Oh you eight Chaos-gods who are in charge of the chambers of the sky, whom Shu has made from the efflux of his members, who tied together the ladder of Shu (or Atum), come and meet your father in me...., for I am he who created and made you just as I was created /II,2/ by your father Atum. I am weary because of the Supports-of-Shu since I [as Shu] lifted up my daughter Nut upon me so that I might give her to my father Atum in his boundaries (i.e. in his celestial regions); I have set Geb under my feet... /II,3/ .... Indeed I am Shu whom Atum created, from whom Re came into being. I was not fashioned in a womb, nor knit together in an egg. I was not conceived, but Atum /II,4/ spat me out in the saliva of his mouth along with my sister Tefenet....The phoenix of Re was the form of Atum which came into being in *ḥḥ* (indefinite space), in *nw* (the Watery Abyss), in *kk* (darkness), and in *tnmw* (gloom).[1] /II,5/ I am Shu, the father of the gods .... It was I who once again begot the Chaos-gods in *ḥḥ*, in *nw*, in /II,6/ *kk*, and in *tnmw*. I am indeed Shu who begot the gods. /II.7/ O you eight Chaos-gods whom I made from the efflux of my flesh, whose names Atum made when *nw* was created, on the day when Atum spoke in it with Nu /II,8/ in *ḥḥ*, in *tnmw*, and in *kk* ....

Spell 79

/II,23/ O you eight Chaos-gods who came forth from Shu, whose names the flesh of Atum created /II,24/ in accordance with the words of Nu in *ḥḥ*, in *nw*, in *tnmw*, and in *kk*....[2]

Spell 80

/II,27/ O you eight Chaos-gods, being Chaos-gods, who encircle the sky with your arms, /II,28/ who gather together the sky and earth for Geb, Shu fashioned you in *ḥḥ*, in *nw*, in *kk*, and in *tnmw*, and he allots you to Geb and Nut, for Shu is eternity and Tefenet is everlastingness[3] .... /II,32/ Thus said Atum: Tefenet is my living daughter, and she shall be with her brother Shu, whose name is 'The Living One'. Her name is Maat (Order or Righteousness)[4] .... /II,35/ .... Nu said to Atum: Kiss your daughter Maat, put her at your nose that your heart may live, for she will not be far from you. Maat is your daughter and your son is Shu, whose name is 'The Living One'. Eat of your daughter Maat (i.e., thrive on her); it is your son Shu who will raise you up [in the sky] ....

Spell 83

/II,46/ To be said over the forepart of a lion[-charm] made from carnelian (?) or from the bone of a vulture, to be given to a man for his neck when he descends to the necropolis as a protection from the soul *(bꜣ)* of Shu, so that he may have power over the winds of heaven /II,47/ .... As for anyone who knows this spell, he will not die again, his foes will not have power over him, no magic *(ḥkꜣ)* will restrain him on earth forever....

Spell 223

/III,208/ Spell for breathing air in god's land (i.e. the realm of the dead).[5] O Atum, give to N. (the deceased) this sweet air which is in your nostrils, for I am the egg which is in the Great Cackler.[6]...

Spell 245

/III,334/ Hail to you who arises and comes into being in this your name of Khepri. This is what is said to you: ... See, I have come and I bring you adoration and purification in the Pure Land, because I am the seed of my conception [achieved] on your behalf by means of your mouth, /III,335/ which bore me for you by means of your grasp (fist) in passion (i.e. in the orgasm of masturbation) .... /III,36/ .... I have arisen [on] my throne by means of myself, without a father and a mother contributing to me (i.e. to my conception and birth) ....

Spell 307

/IV,63/ .... I am the soul *(bꜣ)* who created the Watery Abyss and made a place in god's land (i.e. the realm of the dead); [my] nest will not be seen nor [my] egg broken,[7] for I am the lord of those who are on high, and I have made a nest in the limits of the sky....

Spell 335[8]

/IV,184/ Going out by day from god's land (i.e. the realm of the dead) [by] the revered N., deceased. My speech to you has come into being, Atum. /IV,186/ I was alone. I am Re at his first rising from the horizon. /IV,188/ I am the Great One who created himself. *Who is the Great One who created himself? He is Nu (i.e. the Watery Abyss).* [I am he] /IV,190/ who created his names, Lord of the Ennead. /IV,191/ *Who is he? He is*

Atum who is in his sun disk (aten). /IV,192/ Yesterday is mine and I know tomorrow .... /IV,194/ The warship of the gods is according to my command *(dd)* .... /IV,199/ I am the Great Phoenix who is in On (i.e. Heliopolis). *Who is he? He is Osiris.*[9] /IV,200/ As for me, I am the examiner of what exists. *Who is he? He is Osiris. As for what exists, it is eternity and everlastingness.* /IV,202/ *As for eternity, it is day. As for everlastingness, it is night* .... /IV,244/ I saw Re being born yesterday from the buttocks of the Celestial Cow[10] .... /IV,276/ I am his twin souls dwelling in his two progeny. *What does it mean? As for his twin souls dwelling in his two progeny, they are Osiris* /IV,278/ *when he entered into Djedut (Mendes) and found the soul of Re there, and they embraced.* /IV,280/ *They* (or *he) became his twin souls* ....

Spell 714

/VI,343/ I am Nu the Sole One who has no equal and who was born on the great occasion of my flood when I came into being. I am the one who flew up and whose form is that of Djebenen who is in his egg.[11] I am the one who originated in the Abyss, and see, the Chaos-gods came out to me. /VI,344/ See, I am flourishing. I brought my body into being with my own might. I am the one who made myself. I fashioned myself at my will according to my desire. That which came forth from me was in my charge....

Spell 1030

/VII,258/ See, its starry sky is in On, the sun-folk are in Kheraha because its thousands of souls are born, because their bandages are donned, because their oars are grasped.[12] /VII,259/ I will go with them aboard the lotus-bark at the dockyard of the gods. I will take possession of the bark which has lotus leaves at both ends. I will ascend in her to the sky. /VII,260/ I will sail in her in the company of Re. I will sail in her with Megef-ib. /VII,261/ I will act as pilot in her to the polar region of the sky, to the stairway of the bark of Mercury *(Sbg)*. Spell for sailing in the great bark of Re.

Spell 1034

/VII,278/ Upon your faces, you *Aftet*-snakes! Let me pass, /VII,279/ for I am a powerful one, Lord of Powerful Ones. I am a dignitary of Re, Lord of Maat, whom Edjo made. My protection is /VII,280/ the protection of Re. See, he has gone all around the Field of Offerings, which belongs to me, for I am Re, a greater god than you. /VII,281/ .... Guidance to the Paths of Rostau.[13]

Spell 1035

/VII,282/ I have passed over the paths of Rostau on water and on land; these are the paths of Osiris and are in the limits of the sky. As for anyone who knows this spell for going down into them, he himself will be a god in the Following of Thoth .... /VII,283/ .... But as for him who does not know this spell for passing over these paths, he will be counted as having been taken by the infliction (?) of the [truly] dead, ordained as one who is nonexistent, who shall never have maat.[14]

Spell 1130

/VII,461/ Utterance by him whose names are secret, the Lord of All (*lit.* Lord to the Limit) .... /VII,462/ Go in peace! I will relate to you the four (*var.* two) good deeds which my heart did for me when I was within the body of the Coiled One (i.e., I had the form of a snake) in order that wrong *(isft)* might be silenced (i.e., disorder or chaos be quieted). I have done four good deeds within the portal of the horizon.[15] I made the four winds /VII,463/ that everyone might breathe in his time (i.e. his lifetime). Such was my deed there (i.e. in the matter). I made the great [annual] flood so that the poor as well as the grandee might have power. Such was my deed in the matter. I made every man like his fellows /VII,464/ and I commanded them not to do wrong, but their hearts disobeyed what I had said. Such was my deed in the matter. I made their hearts not to forget the West by making god's-offerings to the regional gods. Such was my deed in the matter. I created the gods from my sweat /VII,465/ and mankind from the tears of my eye …. I shall sail correctly in my bark, for I am Lord of Waters (*var.* of Eternity) when crossing the sky. /VII,466/ …. Hu is with Heka felling that person over there who is ill-disposed toward me[16] ….

Spell 1131

/VII,473/ …. "Keeper to the Double Doors of the Horizon When They are Locked on Behalf of the Gods"--this is the name of their keepers, which is in writing …. As for anyone who does not know what they say, he shall fall into the nets…./VII,474/ But anyone who knows what they say shall pass through them and he shall sit beside the Great God …. [and] he will never perish ….

Spell 1136

/VII,481/ …. I have come here on the wind (*var.* the north-wind), for I am the leg(?) of Shu with which the

Abyss was filled .... Hu who speaks in darkness belongs to me[17]....

Spell 1146

/VII,496/ The image of a mansion .... The place of a spirit (*ꜣḥ*), the place of excellent magic (*ḥkꜣ*) ....

Spell 1162

/VII,506/ Spell for being in the Field of Offerings among the Following of Osiris and among the Following of Thoth every day. They will eat bread among the living. They will never die, breath being in their nostrils.

{For the notes to this document, see page 445.}

Notes to Document II.2

1.   These are characteristics of the primordial watery space in which creation took place. They were personalized and paired with feminine counterparts to become the Ogdoad of Hermopolis. The fourth condition listed here is *tnmw*, which I have translated as "gloom". We have already seen in Chapter Two that Amun and his consort Amaunet sometimes constituted the fourth pair of the Ogdoad. This pair embraced the concept of invisibility. Notice here how these conditions of chaos are immersed in the Heliopolitan doctrines of creation, for we have just seen expressed here the presumably older doctrine of the creation of Shu and Tefenet from Atum's spittle. Atum is himself described as having come into being in the primordial chaos delineated in those negative terms of *nw*, *ḥḥ*, *kk*, and *tnmw*. At this point the Chaos-gods are said to have been created by Shu in the primordial darkness and waters, and indeed the important function of Shu is evident from the fact that he is called the father of the gods.

2.   Again it is stressed that Shu has created the Chaos-gods in the Chaos under the same four conditions mentioned in Spell 76. The flesh of Atum is here held to have created the names of the Chaos-gods.

3.   Here the Chaos-gods are the supporters of the sky. Presumably they are the so-called Supports-of-Shu, which lifted Nut to the sky and kept her aloft. Again we notice that Shu is said to have fashioned them in the primordial chaos described by the four negative conditions mentioned in the earlier spells. Shu is called here "eternity" (perhaps the long stretch of past time)

and Tefenet "everlastingness" (perhaps the indefinitely long future), so that together they represent all time. In Spell 335, given below, a gloss identifies "what exists" (i.e. the cosmos) with "eternity and everlastingness". These two then are identified with "day" and "night". Here then there is some confusion concerning the role of "day" and "night" in the scheme of things. Presumably, at the end of each day an "eternity" has passed and the night represents the coming recreation of the successive elements of the future "everlastingness". Now the night obviously includes the period when chaos is present and the sun-god is overcoming chaos to recreate the cosmos of the day. I hardly need say we should not seek in "night" any notion of a mathematical limit separating "eternity" and "everlastingness".

4. Here we see, mythologically expressed, two aspects of creation, that is, of the coming into being of the cosmos. By creating Shu, his son, as light and air, Atum has brought forth life, and by creating his sister Tefenet in her name of Maat, the great creator has substituted the essential rightness or order (m³ᶜt) of the cosmos for the essential wrongness or disorder (isft) of the chaos (see below, Spell 1130, VII,462).

5. Needless to say, the most important things for the deceased to have at his resurrection in the realm of the dead were the ability to breathe (hence the ceremony of the Opening of the Mouth) and the air to breathe. This spell, it was hoped, would provide the latter.

6. I have described the Hermopolitan doctrine of the Great Cackler above in Chapter Two.

7. Again, we see here possible vestiges of the Hermopolitan conception of a primeval egg, a doctrine I

have discussed above in Chapter Two.

8. This is a long spell, of which I have included but a brief extract. The additions in italics are glosses added to certain coffins. In the glosses Osiris has been prominently inserted into what are clearly Heliopolitan doctrines. In the main text we may have a trace of the doctrine of the creative word in the second sentence where the deceased says that his speech to Atum has come into being.

9. Both in this gloss and the next we find interpretations involving Osiris.

10. This birth is shown pictorially in Fig. II.2a, where the solar bark is seen to emerge from between the back legs or buttocks of the celestial cow. Recall that the cow faces west and the bark moves across her stomach amidst the stars.

11. Here is one more trace of the doctrine of the primordial egg which was discussed in Chapter Two. It appears here mixed with Heliopolitan ideas.

12. This is a picturesque account of the gods, and of the newly deceased in their bandages, i.e. embalmed, boarding the solar bark at the dockyard of the gods in the district of Heliopolis known as Kheraha (see Doc. I.1, n. 114). Recall that the thousands of stars, which, by euphemism, become the starry sky, are thought of either as the gods or as the habitations of the gods. At any rate, here the gods and the deceased, as they board the bark, take up their oars. There is a nice bit of detail in the passage in which the bark is described as being shaped in the form of lotus leaves at both ends.

13. This is a rubric or title that appears more appropriate for the next spell. I have already mentioned in Chapter Two that the earliest view of the location of

Rostau was that it was near or at the necropolis of Saqqara.

14. This is good evidence for the view held by some Egyptians that if the traveler loses his way and does not find the true path, he becomes truly dead, that is, he is in the state of the nonexistent, and one who is nonexistent is one who never will experience maat. In short, he is a part not of the order of the cosmos but rather of the disorder of chaos. I have discussed these various ideas briefly in Chapter Two above.

15. Here the Lord of All gives a remarkable account of what he has done for man by outlining four deeds. In the first place he created the winds that man might be able to breathe. Then he created the annual floods which make the growing of crops so easy that the poor can be sustained as well as the rich. Thirdly, he made every one like his fellows and commanded or perhaps instructed them not to do wrong. However, the great god confesses, men's hearts have led them to disobey him. Hence we seem to have a doctrine of man's moral freedom expressed here.

16. I have already mentioned in Chapter Two that Hū (the personified authoritative or creative command) and Sia (personified creative intelligence) were often found accompanying the solar god in his bark, and that on occasion Heka (personified magical power) appeared instead of Hu (see above, Chapter Two, note 44). Now here we have a case where Hu and Heka (presumably in the bark with the deceased) are together protecting the solar god from some one who is hostile to him.

17. This is probably a veiled reference to Hu as the creative word working in the primordial darkness. Thus, the deceased, by identifying himself with Hu, that

is, by claiming that Hu (creative word) belongs to him, hopes to absorb some of the powers of the creator gods.

Document II.3: Introduction

The *Book of the Dead*

The third stage of the development of funerary documents in ancient Egypt is represented by the collections on papyrus of spells that are known by the title of *Book of the Dead*. Though there is no single Egyptian title for the various versions of the collections, the title most often found is *Spells for Going Forth by Day*, a title that indicates that a main purpose of the spells was to allow the deceased to leave the tomb in any form in which he wished to leave it. As one can readily see by examining the editions and the translations listed below, there is no single canonical collection. Each collection was commissioned or simply bought already prepared for a particular deceased, and each collection contained whatever number of spells was thought desirable and affordable.

The practice of making these collections on papyrus so that the papyrus could be placed in the tomb of the deceased began in the eighteenth dynasty, though the spells themselves were often modified versions of those already used on the coffins of the Middle Kingdom. Just as the number of spells was by no means fixed, so the order of the spells (and thus their ordinal number) was not always precisely fixed, though there was a generally accepted order later. However, since the beginning of the editing of these collections by Lepsius in the middle of the nineteenth century, a general system of numbering has been assumed by the scholars working on this material. Lepsius numbered 165 spells and in Allen's recent translation (see the bibliographic section below) we find the numbers extended to Spell 192 plus Pleyte Spells 166-74. These numbers are not truly indicative of the great variety that exists among the versions, for many spells occur in diverse forms. For example, Spell 15 exists in fourteen different versions, and these versions themselves vary in length and thus in richness of content. One can get a better sense of the complexity of the situation by going through Allen's translation carefully, since he tells us the sources and their dates for each of the variant versions.

Allen's remarks about the general history of the collections are pertinent:[1]

> The content and length of any given Book of the Dead depend on various factors: quantity and quality of scribes and spells available, wealth of the deceased, local usage, etc. During the Empire, only a few spells had begun to occur together in small groups, and to find any two manuscripts of exactly similar layout would be unusual. Before Ptolemaic times, however, perhaps by the fifth century B.C., a definite set of spells in a definite order had become fairly standard. In 1842 the great German Egyptologist Richard Lepsius published in facsimile a late Ptolemaic document of this type belonging to the Turin Museum. The

numbers he assigned to its successive spells are those which we still use, plus further numbers for spells there absent but found elsewhere and added by William Pleyte, Édouard Naville, E. A. Wallis Budge, and the present writer.

The present volume deals with all the spells yet numbered, which total 192, plus some insertions. In following the established order we vary widely, of course, from the diverse orders found within the documents themselves, for the texts here translated are those of the Empire rather than later times whenever available. Even so, earlier versions of many spells occur. All but 79 have a Middle Kingdom background in the Coffin Texts, written mostly on the insides of wooden coffins of the 19th or 20th century B.C. A few even go back to the Pyramid Texts of the 23rd or 24th century B.C., composed then or even earlier and used in the Old Kingdom for royalty alone.

In choosing the extracts given below, I have attempted to illustrate some of the more important cosmogonical and cosmological doctrines which I have discussed in Chapter Two. I have also given a good example of a spell in the *Book of the Dead* (Spell 17) which had its origin in the *Coffin Texts* (Spell 335). Comparison of the two will illustrate the types of changes made. It will also show how the glosses began to expand, so that virtually nothing of the past was abandoned when something new was added.

The same doctrinal tendencies that we saw in the earlier funerary material are still present in the *Book of the Dead*,[2] such as the great importance of the solar god and his creative activities (see below, Spells 15*A*3, 15*A*4, 15*B* 1, 15*B* 2, 15*B* 2 variant, 17); the vestiges of Hermopolitan concepts like the Lake of the Twin Knives (Spell 15*A*3), the Isle of Flame (Spell 15*B* 1), the emergence of the young sun from the primordial lotus (*ibid.*), the birth of Shu (as light) on the primitive hill of Hermopolis (Spell 17), and the primordial egg and the Great Cackler (Spells 54, 56); and further elaboration of the hereafter but with emphasis on Osiris in the Netherworld (Spell 175 and Spell 30*A* [note 3 below]), the Osirian judgment at the Hall of the Two Truths (Spell 125 and Spell 30*A* [again see note 3 below]), and accounts of the Fields of Offerings and of Rushes (apparently by this time a part of Osiris's realm in the Netherworld) which describe the familiar agricultural and pleasurable activities that the deceased had enjoyed before death and burial (Spells 99, 110).

One remarkable spell (175) recounts a conversation between Atum and Osiris. Osiris complains about the conditions of the Netherworld (too silent, no water, no air, very deep, very dark, no sexual pleasures), and Atum offers Osiris in recompense blessedness, quietness of heart, the earthly throne for his son Horus, and above all survival along with Atum at the end of time when Atum will destroy everything he made at the time of creation, i.e. everything and everybody that make up the cosmos except Osiris and himself, and even he (Atum) will return to his primitive serpent-forms.

In one of the most important and well-known spells (Spell 125) we see the wholesale elaboration of an earlier tendency to declare in a tomb biography the freedom of the deceased from wrongdoing. In the deceased's denial of a long list of sinful actions, we see clearly what constituted the proper code of conduct at the time of the New Kingdom. The importance of this for our cosmological study is simply that the creator not only set out maat or order for the cosmos but simultaneously set out with it maat or truth for individuals, i.e. right conduct. It would of course be surprising if very many people of power could truly assert that they had led the blameless life outlined in Spell 125. Indeed there is evident recognition among the Egyptians of the fact that men will often say anything to achieve resurrection in the Netherworld, for there exists a spell whose purpose is to make sure that the deceased's heart does not speak out against him in god's domain.[3] But perhaps they thought that if the statements of innocence were solemnly and formally spoken as a spell, then by the doctrine of the creative word, i.e. the bringing into existence of something by magical word, the assertion could somehow produce the truth. We are reminded that this doctrine is evident in Spell 17, when at the very beginning it is said "My words come to pass" (*ḫpr*, lit., "are created").

I mentioned in the introduction to Document II.2 that the spells began to acquire titles in the Middle Kingdom, titles often written in red ink. These rubrics multiplied as time went on, so that most of the spells of the various versions of the *Book of the Dead* had their special titles. Allen distinguishes the preliminary materials connected with titles from the spells proper by using a marginal P for the preliminary material and a marginal S for the spells themselves. On occasion there are also separately distinguishable terminal comments, which Allen identifies with a marginal T.

Texts and Studies of the *Book of the Dead*

The earliest edition of any version of the *Book of the Dead* was that of C. R. Lepsius, *Das Todtenbuch der Ägypter nach dem hieroglyphischen Papyrus in Turin* (Turin and Leipzig, 1842; repr. Osnabruck, 1969). The papyrus produced in facsimile in Lepsius's work dated from Ptolemaic times and thus contained material not in the earlier versions.

Almost a half-century later E. Naville produced the standard version of the text: *Das ägyptische Todtenbuch der XVII. bis XX. Dynastie aus verschiedenen Urkunden zusammengestellt und herausgegeben*, 3 vols. (Berlin, 1886). This text includes the pictorial vignettes found in various manuscripts.

Convenient, too, is the text produced by E.A.W. Budge, *The Chapters of Coming Forth by Day or the Theban Version of the Book of the Dead. The Egyptian Hieroglyphic Text Edited from Numerous Papyri*, 3 vols. (London, 1910). Budge also prepared a now-out-of-date English translation: *The Book of the Dead. An English Translation of the Chapters, Hymns, etc., of the Theban Recension*, 2nd ed., rev. and enl., 3 vols. (London, 1909).

See the selected spells edited by H. Grapow, *Religiöse Urkunden (Urkunden des ägyptischen Altertumskunde, Abt. V)*, 3 Heften (Leipzig, 1915-17).

By far the most important of the recent studies and translations are those of T.G. Allen: *The Egyptian*

Book of the Dead. Documents in the Oriental Institute Museum at the University of Chicago (The University of Chicago Oriental Institute Publications, Vol. 82) (Chicago, 1969), and *The Book of the Dead or Going Forth by Day. Ideas of the Ancient Egyptians Concerning the Hereafter as Expressed in Their Own Terms (The Oriental Institute of the University of Chicago Studies in Ancient Oriental Civilization*, No. 37) (Chicago, 1974).

Also useful (but not nearly so comprehensive as Allen's translation) is the translation by R. O. Faulkner, *The Ancient Egyptian Book of the Dead*, revised edition by C. Andrews (London, 1985). This revision contains beautiful reproductions of numerous vignettes. Miss Andrews has translated some spells and parts of spells omitted in the earlier edition of Faulkner's translation published by the Limited Editions Club in New York, 1972 (see Bibliography below).

The English Translation

I have followed Allen's translation with little change, except to keep my earlier translations of some of the doctrinal terms, like "eternity" for $n\hbar\hbar$ and "everlastingness" for $\underline{d}t$. I have also retained the marginal letters and numbers used by Allen, which appear here as bold-faced letters at the beginning of the sections they mark. As in the case of my extracts from Documents II.1 and II.2, the extracts for this third document present only a pale picture of the whole complex structure of the various versions of the *Book of the Dead.*

Notes to the Introduction of Document II.3

1. See the translation by Allen, *The Book of the Dead*, pp. 1-2.

2. For a brief but illuminating description and comparison of the spells of both the *Coffin Texts* and the *Book of the Dead*, treated together as one form of funerary texts, see H. Kees, "Toten-Literatur," *Handbuch der Orientalistik*, Abt. 1, Vol. 1, Part 2, pp. 61-69.

3. See Allen, *The Book of the Dead*, Spell 30*A*, p. 40: "Spell for not letting N.'s heart oppose him in the god's domain. He says: My heart of my mother, my heart of my mother, my breast that I had on earth, stand not against me as a witness before the Lords of Offerings. Say not against me 'He really did it' concerning what I have done. Bring no charges against me before the great god the lord of the west (i.e. Osiris)." I have not included this spell in my extracts below.

Document II.3

The *Book of the Dead*

Spell 15 A3[1]

a

/P,1/ Adoring Re as he rises in the eastern horizon of the sky. /2/ Osiris N. shall say:

/S,1/ Hail to thee, Re at his rising, Atum at his setting.[2] Thou risest, thou risest, thou shinest, thou shinest, having dawned as king of the gods. /2/ Thou art lord of the sky and earth, who made the stars above and mankind below, sole god who came into being at the beginning of time, who made the lands and created common people, who made the deep *(nw)* [3] and created the inundation, who made the water and gave life to what is in it, who fashioned the mountains and brought into being man and beast. /3/ Sky and earth greet thee with libations; Truth *(Mꜣꜥt)* embraces thee day and night. /4/ Thou traversest the sky in gladness, the Lake of the Twin Knives having grown calm.... /6/ (As for) him who is in his shrine, his heart is pleased, for he has dawned as dominator of the sky, sole one, keen, who came forth from the deep, Re triumphant, divine youth, heir of eternity *(nḥḥ)* [4] who begot himself and bore himself, sole one, great in number of forms, King of the Two Lands, Ruler of Heliopolis, lord of eternity, familiar with everlastingness *(ḏt)*. /7/ The Ennead is in joy over thy rising. (They) who are in the horizon paddle (thee); (they) who are in the night bark exalt thee....

b

/**S**,3/ No tongue could understand its fellow except for thee alone....

Spell 15*A*4[5]

/**P**/ Adoring Re at his rising in the eastern horizon of the sky.

/**S**,1/ Hail to (thee) child in ⌐the ⌐same way as yesterday⌐,[6] rising from the lotus, goodly youth who has ascended from the horizon and illumines [the ⌐Two⌐ Land⌐s⌐ with] his light....

Spell 15*B* 1[7]

b

/**S**,2/ (Then) I shall see the great god who lives yonder in the Isle of Flame, the youth (born) of gold who came forth from the lotus....[8]

Spell 15*B* 2[9]

/**S**,1/ Hail to thee Re [maker of] all mankind, Atum-Harakhti, sole god, living on truth, maker of what is and creator of what exists of animals and human beings that came forth from his eye, lord of the sky and earth, maker of mankind below and (the stars) above, Lord of the Universe, bull of the Ennead,[10] King of the sky, lord of the gods, Sovereign at the head of the Ennead, divine God who came into being of himself, Primeval One, who came into being at the beginning....

Spell 15*B* 2 variant[11]

/**S**,2/ Joy to thee, maker of the gods, who lifted high the sky to be the pathway of his eyes,[12] who made the earth to be the broad realm of his Sunlight, that every man might perceive his fellow....

Spell 17[13] (*cf.* Doc. II.2, Spell 335)

a

/**P**,3/ (Spell for) going forth by day, assuming whatever form one will, playing senet, sitting in a pavilion, going forth as a living soul by N., after he moors (i.e. dies).[14]
/**S**,1/ My words come to pass. ⟨All[15] was⟩ mine when [I] existed in the Deep; (I was) Re at ⟨his⟩ dawnings when he began his reign.

> What does it mean, that is "Re when he began his reign"? It means when Re began dawning in the kingdom he had created before the uplifting of Shu had come into being, while ⟨he was on⟩ the hill that was in Hermopolis. Now the children [of the feeble one] had been given ⸢with⸣ them that were in Hermopolis.

/2/ I am the great god who came into being of himself,

> Who is he, "the [great] god who came into being of himself"? (He is) water; he is the Deep (i.e. *Nw*), the Father of the gods. Variant: He is Re.

who created himself, lord ⟨of the Ennead⟩,

> Who is he? He is Re when he created the names of his members. So came into being these gods who are in his Following.

(most) irresistible of gods.

> Who is he? He is Atum who is in his Disk. [Variant:] He is Re when he rises from the eastern horizon of the sky.

/3/ Mine is yesterday, and I know tomorrow.

> Who is He? "Yesterday" is Osiris; "tomorrow" is Re. That is the day when the

enemies of the Lord of the Universe were
annihilated and his son Horus was caused
to reign. Variant: That is the day of the
Festival (called) We Abide, that is (the
day) when the burial of Osiris was directed
by his Father ⟨Re⟩ ....

/5/ I am this (great) phoenix that is in Heliopolis, the
examiner of what exists.

Who is he? He is Osiris. As for "what
exists", (that means) the great god. Variant:
it means eternity and everlastingness. As
for "eternity", that is day; as for
"everlastingness", that is night...

Spell 42[16]

/P,l/ Spell for warding off the harm that is done in
Heracleopolis....

/S,2/ My hair is (that of) the Deep; my face is (that of)
the Disk. My eyes are (those of) Hathor; my ears are
(those of) Wepwawet. My nose is (that of) the Presider
over ⌐Xois⌐; my lips are (those of) Anubis [and so on
through other parts of the body] .... /4/ I am one who
has ascended sound, whose name [is not known]. I am
yesterday; my [name] is He Who Sees a Million Years,
who has gone, who has gone along the roads of the
Chief Examiners. [I am lord of everlastingness .... /5/ I
am your protector for millions ⌐(of years)⌐ .... I shall not
die again.... /6/ .... I am the blossom that came forth
from the Deep, and Nut is my mother, O thou who didst
create me, I am one who strides not, the great
commander within yesterday, the commander's portion
being within my hand. There is none that knows me or
shall know me; there is none who grasps me or shall
grasp me. O thou of the egg, thou of the egg, I am

Horus presiding over millions ....

Spell 54[17]

/P,l/ Spell for giving breath to N. in the god's domain.
/S,l/ O Atum, give me the refreshing breath that is in my nose. I am this egg that was in the Great Cackler. (I am) this great magical protection that came into being and separated Geb from the earth. If I live, it lives; if I grow old, it (grows old). If I breathe air, (it breathes air) ....

Spell 56[18]

/S,l/ O Atum, mayest thou give me the refreshing breath that is in thy nose. It is I who occupy this (great) seat in the midst of Hermopolis. I have guarded this egg of the Great Cackler. If I flourish, it flourishes. If I live, it lives; if I breathe air, it breathes air....

Spell 99[19]

c

/S,3/ .... Your offerings to me are barley and wheat; your offerings to me (are) myrrh and clothing. (Your) offerings to me (are) oxen and fowl; (your) offerings to me (are) life, soundness, and health. Your offerings to me (include the right) to go forth (by) day in any form in which I may wish to go forth (from) the Field of Rushes.
/T/   If one recites this spell, he goes forth (by) day from the Field (of Rushes). Given him are a cake, a jar, a *pz(n)*-loaf, a chunk of meat, and (fields of) barley and ⌐Upper Egyptian⌐ wheat 7 cubits (high). It is the Followers of Horus who reap them for him. Then he shall chew (on) this barley and wheat and shall wipe his body therewith, and his body shall be as (those of)

these gods. (So) he goes forth from the Field of Rushes in any form in which he may wish to go forth.

/T var. 2/ A truly excellent spell (proved) a million times.

Spell 110[20]
a 1

/P,1/ Beginning of the spells for the Field of Offerings, the spells for going forth by day, going in and out of the god's domain, attaining the Field of Rushes, existing in the Field of Offerings, the great settlement, ... gaining control there, becoming a blessed one there, plowing there, reaping (there), eating there, drinking there, copulating there, doing everything that is done upon earth....

/S,2/ Lo, I paddle (in) this great bark in the lake(s) of Hotep; it is I who took it (i.e. the bark) from ⌜the limbs⌝ of Shu. (His) limbs and his stars are ⌜years and seasons⌝ .... /3/ I prevail over her (i.e. the field), (for) I am the one who knows her. I paddle in her lake(s), so that I arrive at her settlements. My mouth becomes powerful, and I become sharper than the blessed .... I become a blessed one therein; I eat (there)in, I drink therein. I plow therein, I reap therein, I grind therein. I copulate therein; my ⟨magic⟩ becomes powerful therein ....

Spell 125[21]
a

/P/ What to say on arriving at the broad hall of the Two Truths, cutting N. off from all the forbidden things he has done, and seeing the faces of all the gods....

/S,2/ I have not sinned against anyone. I have not mistreated people. I have not done evil instead of righteousness. I know not what is not (i.e. the nonexistent); I have not done anything bad.... I have not reviled the god. I have not laid violent hands on an orphan. I have not done what the god abominates. I have not slandered a servant to his superior. I have not made (anyone) grieve; I have not made (anyone) weep. I have not killed; I have not turned (anyone) over to a killer; I have not caused anyone's suffering. I have not diminished the food(-offerings) in the temples. I have not debased the offering-cakes of the gods. I have not taken the cakes of the blessed. I have not copulated (illicitly); I have not been unchaste. I have not increased or diminished the measure, I have not diminished the palm(-measure); I have not encroached upon fields. I have not added to the balance weights; I have not tampered with the plumb bob of the balance....

c

/**S**,6/ I (have) purified myself in the southern site. I have gone to rest in the northern settlement, (in) the field of grasshoppers wherein I purify myself at this hour of night or day (for soothing the hearts of the gods when I pass through it by night or by day). "Let him come," say they of me. "Who art thou?" say they to me. "What is thy name?" say they to me. I am the lord of the undergrowth (of) a papyrus clump; He Who is in the Moringa is my name. "What did(st thou) pass through?" say they to me. I passed through a settlement north of a thicket.... [With further answers given correctly, the guardians say] "Come thou, enter through this gate of (this) broad hall of the Two Truths, for thou knowest us." [Similar inquisitions follow, in which the deceased has to produce the names of the jambs of the gates, the leaves of the gate, the floor of the gate, and so on.][22]

Spell 175[23]
b

/P/ To be said by Osiris N.:

/S,l/ O Atum, what means it that I proceed to the necropolis, the silent land, which has no water and no air and is very deep and very dark and (all) is lacking, wherein one lives in quietness of heart and without any sexual pleasures available? "I have given blessedness instead of water, air, and sexual pleasure, quietness of heart instead of bread and beer", says Atum.... (But) every (other) god has mounted his throne in ⟨the bark of⟩ millions (of years). "Thy throne (belongs) to thy son Horus", says Atum .... /2/ "What is a lifetime of life?" says (Osiris). "Thou art (destined) for millions of millions (of years), a lifetime of millions (of years).... And [then] I will destroy all that I have made. This land shall return into the Deep, into the flood, as it was aforetime. (Only) I shall survive together with Osiris, after I have assumed my forms of other (snakes) which men know not and gods see not...."

Notes to Document II.3

1. This is based on a 19th-dynasty papyrus (Ag in Allen's list).

2. This is a variation of the usual formula: the solar god is called Khepri at rising, Re in midday, and Atum in the evening. I have retained here Allen's respectful forms in using his translation of this hymn to Re.

3. Allen usually translates nw as "deep", while ordinarily I have translated it by "Abyss". This is somewhat different from the usual statement that Atum created himself in the Abyss, for here we have him creating it.

4. I stick with my previous translations of $nḥḥ$ and $ḏt$ by "eternity" and "everlastingness". Allen translates them by "perpetuity" and "eternity". As I have noted before, the two words together constitute eternal time as an aspect of creation, i.e. of the cosmos.

5. This version is found in Papyrus Ba in Allen's list (also of the 19th dynasty).

6. This is one of the sections that mention the Hermopolitan doctrine of the sun's emergence from a lotus blossom as a "goodly youth".

7. This version was also found in Papyrus Ba from Allen's list.

8. Two Hermopolitan doctrines are mentioned here: the Isle of Flame or Fire and the primordial lotus blossom.

9. This spell is known from two documents: one from the 18th and one from the 19th dynasty. See Allen's translation. Notice the allusion to mankind's creation from the eye of Re, needless to say from his

tears.

10. The expression "bull of the Ennead" simply means that Re is the progenitor of the Ennead. This title is coupled with the usual epithets of a creator god.

11. From the 19th-dynasty papyrus Da in Allen's list.

12. This is no doubt a reference to the sun-disk and the moon-disk. The statement that follows is the usual one that says that Re created his light in order that man might see.

13. This is based largely on Aa of the 18th dynasty but was restored or added to from Papyrus Ce and occasionally from the version in Spell 335 of the *Coffin Texts*. I have included this spell so that the reader may compare the two versions. I have found it useful to consult the edition of this spell constructed by Grapow from versions of it ranging from the time of the Middle Kingdom through the Late Period (see the text noted above in the introduction to the document).

14. I have included only part of the preliminary material; for the rest of it (i.e. for sections /P,1/ and /P,2/), see Allen's translation. For the vignettes which show the deceased in a pavilion playing senet, see Fig. II.26.

15. In some versions of Spell 335 of the *Coffin Texts*, the *tm* here rendered as "All" was read as "Atum" and was attached to the preceding sentence.

16. Based on two versions of the 18th dynasty (see Allen's translation). The deceased's effort to become a god reflects the similar objective of the part-by-part identification of the deceased with gods which was found in the *Pyramid Texts* (Doc. II.1, Sects. 1303-05). Again note traces of Hermopolitan doctrines in this spell.

17. Based on 18th-dynasty papyri (see Allen's translation). Allen calls the Great Cackler "the Great Honker". This and the next extract (Spell 56) demonstrate the continued interest in the Hermopolitan doctrine of the Great Cackler which was present in the *Coffin Texts*.

18. This spell is from 18th-dynasty papyri (see the translation of Allen). It specifically ties the doctrine of the Great Cackler to Hermopolis.

19. From the 18th dynasty. See Allen's translation. In Chapter Two, I have commented on the details given here concerning the agricultural benefits of being in the Field of Rushes.

20. More details on the good life followed in the Field of Rushes in this spell found in 18th-dynasty papyri (see Allen's translation). This chapter attracted pictorial vignettes (see Figs. II.25-26).

21. This spell from 18th-dynasty papyri (see Allen's translation) is one of the most celebrated of all spells in the *Book of the Dead*. There are two versions of the somewhat misnamed "Negative confession", i.e. lists of sins or wrongful acts which the deceased affirms that he has not committed. I have given much of the shorter list. For the longer list, see Allen's translation, part *b*.

22. The reader will notice here just one more instance of the importance of knowing names for achieving desirable objectives. It is a part of the ancient Egyptian interest in magical words that persisted from the very beginning of their written literature. I have often commented on this in Chapter Two above and have particularly stressed the significance of it for the doctrine of the creative word.

23. Based on both 18th- and 19th-dynasty copies

(see Allen's translation). In Chapter Two I mentioned the importance of this spell for the doctrine of the cessation of the cosmos and the return to the condition of chaos that existed prior to creation.

Document II.4: Introduction

The *Book of Amduat*

The *Book of Amduat* or *Amdat*, which may be translated as the *Book of What Is in the Netherworld*, is one of several works whose earliest copies appear in the tombs of the kings of the New Kingdom. Two other significant works of a similar nature are the *Book of Gates* and the *Book of Caverns*. Though we know nothing about the origins of these works, we can suppose on the basis of extant copies that the *Amduat* is the earliest of them.

I have given a brief account of the *Amduat* in Chapter Two, but certain other details ought to be mentioned here. In the first place, it exists in two versions, a discursive one that is long and a poetic one that is short. Both of the versions are found in the tombs of the kings in the Valley of the Kings in Western Thebes, sometimes the shorter version following at the end of the longer one. The extracts given here are from the long version. Its earliest extant copy is fragmentary and from the tomb of Tuthmosis I (1504-1491 B.C.). The copy on the walls of the tomb of his grandson Amenhotep II is the most complete. Details of the various other copies from the New Kingdom, some of which are in the tombs of nobles (in fact the second-earliest copy is found in the tomb of the vizier Weseramon in the time of Tuthmosis III, where also is found the earliest copy of the short version), are given in the edition of Hornung mentioned below.

The name by which this work is known, *t' md't imit dw't*, is one sometimes applied generally to such underworld-books, and indeed it appears in some later copies of this work.[1] The title given in the earliest version is that which we have included in the extracts below, "The Writing of the Hidden Place (or Chamber)". This is followed by a kind of table of contents of what is included in the work: the locations and activities of the various souls, gods, and shades that occupy the twelve hours or regions of the Netherworld, the speeches of Re to them, the various gates, the course of the hours and their gods, and so on. When we first see this book in the tombs of the kings, it clearly has become a substantial element of the royal (and occasionally noble) funerary literature. The major objective of placing the book in the royal tomb is to ensure the king's rebirth like that of his father Re. There are passages in it that also have a ritual significance, as for example in the introduction to the second hour where we read that offerings will be made on earth in the names of the gods of this region. They will be effective for the celebrant on earth, "as has been proved true a million times".

In the royal tombs the work is laid out on the walls like a gigantic papyrus. It is divided into twelve sections, each representing one of the hours of the night, and it is organized about paintings in each section (see Figs. II.29-40). The character of the work is epitomized by Piankoff in the following way:[2]

> The *Book of Am-Duat* or *Book of What Is in the Netherworld*, is a composition very similar to the *Book of Gates*. It contains twelve

divisions...,each division (except the introductory division) is formed of three registers, the sun barge being in the central one. But there are no gates [depicted], and each division is usually preceded by a short introduction giving the name of the hour. In the *Book of Gates* the barge is the same in all twelve divisions; in the *Book of Am-Duat*, however, the number of the crew and the aspect of the barge change in almost every division. The crew is usually composed of the Opener of the Ways [Wepwawet], Mind [Sia], Mistress of the Barge, [Re, i.e. Flesh of Re,] Horus the Praiser, the Bull of Truth, the Watchful One, Will [Hu], and the Guide of the Barge. In the second division Isis and Nephthys also appear on the barge, in the shape of two serpents. In the first division the prow of the barge is covered by a mat of reeds. In the second, sixth, and all other divisions with the exception of the fourth and fifth, the prow and the stern both terminate in a lotus flower. In the eleventh division a disk rests on the lotus flower of the prow; in the twelfth a scarab. In divisions four and five the prow and the stern of the barge both terminate in the head of a serpent. The bow is towed in divisions four,

> five, eight, and twelve. The serpent which envelops the god appears only from the seventh division on. In the second and third divisions the barge is seen in the great waters, in all others it is placed on a small section marked as water.

As Piankoff notes, this description applies to the older, more traditional versions of the book.

Though, as I have said, the work was clearly a funerary work by the time of its earliest known copy from the tomb of Tuthmosis I, there is some reason to believe that it may have originated as an early topography of, and a guide to, the Netherworld, modeled after works like the Onomastica that describe the entities of this world. This seems to be borne out by the initial table of contents and by the fact that at each hour the work catalogues the peculiar features of the region and the inhabitants of that hour. Furthermore, the insistence again and again that the gods represented are "like this in the Netherworld" seems to reflect a style that emphasizes the veracity of accounts, a style that seems to give the book a quasi-scientific or empirical air of the sort found in the Edwin Smith Surgical Papyrus, which we shall study in Chapter Five of Volume Three.

Altenmüller has suggested that the work may go back to the fourth or fifth dynasty,[3] a time when some sort of medical works like the surgical papyrus seem to have been composed. Whether such speculations are true or not, the purpose of the inclusion of this document in my account is to give the reader some idea of standard views of the Otherworld in the New Kingdom. I have also included it to show the persistence of the cosmogonic and cosmological ideas in works other than collections of spells or hymns.

Generally the cosmogonic ideas represented in the *Book of Amduat* are those associated with the Heliopolitan system and its modifications by the time of the growing dominance of Amon-Re. The whole drama confirms the primacy of the creation of Re by himself as Khepri. It surely is significant that in the second register of the twelfth hour Re is towed through the body of a snake and "comes out of its mouth, being born in his forms of Khepri, and of the gods who are in his bark as well". Hence as the creator creates his new emergent forms of Khepri he also creates the forms of the other gods of his crew as forms of himself. Presumably the same kind of creation by the multiplication of his forms took place when he brought the Ennead into creation, though he seems to say in the first register of the fifth hour that such a creation of forms took place after the creation of their bodies: "May you breathe, O Ennead of gods, who came into being from my flesh when your forms were not yet created."

The comparison of the rebirth of Re with the birth process of the beetle as he pushes his egg along is alluded to in the first register of the tenth hour when we are told that the gods of the Cavern of the West ("where Khepri rests with Re") are "in the forms and births of Khepri when he carries his egg toward this city in order to come forth afterwards in the Eastern Horizon of Heaven". As we have seen, snake forms like that of Neheb-kau ("The Provider of Forms") were important in the early stages of creation, and that importance is reflected throughout this book, particularly in the last hours of the night that precede the actual rebirth of Re.

But following his self-generation the creator god created the gods and mankind. That role of creation is reflected in this work by continual references to his gift of life and breath to all the gods and spirits of the Netherworld. For example, in the third register of the second hour, the gods who have already confirmed his lordship of time by presenting him with year- and age-signs are said to "live through the voice of this Great God. Their throats breathe when he calls them and assigns them their duties." Clearly the doctrine of the creative word that is a part of the Heliopolitan system in the form of the divinities Sia and Hu as personalized Conception and Command is evident in the temporary revivification of the underworld gods when they hear Re's voice each night. Indeed I have stressed here and in the notes below the fact that Sia and Hu accompany the solar god as part of the crew of his bark, Heka (Magic) being substituted for Hu in one instance (see Doc. II.4, notes 2 and 4). The importance of magical utterances is stressed in the second register of the seventh hour where we read that the solar bark "is towed by the magical utterances of Isis and the Elder Magician". But here the reference is not so much to creative words as it is to destructive ones, since the purpose of the magic of Isis and the Elder Magician is to fend off the serpent Neha-her or the serpent Apep.

The pre-existent Abyss from which the Great God brought himself into being lurks everywhere in the Netherworld of this book and it is seen to constitute a constant danger to the creator god. Similarly, those creatures of the Abyss, the nonexistent ones, are said, in the third register of the third hour, to be annihilated by the gods of that hour, and it is the nonexistent ones who live in a special "Place of Annihilation-houses".

It is also abundantly clear from the *Amduat* that the concept of the birth and death of gods has been significantly elaborated beyond (1) the earlier Heliopolitan idea of successive generations of gods, starting with Atum, then proceeding to his children Shu and Tefenet, then to Geb and Nut, followed by their children Osiris, Isis, Seth and Nephthys, and then to Horus, the son of Osiris and Isis: in all, five generations; and (2) the death of Osiris and his resurrection in the Netherworld. Now we find mentioned in this work not only the corpse of Osiris (second register of the fifth hour), but also those of Sokar and Khepri himself. In the fifth hour we see Sokar's head sticking out from his pyramidal burial mound of sand, while the body of Khepri "in his own flesh" is mentioned in the second register of the fifth hour. Indeed, the solar god when traveling as the "Flesh of Re" is a corpse himself, though admittedly a very powerful one capable of bringing about his own rebirth.

It is surely of interest that, upon this rebirth, Re leaves behind his mummified "Image of Flesh" leaning against the extreme boundary of the twelfth hour. Further, the corpses of the many anonymous gods residing in the regions of the Netherworld lie inert until revived for a time by the voice of Re as he talks to them. In the first register of the fifth hour the Great God himself confirms the death of gods by saying: "How beautiful is the great way inside the earth, the way to the grave and resting place of my gods," and the very name of the city of the eighth hour "Sarcophagi of her Gods" is further evidence that we have divine corpses. Presumably these gods who live in the Netherworld and who have corpses are distinguished in some fashion from the gods who live in the sky, the star-gods. Such gods are spoken to by Horus in the third register of the seventh hour, and he says to them: "May your flesh be right, may your forms come into being, so that you might be at rest in your stars." But even here a kind of resurrection from corpses is implied.

The doctrine of the secret names and forms of the gods is also briefly reflected in this book, when the gods who sit upon their hieroglyphic signs of clothing in the third register of the eighth hour are said to be like this, "as the secret forms of Horus, the heir of Osiris". In the ninth hour forms and creations or transformations (hprw) seem to lead a kind of disembodied and separate existence, for the city of the ninth cavern is named "That Which Springs Forth for Forms, That Which Lives for Transformations". Similarly the gate of the tenth hour is called "Great of Creations (or Transformations), The One Bearing of Forms." The images of the gods, presumably their statues or representations, are constantly mentioned throughout the work and obviously have special significance for the Egyptian view of the gods. While we may not know the secret forms of the gods, we can at least perceive their images, though the meaning of the images or their true nature is not knowable.

As in the case of human beings who have bas or souls that allow them after death to leave their corpses, to go forth and to return, so too the gods and deceased spirits in the Netherworld of this book have such additional spiritual personalities, and indeed the bas are included in the list of beings in the introductory table of contents whose locations in the Netherworld are to be described in the work.

Other Works Concerning the Solar Bark

Since I have treated the cosmological aspects of the *Book of Amduat* in some detail, I can limit myself here to a briefer consideration of the other two comparable funerary books that describe the night journey of the solar bark. The work most like the *Amduat* is the *Book of Gates.*[4] Its earliest copy dates from the time of King Haremhab (1319-1307 B.C.) and is on the walls of his tomb in Thebes. It contains less than half of the work. The first complete copy is found on the sarcophagus of Seti I (1306-1290 B.C.), originally in his tomb, but now in the Museum of Sir John Soane in London. A number of its features closely resemble those of the *Amduat*: the twelve hours, the division of each hour into three registers with the middle register representing a river and the upper and lower registers its banks, the passage along the middle register by the sun bark, the form of the sun-god as a ram-headed human figure (standing in a pavilion about which the Enveloper-snake is draped) who is called The Flesh of Re. The most distinctive features of the book are the great gates at the end (according to Hornung but at the beginning according to Piankoff) of each hour, with doors guarded by snakes.

The sun-god begins the journey at the Western Desert Mountain in front of the first gate, arriving there in the beetle form, i.e. as a beetle imposed upon the sun disk and surrounded by a snake with its tail in its mouth (see Fig. II.41). This is somewhat surprising since the Khepri form of the sun-god is usually confined to its first appearance at dawn. But perhaps here it indicates that the sun begins the journey with its Khepri form gestating within the disk (fetuslike) and

kept there by the surrounding snake, which, as we have
seen (Chap. II, n. 27), represents the surrounding chaos
out of which the sun must emerge at the end of his
night journey. Also of interest to our concern with
cosmogony is the fact that the crew of two that is in
the bark here and throughout the night journey consists
of Sia and Heka, the two gods concerned with the
creative word, Heka having replaced Hu, a not
uncommon substitution. Further of interest in the first
hour is a brief statement over the Gods of the Western
Mountain epitomizing the Great God's creation:[5]

> The ones who come into being from
> Re, from the brilliance of his eye,
> who came forth from his eye. Re
> grants them the hidden place, to
> which are brought men and gods, all
> cattle, all worms, which this Great
> God has created.

Moving to the first gate, we see a statement of
Sia to the Guardian of the Desert that is of interest for
the doctrine of Re's autogenesis:[6] "Open your gate to
Re, You upon your door, Akhti. The Hidden Place (or
Chamber) is in darkness until the creation-forms (hprw)
of this god are created." After the bark with Re in his
ram-headed form and his two crew members, Sia and
Heka, has passed through the gate, gods come to meet
the bark, and we read over them in the middle register
of the second hour a statement that epitomizes the
purposes of the voyage:[7]

> The sailing of this Great God on the
> ways of the Netherworld. The
> towing of this god by the gods of
> the Netherworld in order to diversify
> that which is on the earth, to take

care of those who are in it (the
earth), to render judgments in the
West, to make the great into the
small (i.e. to equalize the differences)
among the gods who are in the
Netherworld, to set the spirits (*ḫw*)
in their places, to deliver the damned
to their judgment, to annihilate the
bodies of the wicked, to confine the
bas (of whom?). Re says:....Sia and
Heka associate themselves with me
in order to take care of you and to
bring into being (*sḫpr*) your forms
(*irw*).

Nothing could be clearer concerning the role that
Sia and Heka (Zandee has Sia, Hu, and Heka in his
translation) were to have in the voyage than this
statement: they were present to help the Great God
create the forms of the gods, presumably by conception
and magical utterance. Incidentally, in the third register,
the representation of the elderly Atum leaning on his
staff, which is ordinarily the evening form of Re, seems
to be assisting Re, but after calling Re his father he
makes it clear that they are in effect one:[8] "I am the
son who comes forth from his father, and I am the
father who comes forth from his son."

The gate at the end of the third hour (Fig. II.42)
provides the model for the remaining gates. Like it, all
contain guardians, uraei, elaborate collections of insignia
or adornments (*ḫkrw*), and erect snakes in the
doorways. I note among its features that nine
mummified gods are placed one upon the other in front
of the gate. They are labeled as the Second Ennead and
show once more the firm belief in the death of even

the great gods and the application to them of the funerary practices provided for the human deceased. Similar groups of nine gods appear before the succeeding gates (the Third Ennead, the Fourth Ennead, and so on through the Ninth Ennead). One distinctive gate arrangement is found in the fifth gate (see Fig. II.43) where the Judgment Hall of Osiris is inserted between the gateway and the door, illustrating the great popularity of the imagined ceremony in which Osiris judges the dead. Not only is the Osirian judgment procedure woven into this work, but the Osirian funerary ritual as well. This is evident from the inscription that appears in the third register of the sixth hour where an address is made to mummified gods that lie on their biers, an address in which they are urged to take up their flesh, put together their bones and knit together the parts of their bodies,[9] urgings that were a part of the conventional Osirian ritual for the dead. Indeed resurrection procedures abound in this work.

In the middle register of the eleventh hour we see a cobra with one human head facing forward, another backward, and a snake's head in between (Fig. II.44). She is called "The One Who Establishes Lifespans $(^ch^cw)$ and Writes them down as Years on this Uraeus".[10] I note this now only to show once more the Egyptian image of the two-headed nature of time, one head looking toward the past and the other toward the future. In the third register we see twelve goddesses with stars on their heads, who have a tow rope in their hands (see Fig. II.45). They are the hour-goddesses who guide the Great God, and Re says to them that his birth is their birth and his creation is their creation, and he tells them that they establish lifespans and give years to those who are among them.[11]

My final comments on the *Book of Gates* concern the climactic events at the end of the twelfth hour. Before the twelfth gate are two pillars with the heads of Khepri and Atum (Fig. II.46). Behind the gate we see two doors each with its guarding snake. Just beyond the doors are two figures of uraei, the one at the top of the register called Isis and the one at the bottom Nephthys. Notice that we have only one register beyond the last gate and it is devoted entirely to the rising of the sun-god in his form of the Khepri beetle. Nun, the personalized Abyss, lifts the bark (now the Daybark) up out of the abysmal waters, repeating thereby the initial creation of the Great God from the Primordial dark waters. The beetle is between the goddesses Isis and Nephthys. Above the beetle is the sun disk being received by the sky goddess Nut. She stands on the head of Osiris, "who encircles the Netherworld".[12] The crew of the bark, which, along with Re, had only contained Sia and Heka during the night journey, now has additional gods so that the complete crew includes three unnamed gatekeepers (with door-signs over their heads), Nephthys, Khepri, Isis, Geb, Shu (*or* Maat), Heka, Hu, and Sia.[13] The one point of interest for us is that now both Hu and Heka are together on the crew, thus intensifying the capacity of creative utterance.

The third and last book of the night voyage, which I shall discuss even more briefly than the *Book of Gates*, is the *Book of Caverns*.[14] The earliest copy is that in the Osireion of Seti I's temple at Abydos, built by his grandson Merneptah (1224-14 B.C.). This work differs considerably from the two works already examined. In the first place the *Book of Caverns* has only six divisions (reproduced on six tableaux). As for the registers used in these tableaux, there is no essential uniformity in the number of registers, as there was in the two other works. Actually we have only a series of somewhat disconnected scenes. As a result, there is no middle register that runs through the whole and represents the underworld river on which the solar bark moves. In fact in the six tableaux we find only one small representation of the front section of the solar bark, namely that occurring in the so-called final picture of the work (Fig. II.47), where we see the Khepri beetle rising after the front of the bark has been towed to the edge of the Abyss. In that bark we see three forms of Re, the first being a bird standing on an oval, who is Re as Osiris (it is labeled as Osiris), the second is the Khepri form, and the third is Re's night form as a human figure with a ram's head. The rising of Khepri is shown twice more in the sixth tableau.

A particularly distinctive feature of the *Book of Caverns* is the manifold use of ovals which most often are laid on their sides so that the enclosed figures of the gods simulate reclining figures. This oval is called *ḏb't*, which is ordinarily translated as "coffin" but, as Piankoff has suggested,[15] here seems to have "the sense of an envelope or cocoon in which new life is being formed". The constant interest in revivification of the gods in this work dovetails with the paramount attention given in it to Osiris. That attention is immediately evident in the very first speech of Re in the first register of the first division:[16]

> O gods who are in the Netherworld,
> (in) the first cavern of the West,
> doorkeepers of the districts of the
> Silent Region, Ennead of the Regent

of the West (i.e. Osiris), I am Re who is in the Heaven. I enter into the utter darkness. I open the gate of the sky in the West. Behold I enter into the Land of the West. Receive me, your arms toward me. Behold, I know your place in the Netherworld. Behold, I know your names, your caverns, your secrets. I know from what you live, when the One of the Netherworld (i.e. Osiris) orders you to live. Your throats breathe when you hear the words of Osiris .... Re calls out to the gods who are in the first cavern of the Netherworld .... Behold, I enter into the beautiful West in order to take care of Osiris, to greet those who are in him. I set his enemies in their place of execution.

As Re addresses other gods he repeats that he has come to take care of Osiris. Furthermore, as we saw above, there was an identification of Re with Osiris, when he appears as a bird along with two other forms of Re in the solar bark at the end of its journey. In fact, it was not uncommon in the later period to see the mixing of Osiris and Re.[17]

Texts and Studies of the *Book of Amduat*

The most important single work treating of the *Book of Amduat* is the edition, translation, and commentary included in E. Hornung's *Das Amduat: Die Schrift des Verborgenen Raumes* (*Ägyptologische*

ANCIENT EGYPTIAN SCIENCE

Abhandlungen, Vols. 7, 13), 3 Parts (Wiesbaden, 1963, 1967). The first two parts are concerned with the longer version, Part I being the text and Part II the translation and commentary. Part III covers all aspects of the shorter version. As the result of his careful preparation of the text, Hornung's German translation is considerably more accurate than the English translation made earlier by A. Piankoff, *The Tomb of Ramesses VI*, ed. by N. Rambova (*Bollingen Series*, Vol. 40.1) (New York, 1954), pp. 227-318. A separate collection of plates appeared with this volume. Since Piankoff's work contained much which was new and ground-breaking in the study of books regarding Egyptian views of the underworld, it remains of great value to the student of Egyptian religion.

An improved German translation by Hornung, with a new metric arrangement, appeared in his invaluable *Ägyptische Unterweltsbücher* (Munich, 1972, 2nd ed. 1984), pp. 57-194.

The English Translation

I have kept a careful eye on Hornung's German translations, but I have also followed Piankoff, who on many occasions presents an apt or picturesque rendition. We often find that the text does not contain whole sentences but only labels of what is pictured, and hence the result is a rather staccato form of expression. I have given all the pictorial representations of the hours in the form denuded of text found in Hornung's edition and in the various translations.

Notes to the Introduction of Document II.4

1. See H. Altenmüller, "Toten-Literatur, 22. Jenseitsbücher, Jenseitsführer," *Handbuch der Orientalistik*, Abt. 1, Vol. 1, Part 2, pp. 70-72.
2. Piankoff, *The Tomb of Ramesses VI*, p. 227.
3. Altenmüller, *op. cit.* in n. 1, p. 72.
4. The text was published by C. Maystre and A. Piankoff, *Le livre des portes*, 3 vols.(Cairo, 1939-62). Volumes 2-3 were published by Piankoff alone. Piankoff also published an English translation in his *The Tomb of Ramesses VI*, pp. 137-224. Hornung was highly critical of Piankoff's text and prepared a text of his own (edited with the assistance of A. Brodbeck and E. Staehelin): *Das Buch von den Pforten des Jenseits (Aegyptiaca Helvetica*, Vol. 7), 2 parts (Basel and Geneva, 1979-84). The first volume contains the text and the second a German translation and commentary. See also his translation in *Ägyptische Unterweltsbücher*, pp. 54-55, 195-308. Another English version was made by J. Zandee in *Liber amicorum. Studies in Honour of Professor Dr. C. J. Bleeker* (Leiden, 1969), pp. 282-324. As I have indicated in the text, I have followed Hornung in his assumption that the gates come at the end of the hours, not at the beginning. Hence the treatment of the sun-god at the Western Mountain, which Piankoff labels as a prologue, becomes for Hornung the first hour. Hence the numbers of the hours in Hornung's version are each one number higher than in Piankoff's text and translation.
5. See Maystre-Piankoff, *Le livre des portes*, Vol. 1, pp. 15-17; Hornung, *Das Buch von den Pforten*,

Part 1, p. 1, Part 2, pp. 32-33; Hornung, *Ägyptische Unterweltsbücher*, p. 197; Piankoff, *The Tomb of Ramesses VI*, p. 141.

6. Maystre-Piankoff, *Le livre des portes*, Vol. 1, p. 25; Hornung, *Das Buch von den Pforten*, Part 1, p. 13, Part 2, p. 43; Hornung, *Ägyptische Unterweltsbücher*, p. 201; Piankoff, *The Tomb of Ramesses VI*, p. 144.

7. See the works of the preceding note: Maystre-Piankoff, pp. 30-34, 34-35; Hornung, *B.v.d.P.*, Part 1, pp. 25-30, Part 2, p. 58; Hornung, *Ä. U.*, p. 204; Piankoff, p. 146.

8. Maystre-Piankoff, Vol. 1, p. 60; Hornung, *B.v.d.P.*, Part 1, pp. 36-37, Part 2, p. 68; Hornung, *Ä.U.*, p. 206; Piankoff, p. 147.

9. Maystre-Piankoff, Vol. 2, pp. 69-70; Hornung, *B.v.d.P.*, Part 1, pp. 236-37, Part 2, p. 168; Hornung, *Ä.U.*, pp. 247-48; Piankoff, p. 176.

10. Maystre-Piankoff, Vol. 3, pp. 66-67; Hornung, *B.v.d.P.*, Part 1, pp. 365-66, Part 2, p. 257; Hornung, *Ä.U.*, p. 289; Piankoff, p. 209.

11. Maystre-Piankoff, Vol. 3, pp. 87-88; Hornung, *B.v.d.P.*, Part 1, pp. 370-71, Part 2, p. 262; Hornung, *Ä.U.*, pp. 291-92; Piankoff, p. 211.

12. Maystre-Piankoff, Vol. 3, p. 168; Hornung, *B.v.d.P.*, Part 1, p. 410, Part 2, p. 290; Piankoff, p. 224.

13. This is the order of the gods in Maystre-Piankoff, Vol. 3, p. 179, and Hornung, *B.v.d.P.*, Part 1, p. 410, Part 2, p. 290. Hornung, *Ä.U.*, p. 308, gives the following names, ordered from the bow: Sia, Hu, Heka, Maat, Geb, Isis, Nephthys, and the three Gatekeepers. Piankoff, p. 224, has a similar order except that Thoth replaces Maat.

14. A. Piankoff, *Le livre des quererts* (Cairo, 1946). Published earlier in *Bulletin de l'Institut Français*

d'Archéologie Orientale, Vols. 41-45 (1942-47). The latter is the version I have used. Included is the text and a French translation. Piankoff also produced an English translation in his *The Tomb of Ramesses VI*, pp. 45-133. Finally, consult the German translation of Hornung, *Ägyptische Unterweltsbücher*, pp. 311-424.

15. Piankoff, *The Tomb of Ramesses VI*, p. 47, n. 2.

16. Piankoff, *Le livre des quererts*, *Bulletin*, Vol. 41, pp. 7-8, plate III; Piankoff, *The Tomb of Ramesses VI*, p. 49; Hornung, *Ägyptische Unterweltsbücher*, pp. 311-12.

17. P. Derchain, *Le Papyrus Salt 825* (Brussels, 1957), pp. 35-37. See the Introduction to Doc. II.5 below.

Document II.4

The *Book of Amduat (Amdat),* i. e. The *Book of What Is in the Netherworld*

THE TITLE AND INTRODUCTION

The Writing of the Hidden Place (*or* Chamber), which [concerns] the places where the souls, the gods, the shades, and the spirits stand and what [they] do. The beginning is the Horn of the West, the Gate of the Western Horizon, the end is [The End of] Utter Darkness, the Gate of the Western (! Eastern?) Horizon. [It gives us] knowledge of the underworld souls, knowledge of what they do, knowledge of their glorifications of Re, knowledge of the mysterious souls, knowledge of what is in the hours (i.e. the regions of the hours) and their gods, knowledge of what he (Re) says to them, knowledge of the gates and the path over which the Great God passes, knowledge of the course of the hours and their gods, knowledge of those who thrive and those who are annihilated.

TITLE OF THE FIRST HOUR

This god enters into the Western Gate of the Horizon. Seth stands on the bank of the river. There are 120 *itrw*[1] through this gateway until the bark reaches those residing in the Netherworld and passes through to Wernes.

THE FIRST HOUR (Fig. II.29)

[*The first register, above the baboons*.] The names of the gods who open the doors for the Great Soul: Benti, Ifi, Dehdeh, The Heart of the Earth, The Sweetheart of the Earth, The One Who Praises, The One Who Opens the Earth, The Soul of the Earth, The One Whom Re Has Seen ....

[*The second register, over the first part.*] The two goddesses Truth tow this God in the Nightbark, which passes through the gateway of this city. It is 120 *itrw* (*mistakenly*, 200) after that before he reaches Wernes, which is 300 (! 309?) *itrw* in length. He grants plots of land to the gods who are in his Following. The name of this field (i.e. region) is the Water of Re. The One Who Belongs to Both Flames is the name of its guardian. This god (Re) begins to give orders and to care for those who are in the Netherworld in this field. [*Gods in the Bark of Re.*] Wepwawet (i.e. the Opener of the Ways), Sia (Understanding),[2] Lady of the Bark, The Flesh [of Re], Horus the Praiser, The Bull of Truth, The Watchful One, Hu (Authoritative Utterance), The Guider (Helmsman) of the Bark ....

[*Above the third register.*]... and this gateway, along which this god is towed in the form of a ram. He is changed after he passes this gateway, without the dead following behind him. They stand in the gateway. He gives orders to these gods whom he finds in this gateway. This is done in the secret places of the Netherworld represented like this, holy and hidden, for the few who know this.

[*Above the boat with the scarab beetle.*] Osiris, Khepri (?), Osiris .... [followed in the fourth register by the names of the baboons, twelve serpents, nine gods with lifted arms, and the names of twelve goddesses, and at the end of the division or hour there is a long passage which ends as follows:] This god passes by them (i.e. the residents at the gateway) and they wail after he has gone by them toward Wernes. It is done like this in the secret part of the Netherworld .... There are 120 *itrw* to travel to this gate. The Hour [goddess] who is the guide of this gateway is She Who Cleaves the Brows of the Enemies of Re.

THE SECOND HOUR (Fig. II.30)
INTRODUCTION

To rest in Wernes by the majesty of this god, to sail the fields on the Waters of Re. This field (i.e. region) is 309 *itrw* in length and 120 in width. This Great God makes grants of plots of land to the gods in this region. The name of the Night-hour [goddess] who guides this god is The Learned One, She Who Protects her Lord. The name of the gate of this city is The Devourer of All ....

This Great God makes grants of plots of land to the gods of the Netherworld. He cares for those who are in this field (region) .... These representations of the souls of the Netherworld are done in painting .... There will be offerings to them on earth in their names, which will be effective for a man on earth, as has been proved true a million times.[3] [*Above the upper register.*] They are like this. They adore this Great God after he has reached them. Their voices lead him to them. Their wailing trails him after he has given them orders [and left] .... [The gods in the first register are named.]

[*Above the solar bark in the middle register is a corrupt passage and over the bark of Re the names of the gods, where* Heka (Magic) *is substituted for* Hu.][4]

[*Above the gods of the third register we read:*]
These [gods] are like this. They present this Great God
with the year-signs.[5]   They present him with
ages-branches (i.e. those representing great periods)
which they hold in their hands. This Great God gives
them orders. They call him and they live through the
voice of this Great God. Their throats breathe when he
calls them and assigns them their duties. He bestows on
them the plants which are in their fields. It is they who
give the green plants of Wernes as food to the gods
who are in the Following of Re ....

THE THIRD HOUR (Fig. II.31)
INTRODUCTION

[*The introduction is at the end of the second
hour:*] This Great God rests in the Field of the
Offering-dwellers and sails the Water of Osiris. This
field is 309 *itrw* in length. This Great God gives orders
to the souls who are in the Following of Osiris in this
city. The name of the Night-hour [goddess] who leads
this Great God [in this cavern] is She Who Cuts Up
Souls. The name of the gate of this city is The Seizer ....

[*Above the representations of the first register:*]
They are like this in the Netherworld in the flesh of
their own bodies. Their souls (bas) speak for them and
their shades (or shadows) rest on them .... This is what
they have to do in the West, to crush the adversary, to
cause Nun to come into being to produce the
inundation, for under them comes forth wind from the
earth ....

[*Above the second register:*] This Great God
passes along the Water of the Sole God, he who creates
food-offerings .... He travels by Osiris in this city. This
Great God rests for a time in this city and gives orders

to Osiris and to those who are in his Following ....

[*Above the third register.*] They are like this. They worship this Great God while this Great God gives them orders. They live (i.e. come alive) while he calls them .... This is what they have to do in the West: to roast and cut up the souls, to imprison the shades, to annihilate those who are the nonexistent,[6] who belong to their Place of Annihilation-houses ....

THE FOURTH HOUR (Fig. II.32)
INTRODUCTION

The towing of the majesty of this Great God is arrested in the mysterious Cavern of the West whose form is holy. Taking care of those who are in it without being seen by them. The name of this cavern is Life of Forms. The name of the gate of this cavern is She Who Hides the Towing. The name of the Night-hour [goddess] who leads this Great God [in this cavern] is She Whose Power Is Great. He who knows these images will eat bread beside the living in the Mansion of Atum.

The mysterious ways of Rostau, the holy roads of the Imhet Necropolis, the hidden gates in them, the Land of Sokar, He Who is on his Sand .... [Among the divinities are a number of serpents and over the one with two heads we find the name:] Neheb-kau (The Provider of Kas)[7] ....

[*Second register, above the solar bark:*] Wepwawet, Sia, Lady of the Bark, The Flesh of Re, Horus Who Praises, The Bull of Truth, The Watchful One, Hu ....

[*Third register, above the bark with the serpent:*] He is like this in his boat. He guards the Imhet Necropolis .... [*To the right of the sloping passage:*] This is the mysterious image of Imhet Necropolis. Light is on her every day at the birth of Khepri, who comes forth from the faces of the *mnmnw*-snake; then Khepri distances himself....

THE FIFTH HOUR (Fig. II.33)
INTRODUCTION

This Great God is towed along the right ways of the Netherworld upon the upper half of the Mysterious Cavern of Sokar, He Who Is on his Sand. Invisible and not perceptible is the image of the earth which covers the flesh of this god. The gods, among whom this god is, hear the voice of Re when he calls in the vicinity of this god. The name of the gate of this city is Stopping-point of the Gods. The name of the cavern of this god is The West.[8] The name of the Night-hour [goddess] who guides this Great God [in this cavern] is She Who Guides within Her Bark .... The secret ways of the West, ... The Holy Place at the Land of Sokar, the Flesh, the body in (its) first manifestation-forms *(ḫprw)*, ... [The diagram of the road to Sokar] is made according to the plan which is drawn in the hidden region of the Netherworld ....

[*First register, above the first group of gods.*] Words said by this Great God: O Goddess of the West, give thy arm. How beautiful is the great way inside the earth, the way to the grave and resting place of my gods. May you breathe, O Ennead of gods, who came into being from my flesh when your forms *(irw)* were not yet created. May your provisions be stable. I shall protect you if you protect me. It is you whom I have decreed to be holy for protecting me in the Land of the West.

[*Second register above the solar barge.*] This

Great God travels along, being towed over this cavern in his bark [called] Life of Souls, which is in the earth. The gods of the Netherworld say to this Great God: .... Speak to Osiris, O Re. Call out, O Re, to the Land of Sokar, so that Horus on his Sand may live. Come to Khepri, O Re, Come to Re, O Khepri .... [The solar bark contains the same gods as before, once more including Sia and Hu] ....

[*Center of the third register, under the head coming out of the pyramidal mound of sand*:] The Flesh of Sokar, He Who Is on his Sand.

[*Right and left of this inscription*:] The image is like this in the thick darkness. The oval which belongs to this god (i.e. Sokar) is lighted up by the eyes in the heads of the Great God. Both legs which are inside coils of the Great God are lighted up while he protects his image ....

THE SIXTH HOUR (Fig. II.34)
INTRODUCTION

This Great God rests in the deep [in this cavern called] Mistress of the Inhabitants of the Netherworld. This god orders that these gods take hold of their gods-offerings in this city .... He grants them plots of land for their offerings .... The name of the gate of this city is The One with Sharp Knives. The name of the Night-hour [goddess] who guides this Great God [in this cavern] is Mesprit (Arrival), She Who Gives the Correct [Way].

[*First register*:] Words spoken by this Great God to the gods in this field :.... May you be strong in your necks (i.e., may you be very strong) and powerful in your scepters (i.e. very powerful) that you may protect Osiris against those who acted against him and who

-497-

robbed him ....

[*Second register.*] This Great God travels in this city upon the water; he rows in this field in the neighborhood of the corpse of Osiris. This Great God gives orders to those gods who are in the field. He moors at those mysterious mansions which contain the images of Osiris .... The majesty of this Great God speaks to the Kings of Upper Egypt, who are provided with offerings, to the Kings of Lower Egypt, and to the Spirits who reside in this city. Your kingdoms are yours, O Kings of Upper Egypt; may your white crowns be consigned to you ...; your red crowns are yours, O Kings of Lower Egypt; your spirits are yours, O Spirits .... The Kings of Upper Egypt, who are provided with offerings, the Kings of Lower Egypt, and the Spirits, who are in the earth, are like this. They stand near their caverns, and they hear the voice of this god daily .... [Referring to the figure surrounded by the five-headed serpent, a text says:] This is the body of Khepri in his own flesh. The snake [called] Many Faces guards him. He (the snake) is like this, his tail is in his mouth[9] ....

[*Third register.*] Words spoken by this Great God to these gods: O Gods at the head of the Netherworld, in the Following of the Lady of the Netherworld, the standing and sitting ones of the Abyss, who are in their fields: You are the gods whose heads shine and whose corpses stand. You are the goddesses who go after Khepri to the place in the Netherworld having his body[10] ....

THE SEVENTH HOUR (Fig. II.35)
INTRODUCTION
The majesty of this Great God rests in the

Cavern of Osiris. The majesty of this god gives orders in this cavern to the gods who are in it. This god takes another form in this cavern.[11] He turns away from Apep by means of the magic of Isis and the Elder Magician.[12] The name of the gate of this city, which this god passes by, is the Portal of Osiris. The name of this city is Mysterious Cave. The name of the Night-hour [goddess] who guides this Great God [in this cavern] is She Who Repulses the Serpent Hiu and Cuts Off the Head of the Serpent Neha-her ....

[*First register, over the gods.*] Words spoken by this Great God: O August One, give me your hand that Horus may come out of your head (*or* loins) .... This Great God speaks to Osiris, He who is in the Enveloper-snake .... [whose name is] The Enveloper, Life of Forms ....

[*Second register, over the solar bark.*] This Great God travels in this city on the road of the Cavern of Osiris, along which he is towed by the magical utterances of Isis and the Elder Magician in order to avoid the serpent Neha-her. These magical utterances of Isis and the Elder Magician are made in order to repulse Apep from Re in the West, in the hidden part of the Netherworld .... The sandbank [later named Sadjau] of Neha-her in the Netherworld is 440 cubits in its length; he fills it with his coils .... It ... is 440 cubits in width ....

[*Third register.*] This image is of Horus on his throne. This image is like this. What he has to do in the Netherworld is to set the stars into motion and to produce the positions of the hours in the Netherworld. The majesty of Horus of the Netherworld speaks to the star-gods. May your flesh be right, may your forms come into being, so that you might be at rest in your stars. May you stand up before this Re of the Horizon,

who is in the Netherworld every day. You are in his
Following and your stars are before him until I have
wandered through the beautiful West in peace ....

THE EIGHTH HOUR (Fig. II.36)
INTRODUCTION

The Majesty of this Great God rests by the
caverns of the mysterious, who are upon their sand
[banks]. He gives them orders from his bark, and his
gods tow him in this city in [his] holy form of the
Enveloper. The name of the gate of this city is That
Which Stands without Becoming Weary. The name of
this city is Sarcophagi of Her Gods. The name of the
Night-hour [goddess] who guides this Great God is Lady
of the Deep Night ....

[*First register, above the first group of three
divinities*:] They are like this in their dress as the secret
[forms] of Horus, the heir of Osiris. This god calls to
their souls after he has entered this city of the gods
who reside on their sand. And there is heard from this
cavern the noise of their voices like [the humming of]
many bees when their souls cry out to Re. The Secret
One is the name of this cavern ....

[*Second register, above the solar bark*:] This Great
God travels in this city, being towed by the gods of the
Netherworld, in his secret image of the Enveloper-snake
.... [Above nine gods in the form of the hieroglyphic
sign for Follower, i.e. in front of the eight gods who are
towing the bark, we read:] It is his voice that the gods
hear after he has called them while the forms of their
bodies remain on their corpses which are under the sand
.... When this Great God calls them, what is in them
comes alive .... and what they do is to put to the sword
the enemies of Re .... [Above the four rams is written:]

The Secret Images of Tatenen, the first appearances *(hprw)*, the rams which are in the earth, near whom Horus has hidden the gods[13] ....

[*Third register, above the first divinities.*] They are like this, upon their clothing [signs], as the secret forms of Horus, the heir of Osiris. This Great God calls to their souls after he has entered the city of the gods who reside on their sand. This god calls out to them on both banks of the land. A cry is heard from this cavern like the caterwaul of a tomcat when their souls cry out to Re .... [and the cries of the various other divinities are compared to sundry sounds, like the roar of the living, the sound of a bank falling into the flooding water, the cry of a divine hawk, and the sound of a nest full of birds].

THE NINTH HOUR (Fig. II.37)
INTRODUCTION

The majesty of this Great God rests in this cavern. He gives orders from his bark to the gods who are in it (the cavern). The crew of this god rests in this city. The name of the gate of this city ... is Guardian of the Flood. The name of this city is That Which Springs Forth for Forms, That Which Lives for Transformations *(hprw)*. The name of the Night-hour [goddess] who guides this Great God is She Who Adores, She Who Protects Her Lord ....

[*First register, above the twelve gods sitting on hieroglyphic signs for clothing.*] They are like this in the Netherworld, established upon their clothing-signs, in their forms and in their images made by Horus. Re speaks to them. You are provided with your clothing, you are made holy by your clothing. Horus has clothed you there just as he hid his father in the Netherworld which hides the gods ....

[*Second register.*] This Great God rests from his rowing in this city and his crew rests with his bark and his secret image of the Enveloper-snake. This Great God gives orders to the gods who are in this city. [The gods in the bark are named again, and then above the twelve gods with paddles we read:] These gods are the crew of the bark of Re, who row He Who Is In the Horizon so that he may rest in the Eastern Gate of the Sky. What they have to do is to row Re every day to this city ....

[*Third register.*] They are like this in the Netherworld, established upon their clothing-signs in their own flesh. It is they who light up the darkness in the chamber which contains Osiris. It is the flames of their mouths which overthrow enemies in the Netherworld ....

THE TENTH HOUR (Fig. II.38)
INTRODUCTION

This Great God rests in this cavern. He gives orders to the gods in it. The name of the gate of the city, by which the Great God enters, is Great of Creations (*ḫprw*), The One Bearing of Forms. The name of this city is She Who Is Deep of Water and High of Banks. The name of the Night-hour [goddess] who guides this Great God along the secret way of this city is The Raging One Who Slaughters Those Left Behind.

The Secret Cavern of the West, where Khepri rests with Re, in which gods, spirits, and the dead lament for the secret images of Igeret (the Silent Region) ....

[*First register, referring to the beetle.*] They are like this in the Netherworld in the forms and births of Khepri when he carries his egg toward this city in order to come forth afterwards in the Eastern Horizon of Heaven ....

[*Second register, above the solar bark.*] This Great God travels like this in his bark in this city. His crew of gods row him along. The gods in this city rest in the water, in which their oars are, and they breathe at the sound of the paddling of this crew of gods ....

[*Third register, referring to Horus and the drowned ones.*] Words spoken by Horus to the drowned ones, to the overturned ones, to those who float on their backs, who are in the Abyss of the Netherworld: O you drowned ones ... may there be air for your souls so that they shall not be choked, may your arms row without being hindered, may you open the right road in the Abyss with your feet, without your knees being hindered ....

THE ELEVENTH HOUR (Fig. II.39)
INTRODUCTION

This Great God rests in this cavern and gives orders to the gods therein. The name of the gate of this city by which this Great God enters is Restingplace of the Inhabitants of the Netherworld. The name of this city is Mouth of the Cavern, The One Who Reckons the Bodies. The name of the Night-hour [goddess], who guides this Great God [in this cavern], is Starry One, Mistress of the Bark, She Who Repulses the Rebel at His Appearance. This secret cavern of the Netherworld, which this Great God passes in order to come forth, forms the Eastern Mountain of Heaven. Everlastingness (*dt*) swallows up her images before the Seeing One,[14] who is in this city, and she gives them back [again] upon the birth of Khepri on earth ....

[*First register, referring to the first figures.*] He is

like this. He rises before Re without leaving his place in the Netherworld. He is like this. When this god calls to him, the figure of Atum appears on his back. He swallows his images afterwards. He lives on the shades of the dead, [namely] his body and his head. [The name of the first god with two heads is] The One Provided with a Face, Lord of Everlastingness.[15] [The text concerning the goddess mounted on a serpent says:] This is her own body. She is above the constellation of Shedu (?). What she has to do is to live [by means of?] the voice of Re every day. She swallows her images in this city. It is the Eleventh Hour, one of the followers of the god (i.e. Re).

[*Second register, above the solar bark*:] This Great God travels like this in this city. His crew of gods rows him toward the Eastern Horizon of Heaven. The serpent Luminous One at the bow of the boat leads this Great God toward the road of darkness, by means of which what is in it and those on earth are lighted. [The text above the twelve gods who carry the serpent reads:] They are like this before this Great God. They carry the Enveloper-snake on their heads toward this city. They pass on after Re toward the Eastern Horizon of Heaven. This god calls to them by their names and decrees for them their duties. Re says to them: May you protect your images, may you lift up your heads, may your arms be strong and your legs firm ....

[*Third register*:] Orders by the majesty of this Great God to cut to pieces those who fought his Father Osiris, the corpses of the enemies, the bodies of the dead, those who are turned upside down and are hindered from walking ....

THE TWELFTH HOUR (Fig. II.40)
INTRODUCTION

The majesty of this Great God rests in this cavern at The End of Utter Darkness.[16] Born will be this Great God in his forms (ḥprw) of Khepri at this cavern. Nun and Nunet, Heh and Hehet arise (ḥpr) in this cavern at the birth of this Great God that he may come forth from the Netherworld, let himself down in the Daybark, and arise from the thighs of Nut. The name of the gate of this city is That Which Praises the Gods.[17] The name of this city is Creation (ḥpr) from the Darkness, Appearance of Births. The name of [the goddess of] the Night-hour in which this Great God has come into being is She Who Sees the Beauty of Re ....

[*First register, above the first set of goddesses.*] They are like this in their own bodies. The uraei (cobras) come out of their shoulders after this Great God has reached this city. They are in the Following of this god. The flames of the mouths of their uraei repel the Apep-snake from Re at the Eastern Gate of the Horizon. They cross the sky in his Following in their places in the Daybark.

These gods return after this Great God has passed by the Secret Sandbank of Heaven, and they [then] rest on their thrones. It is they who gladden the hearts of the gods of the West with Re-Harakhti .... [Above the twelve adoring gods we read:] They are like this. They adore this Great God in the morning when he rests in the Eastern Gate of Heaven. They say to Re: The born is born, the created is created, the revered one of the earth, the soul of the heavenly master. The sky belongs to your soul that it may rest in it. The earth belongs to your body, O Lord of Reverence ....

[*Second register, referring to the solar bark.*] This god travels like this in this city along the backbone of this secret image of the serpent Life of the Gods when his gods tow him.[18] He enters into its tail and comes out of its mouth, being born in his forms of Khepri, and of the gods who are in his bark as well. He rests upon the secret image of Shu, who separates the sky from the earth and the complete darkness .... [Text referring to the twelve gods who tow the bark:] They are like this. They tow this Great God along the backbone of the serpent Life of the Gods .... They enter this secret image of the serpent Life of the Gods as the Honored Ones (i.e. as Elders) and they come out as the Youths of Re every day....They are in their own bodies when they come out after the Great God into the sky. The secret image of the serpent Life of the Gods is in its place in the Netherworld, and it does not any day go to any other place ....

[*Third register, referring to the group of divine couples.*] They are like this in their own bodies. They rest before Re in the sky. It is they who receive this Great God at his coming forth with them in the East of the sky every day. They themselves belong to their gate in the horizon, while their images of the Netherworld belong to this cavern .... [Above the mummy at the end of the register we read:] Image of the Flesh.[19] He is like this as an image which Horus hid in the complete darkness. It is this secret image which Shu supports under Nut, and accordingly the great flood comes forth from the earth and from this image.

Notes to Document II.4

1. The measure of the *itrw* is about 10.5 km. It is often translated by the Latinized Greek term *schoenus* and the English word *league*. It is of interest that even in these mythologized and fanciful accounts the priestly authors attempted to give them an air of verisimilitude by inserting precise land measurements as if they were preparing regular royal itineraries.

2. There are two points of interest to us in this list of the gods accompanying the solar god. The first is that both Sia and Hu are members of his crew. Thus he has with him the two gods who are important for conceiving and uttering creative and magical words. The second is the fact that in Fig. II.39 Re has the ram-headed form which will be called "The Flesh of Re", and which will be discarded at the end of his voyage for his form of the newly created Khepri, the scarabaeus beetle.

3. This is the formula that is often added to spells, namely that they have been thoroughly tested and proved effective a million times. For example, at the end of my extract from Spell 99 of the *Book of the Dead* (Doc. II.3) we read: "A truly excellent spell (proved) a million times".

4. I have commented earlier on the virtual equivalence of Hu as personified authoritative or creative utterance and Heka as personified magical utterance. Hence it is not surprising to see Heka substituted here for Hu as a member of the crew of the solar bark. See Chapter Two, end of note 44.

5. The year-signs that these gods hold in their hands are usually identified as palm branches stripped

of their leaves and notched for counting purposes. The gods in front of them hold the longer branches, which both Piankoff and Hornung simply call measures. There seems to be little doubt, however, that they are the signs of long ages, centuries if you will, that are shown in temple reliefs as being granted to kings by the gods. The great many notches which we see on the stems of these branches are a good indication that I am correct in this assumption.

6. This is a clear reference to creatures of the chaos, those who are described as not among the beings of the cosmos. They are the truly dead, the nonexistent ones. In the very first paragraph of this work these same nonexistent beings are called the annihilated ones.

7. Neheb-kau is one of the principal creator-snakes, whose role in the early stages of creation I have mentioned in Chapter Two.

8. I follow Hornung in his interpretation of *imn* with the hill-country determinative as Amenet, i.e. the West, rather than accepting Piankoff's translation of it as the "Hidden One".

9. For comments by Hornung on the snake with the tail in his mouth, see above, Chapter Two, n. 27.

10. This is but one of many references in this work to the bodies or corpses of gods, thus illustrating how widely it was believed that gods die and their bodies are buried like those of human beings. I already spoke of this belief in Chapter Two above when I was attempting to characterize the nature of gods.

11. His form here is that of the Enveloper-snake, depicted on Fig. II.35 as the ram enveloped by a snake.

12. This illustrates that even the greatest of the gods needs the magical assistance of two such expert

practitioners as Isis and the Elder Magician. The purpose of this magic was to fend off the sun-god's perennial enemy, the snake Apep, whose harassment of Re I have mentioned above in Chapter Two. The snake is symbolic of all the difficulties and dangers to the creator god on his nightly journey of re-creation, and hence he (the snake) must be killed again and again.

13. The first part of this passage speaks of Tatenen and thus of the early stages of creation when the land rose and the first forms appeared. This no doubt reflects the popular view of Ptah as one of the great creators, and especially in his form of Ptah-Tatenen. The reference to Horus hiding the gods surely has something to do with his great struggle with Seth.

14. This is no doubt an obscure reference to the nature of eternal time and its relationship to creation. Here Everlastingness first swallows her images, which perhaps consisted of past time along with what was created, and then spits them forth anew with the rebirth of the sun and the accompanying cosmos. This might give some support to the view that eternity *(nḥḥ)* was the long stretch of time from the initial creation and everlastingness *(ḏt)* the indefinite stretch of future time.

15. Notice that the god here called the Lord of Everlastingness has two heads, one that looks behind him and the other that looks in front of him. This seems to support the comment made about everlastingness in the preceding note.

16. This is almost surely the name of the twelfth cavern. Note that it was given in the introductory paragraph at the very beginning of the work as the eastern terminus, following the name of the western

terminus.

17. The literal meaning is that this is the place that lifts up the gods, i.e. is the final place where Re and his crew are lifted up to come again into this world.

18. We have here a pictorial representation of the sun being drawn through the snake-form in order to abandon it and assume the rejuvenated form of the beetle. This is just one more reference to the important part in creation that snakes were thought to have played.

19. The discarded mummy, "Image of the Flesh", symbolizes that the sun-god has abandoned his dead, bodily form in order to be re-created as Khepri. Like all those who suffer initial death he is depicted as being mummified, thus leading to his resurrection.

Document II.5: Introduction

The *Litany of Re*

In my discussion of the books of the underworld voyage of the sun-god in the introduction to Document II.4, I mentioned that such books became popular additions to the tombs of the kings under the New Kingdom. Like those books, the *Litany of Re* (whose ancient title was the *Book of the Adoration of Re in the West and of the Adoration of the One Joined Together in the West*) also was concerned with the transformations of Re in the course of his daily journey as he rose every morning in a repetition of his initial creation. It is not surprising therefore that all but one of the earliest copies of this work were also found in the royal tombs (or on royal paraphernalia) in Western Thebes, in the apparent hope for the king's resurrection or re-creation like his father Re. According to Piankoff:[1]

> The main theme of the *Litany of Re* is the meeting of opposites, Re and Osiris, who become united and form an entity. This ritual is probably a part of the royal ritual to transform the Osiris-king into a new Re. The whole of the *Litany* is an amplification of a short passage of Chapter 17 of the *Book of the Dead*, describing the merging of the two opposites who become a Twin soul

[see the earlier version of this passage in Spell 335 of the *Coffin Texts* in Doc. II.2, IV, 276, 280] .... It has been suggested that the Litany is a later development of the short text (the so-called *Short Litany*) which accompanies the seventy-four figures of Re and is met with already in the tomb of Thutmosis III. These figures appear, quite apart from the *Litany*, in some later tombs--those of Seti I, Ramesses III, and others [see Figs. II.48-49] .... The seventy-four names embedded in hymnlike invocations became the first section, the beginning of the *Great Litany* .... [Certain changes brought] the total number of the divine names to 75. The oldest complete version of the *Great Litany* known to us so far is on the shroud of Thutmosis III, in the Egyptian Museum, Cairo .... The *Litany of Re* preceded by the title and the representation of the solar disk descending into the Netherworld appears first in the tomb of Seti I. It is also met with in the tombs of Ramesses II, Meneptah [or Merenptah], Amenmes, Siptah, Seti II, Ramesses III, Ramesses IV, and (in a shortened form) Ramesses IX. An abridged version ... accompanies the forms of Re in the mortuary temple of Ramesses II at Abydos. Finally,

> parts ... are engraved on the stone sarcophagi of the Late Period .... After a brief preface, the *Litany* opens with seventy-five invocations to the Forms of Re, followed by a series of prayers and hymns in which the identity of Re and Osiris, of whom the dead king is a manifestation, is constantly stressed .... The invocations of the seventy-five forms of Re allude to the manifestations of the divinity in the cyle of creation, his descent to the Netherworld, his death, and his rebirth. Here the thought does not develop through concepts; it is not a systematic description of a process. We are dealing with myths and allusions to them as symbols.

Since this was written E. Hornung discovered a version of the *Litany* in the tomb of Weseramon, a vizier of the time of Tuthmosis III, which seems to be the oldest known copy of the text.[2] Hornung also thoroughly disposed of the earlier view mentioned by Piankoff (in the quotation above) that a *Great Litany* grew out of a shorter *Litany* composed to go with 74 figures,[3] and he persuasively argued that the figures (76 and not the number 74 mentioned in the text) were in fact added to illustrate the manifold activities and properties of the so-called One Who Is Joined Together (i.e. Re-Osiris) during his night journey. So in fact they are not forms of Re alone that present him as an all-embracing monotheistic god of whom all other gods are merely manifestations.

In my extract below I have included only the 75 invocations to Re. We see reflected in them the main themes of creation which I have mentioned repeatedly in Chapter Two. Most of these I have called attention to in the notes to the document: the autogenesis of Re, the creation of the Ennead, Re's designation as the Weeper (i.e. the creator of man from his tears), his vivification and re-creation of the gods, souls, and spirits in the Netherworld, his granting of the breath of life, his infusion of the primordial darkness with light, his creation by authoritative word, his creation of continuous time, and so on.

Texts and Studies of the *Litany of Re*

The first text and study of the *Litany* is that of E. Naville, *La Litanie du Soleil: Inscriptions recueillies dans les tombeaux des rois à Thèbes* (Leipzig, 1875). It includes a translation (very much out of date), a commentary, and 49 still useful plates that give transcriptions of the texts from various tombs.

The French translation of Naville's text was turned into English and published in S. Birch, ed., *Records of the Past: Being English Translations of the Assyrian and Egyptian Monuments*, Vol. 8 (London, 1876), pp. 103-28.

The first important recent work concerned with the *Litany* was that of A. Piankoff, *The Litany of Re (Bollingen Series*, Vol. 40.4) (New York, 1964). It includes a translation into English, a commentary, and plates depicting the principal copies, as well as a study of parallels to the *Litany*.

But by far the most important work to appear is the text, German translation, and commentary of E.

Hornung, *Das Buch der Anbetung des Re im Westen (Sonnenlitanei) (Aegyptiaca Helvetica*, Vol. 3), 2 parts (Geneva, 1975-76). It contains the texts of all the copies.

I have also found useful Goyon's treatment of the few fragments from the *Litany* that once appeared in Taharqa's edifice near the sacred lake at the temple of Karnak: R. A. Parker, J. Leclant, and J.-C. Goyon, *The Edifice of Taharqa by the Sacred Lake of Karnak (Brown Egyptological Studies*, VIII) (Providence, Rhode Island, and London, 1979), pp. 30-35, plates 12-15.

The English Translation

My translation depends heavily on that of Hornung. The bold-faced numbers are those followed in Hornung's text and in the translations of Piankoff and Hornung, but at the same time they represent successive columns in most of the tomb versions. The copy in the tomb of the vizier Weseramon was abbreviated as U by Hornung. The shroud of Tuthmosis was designated as T III, the references to the other versions used by Hornung are obvious: S I and S II (tombs of Seti I and Seti II), R II, R III, R IV, and R IX (tombs of the various Ramesside kings with the designated numbers). I have generally followed the text of U and T III, as indeed did Hornung, though the other copies were also employed.

I have often added in parentheses the Egyptian words that have significance for our study of cosmogony. Some of the readings I have adopted from Hornung are clearly controversial and the reader should consult the notes to his translation for his arguments. I have occasionally used apt translations of Piankoff. I have not followed Naville and Piankoff in translating the constantly used epithet *ḳ'-sḥm* as "supreme power", for it may incline readers to believe in a monotheism that is really not present in the text. Literally it means "high of power" (or more felicitously "thou with high power", as Hornung's German translation has it). I have settled for "with exalted power". Note also that I have restored the name Tatenen in the two places it appears (Invocations **3** and **66**), thus abandoning Piankoff's quite ambiguous translation "Exalted Earth". The literal meaning is, of course, "The Land Which Rises". It is the personification of a chthonic force, which, as we have seen above in Chapter Two, was joined with the name of Ptah to become Ptah-Tatenen.

Notes to the Introduction of Document II.5

1. Piankoff, *The Litany of Re*, pp. 10-12, 16-17.
2. Hornung, *Das Buch der Anbetung des Re im Westen*, Part 1, p. vii.
3. *Ibid.*, Part 2, pp. 30-53.

Document II.5

The *Litany of Re*

Beginning of the Book of the Adoration of Re in the West and of the Adoration of the One Joined Together[1] in the West. This book should be recited at night after being drawn on the ground in a field.[2] This is the victory of Re over his enemies in the West. It is beneficial to a man upon earth; it is beneficial to him after he has died.

/1/ Praise to thee, Re, with exalted power, Lord of the Caverns, with hidden forms *(irw)*, he who rests in the secret [places] when he makes transformations *(hprw)* of himself into Deba of the One Joined Together.

/2/ Praise to thee, Re, with exalted power, this Khepri (Becoming One) who flutters his wings, who sets (descends) into the Netherworld, as he makes transformations of himself into He Who Comes Forth from His Own Members.[3]

/3/ Praise to thee, Re, with exalted power, Tatenen *(lit.* The Earth Which Rises), who fashions *(ms)* his gods, he who protects those among whom he is, he who makes transformations of himself into He at the Head of His Cavern.

/4/ Praise to thee, Re, with exalted power, who makes the earth visible, who gives light to those in the West, he whose forms *(irw)* are his becomings *(hprw)*, as he makes transformations of himself into his Great Disk.

/5/ Praise to thee, Re, with exalted power, with a ba

who speaks [to those in the Netherworld], he who is content with his speech, who protects the spirits (ʾḫw) of those in the West while they breathe through him.

/6/ Praise to thee, Re, with exalted power, Unique One, with powerful appearance, who is joined to his body, he who calls his gods while passing through his secret caverns.[4]

/7/ Praise to thee, Re, with exalted power, who calls his Eye, who addresses his Head, he who gives air to the bas in their places that they may receive their breaths.

/8/ Praise to thee, Re, with exalted power, he who attains his ba, who annihilates his enemies, he who decreed the punishment of the damned.

/9/ Praise to thee, Re, with exalted power, the one who is dark in his cavern, he who decrees that there be darkness in the cavern which hides those who are in it.

/10/ Praise to thee, Re, with exalted power, who gives light to the bodies, who is on the horizon, he who enters his cavern.[5]

/11/ Praise to thee, Re, with exalted power, who approaches the hidden cavern of He at the West; surely thou art the body[6] of Atum.[7]

/12/ Praise to thee, Re, with exalted power, who comes to what Anubis has interred; surely thou art the body of Khepri (the Becoming One).

/13/ Praise to thee, Re, with exalted power, whose lifetime (i.e. existence) is longer than that of the West and her images; surely thou art the body of Shu.

/14/ Praise to thee, Re, with exalted power, sparkling star for the dead; surely thou art the body of Tefnut (i.e. Tefenet).

/15/ Praise to thee, Re, with exalted power, who gives orders to the nw-gods (i.e. the time-gods or hour-gods) at their times; surely thou art the body of Geb.

/16/ Praise to thee, Re, with exalted power, with great examination of that place where he is; surely thou art the body of Nut.

/17/ Praise to thee, Re, with exalted power, Lord of Journeyings (?$^{\prime CC}w$) for those who are before him; surely thou art the body of Isis.

/18/ Praise to thee, Re, with exalted power, with shining head for that which is before him (or for those who are before him); surely thou art the body of Nephthys.

/19/ Praise to thee, Re, with exalted power, filled with members (i.e. intact in body), the Unique One, with veins joined; surely thou art the body of Horus.[8]

/20/ Praise to thee, Re, with exalted power, shaper (?), he who shines in the flood; surely thou art the body of Nun.

/21/ Praise to thee, Re, with exalted power, whom Nun protects, he who comes forth from that which he has been; surely thou art the body of the Weeper.[9]

/22/ Praise to thee, Re, with exalted power, he of the two cobras, the one ornamented (?) with two plumes; surely thou art the body of the Putrifying One (see n. 17 below).

/23/ Praise to thee, Re, with exalted power, he who enters and comes forth, and vice versa, who belongs to his secret and hidden Cavern; surely thou art the body of the *Adju*-fish.

/24/ Praise to thee, Re, with exalted power, ba to whom is presented his missing Eye; surely thou art the body of the Divine Eye.

/25/ Praise to thee, Re, with exalted power, ba who stands, Unique One, who protects what he has engendered; surely thou art the body of Netuty.[10]

/26/ Praise to thee, Re, with exalted power, with a raised head and great horns; surely thou art the Ram, Great of Forms (or Transformations) (ḫprw).[11]

/27/ Praise to thee, Re, with exalted power, he who shuts off light in the Igeret (the Silent Region); surely thou art the body of the West.

/28/ Praise to thee, Re, with exalted power, with a ba who sees in the West; surely thou art the body of the Cavern-dweller.

/29/ Praise to thee, Re, with exalted power, he of the wailing soul, the Weeper; surely thou art the body of the Mourner.

/30/ Praise to thee, Re, with exalted power, the one with the arm which comes out,[12] who is praised for his Eye; surely thou art the body of the One with Hidden Members.

/31/ Praise to thee, Re, with exalted power, the one who sinks into the Secret Region; surely thou art Khentamenti (i.e. He at the Head of the Westerners).[13]

/32/ Praise to thee, Re, with exalted power, the one rich in forms (or transformations) (ḫprw) in the Holy Chamber (i.e. the Hidden Place); surely thou art the body of the Sacred Beetle (i.e. the One Who Is Becoming) (ḫprr).

/33/ Praise to thee, Re, with exalted power, who gives his enemies to their guard; surely thou art the body of the One Who Is Feline (i.e. cat-headed).

/34/ Praise to thee, Re, with exalted power, who shines in the Secret Place; surely thou art the body of the Ejaculator [of semen].

/35/ Praise to thee, Re, with exalted power, with wrapped body and breathing throat; surely thou art the body of He Who Is in the Coffin.[14]

/36/ Praise to thee, Re, with exalted power, who calls

the bodies who are in the Netherworld; they breathe [accordingly] and their decay is arrested; surely thou art the body of He Who Causes [Bodies] to Breathe.

/37/ Praise to thee, Re, with exalted power, the one with mysterious face and inflamed Divine Eye; surely thou art the body of Shay (Fate).

/38/ Praise to thee, Re, with exalted power, Lord of Rising, he who alights (or comes to rest) in the Netherworld; surely thou art the body of the Ba Who Alights.

/39/ Praise to thee, Re, with exalted power, whose body is more hidden than those among whom he is; surely thou art the body of Those with Hidden Bodies.

/40/ Praise to thee, Re, with exalted power, stouter of heart than those who are in his Following, who orders heat (i.e. flames) into the Place of Annihilation; surely thou art the body of the Flaming One.

/41/ Praise to thee, Re, with exalted power, who decrees annihilation, who creates breath by means of his forms (ḫprw), the one who is in the Netherworld; surely thou art the body of He of the Netherworld.

/42/ Praise to thee, Re, with exalted power, thou with lifted head who presides over his time (or oval, i.e. the Netherworld), shining one in the Secret Region; surely thou art the body of the Shining One.

/43/ Praise to thee, Re, with exalted power, with joined-together members, the body of He Who Is Prominent in the Earth; surely thou art the body of He with Joined-together Members.

/44/ Praise to thee, Re, with exalted power, who creates secret things and generates bodies; surely thou art the body of the Secret One.

/45/ Praise to thee, Re, with exalted power, he who has provided for those in the Netherworld when passing the Secret Caverns; surely thou art the body of He Who Provides for the Earth.

/46/ Praise to thee, Re, with exalted power, the one whose flesh jubilates when seeing his bodies, with ba honored when passing by his members; surely thou art the body of the Jubilating One.

/47/ Praise to thee, Re, with exalted power, elevated one, with drippings from [his] Whole Eye, the ensouled one for whom his Glorious Eye is being filled; surely thou art the body of the Elevated One.

/48/ Praise to thee, Re, with exalted power, he who makes right (i.e. passable) the ways in the Netherworld and opens the roads in the Secret Region; surely thou art the body of He Who Makes Right the Ways.

/49/ Praise to thee, Re, with exalted power, the ba who travels with passing steps; surely thou art the body of the Traveler.

/50/ Praise to thee, Re, with exalted power, who gives orders to his stars when he illuminates the darkness in the Caverns, with secret forms; surely thou art the body of The Illuminating One.

/51/ Praise to thee, Re, with exalted power, who has made the caverns and who causes the bodies to come into being by what he himself has decreed;[15] mayest thou decree, O Re, for those who exist and those who do not exist,[16] for the gods, the spirits, and the dead; surely thou art the body of He Who Causes Bodies to Come into Being (sḫpr).

/52/ Praise to thee, Re, with exalted power, very secret one, this hidden one whose bas of the Head are like his image, who causes those in his Following to move on; surely thou art the body of the Hidden One.

/53/ Praise to thee, Re, with exalted power, with shining (or straightened) horn, Pillar of the West, with darkened locks, who is in the boiling pot; surely thou art the body of the Shining (or Straightened) Horn.

/54/ Praise to thee, Re, with exalted power, with exalted forms when he traverses the Netherworld and causes the bas in their Caverns to jubilate; surely thou art the body of the One with Exalted Forms.

/55/ Praise to thee, Re, with exalted power, who unites himself to the Beautiful West, at whom those of the Netherworld rejoice when seeing him; surely thou art the body of the Jubilating One.

/56/ Praise to thee, Re, with exalted power, Great Cat, who protects the gods, the judger, President of the Tribunal, he at the head of the Holy Cavern; surely thou art the body of the Great Cat.

/57/ Praise to thee, Re, with exalted power, he whose Eye rescues and whose Brilliant Eye speaks while the bodies are in mourning; surely thou art the body of He Whose Brilliant Eye Speaks.

/58/ Praise to thee, Re, with exalted power, whose ba is distant and whose bodies are hidden, he who illuminates when he sees his secret things; surely thou art the body of the Distant Soul.

/59/ Praise to thee, Re, with exalted power, with exalted ba, when he repulses his enemies, and when he decrees the flame against his transgressors; surely thou art the body of the Exalted Ba.

/60/ Praise to thee, Re, with exalted power, Putrifying One,[17] who hides the decomposition, he who has power over the bas of the gods; surely thou art the body of the Putrifying One.

/61/ Praise to thee, Re, with exalted power, the Great Elder in the Netherworld, Khepri who becomes the child (*miswritten as* the Two Children); surely thou art the body of the Child (Two Children?).

/62/ Praise to thee, Re, with exalted power, great traveler, who repeats the travels, Ba with bright body and dark face; surely thou art the body of the Dark-faced One.

/63/ Praise to thee, Re, with exalted power, who protects his body, who judges the gods as the mysterious Blazing One, who is in the Earth; surely thou art the body of the Blazing One Who Is in the Earth.

/64/ Praise to thee, Re, with exalted power, lord of bonds for his enemies, Unique One, Great One, chief of long-tailed monkeys; surely thou art the body of the One Who Binds.

/65/ Praise to thee, Re, with exalted power, who orders fire into his cauldrons, who severs the heads of the annihilated ones; surely thou art the body of He of the Cauldron.

/66/ Praise to thee, Re, with exalted power, generator with completed forms (or fashionings), Unique One, who lifts up the earth by his magical power; surely thou art the body of Tatenen (The Earth Which Rises).

/67/ Praise to thee, Re, with exalted power, for whom the awakened ones rise,[18] [for] those who are on their biers without seeing their secrets; surely thou art the body of the Awakened Ones.

/68/ Praise to thee, Re, with exalted power, Djenty[19] of the sky, star of the Netherworld, who causes [his] mummies to come forth (or whose mummies flow forth); surely thou art the body of the One Who Causes Bodies to Come Forth.

/69/ Praise to thee, Re, with exalted power, the cheering baboon, thou of Wetjenet (i.e. the desert homeland of the sun in the east), Khepri, thou with just forms; surely thou art the Baboon of the Netherworld.

/69/ Praise to thee, Re, with exalted power, the cheering baboon, thou of Wetjenet (i.e. the desert homeland of the sun in the east), Khepri, thou with just forms; surely thou art the Baboon of the Netherworld.

/70/ Praise to thee, Re, with exalted power, he who renews the earth and opens up what is therein, thou with the ba who speaks and extols his members;[20] surely thou art the body of He Who Renews the Earth.

/71/ Praise to thee, Re, with exalted power, Nehi,[21] who burns his enemies, flaming one, with fire-spitting tongue; surely thou art the body of Nehi.

/72/ Praise to thee, Re, with exalted power, traveler with passing glance, one who causes darkness to come into being after his light [passes]; surely thou art the body of the Traveler.

/73/ Praise to thee, Re, with exalted power, Lord of Bas, he who is in his Benben-house,[22] chief of the gods who are in the Forehall; surely thou art the body of the Lord of Bas.

/74/ Praise to thee, Re, with exalted power, glittering light (?), he of the Benben-house, who ties time together;[23] surely thou art the body of the Glittering Light (?).

/75/ Praise to thee, Re, with exalted power, Lord of Darkness, who speaks as a corpse, ba who calls to those who are in the Caverns; surely thou art the body of the Lord of Darkness.

{For the notes to this document, see page 527.}

Notes to Document II.5

1.   Piankoff, *The Litany of Re*, p. 11: "In the *Litany* we meet two terms: 'The One Joined Together' and 'Deba of the Joined Together'. On the representations of the forms of Re [see Fig. II.49. Reg. 1], this last name accompanies a figure of a mummiform Osiris wearing the crown of Upper Egypt. It obviously depicts Osiris infused with the soul of Re." See also Fig. II.50.

2.   *Ibid.*, p. 22, n. 3: "The 75 forms of Re were drawn on the ground and the text was pronounced over them."

3.   This expresses Re's autogenesis as mirrored by his daily rising in the form of Khepri.

4.   We recall that in the *Book of Amduat* the Great God called to the inert gods of the various caverns and revivified them, at least temporarily, with his voice.

5.   Piankoff, *The Litany of Re*, p. 23, n. 16: "The first ten Invocations apply to Re in his two aspects, the solar and the Osirian. In the tomb of Mentuemhat these forms are disposed in such a way that the solar forms are above, the Osirian below. Invocations 1,3,5,7,and 9 are addressed to the Osirian forms (Pl. 2)."

6.   *Ibid.*, n. 17: "*Body* or *bodies* are used without discrimination. T III has 'bodies' in Invocation 11, 'body' in Invocations 12-17, 19, then 'body' seems to prevail." But I follow Hornung and use "body" everywhere.

7.   After mentioning the bodies of Atum and Khepri (the before and the after forms of the Great God during the night journey), the author gives invocations to Re's manifestations as various Gods of the Ennead:

Shu, Tefenet, Geb, Nut, Isis, Nephthys. He then jumps to Horus, omitting Osiris and Seth.

8. Piankoff, *The Litany of Re*, p. 24, n. 26: "All energy is being concentrated for the birth of the new sun in a new cycle. Invocations 11 to 19 represent the act of creation."

9. The reference is probably to Re as the creator of man, since man was created from the tears of Re, as I often remarked in Chapter Two above.

10. For Netuty, see E. Hornung, *Das Buch der Anbetung des Re im Westen*, Part 2, p. 107, n. 73.

11. Piankoff, *The Litany of Re*, p. 24, n. 34: "Re goes down into the Netherworld 'in the form of a ram' .... The ram with the attached head is a symbol indicating that the dying god is going to rise again."

12. We recall that the Disk (Aten) in the Amarna representations has arms coming out of it (see Fig. II.23).

13. The identification here is with Osiris, not with the older god whose epithet Osiris absorbed.

14. Piankoff, *The Litany of Re*, p. 25, n. 39. In the text Piankoff gives his name as Debaty and in the note presents "He of the Coffin" and "He who adorns" as alternative translations.

15. This is surely a reference to creation by the authoritative word.

16. I have remarked on the significance of this phrase for the Egyptian doctrines of creation, the existent things being those of the cosmos, the nonexistent being those of chaos.

17. See Hornung, *Das Buch der Anbetung des Re*, Part 2, p. 116, n. 167.

18. Piankoff. *The Litany of Re*, p. 28, n. 63: "In the *Pyr. Texts* the dead king is received by Geb, who

places him at the head of the spirits, the Imperishable Stars, then: 'They of the secret place adore thee; the great assemble for thee; the watchers stand before thee' (Sect. 656de). The watchers are the gods sitting behind Amon: *Wörterbuch-Belegstellen*, I, 40."

19. See Piankoff, *The Litany of Re*, p. 28, n. 64, for a discussion of the *djent*-jar, which "is the cradle and at the same time a vessel for crossing the celestial waters [*Pyr. Texts*] Sect. 1185;...".

20. Piankoff, *The Litany of Re*, p. 28, has "rearer of his members". The reading I have adopted for *rnn* is in accord with the determinative. Goyon, *The Edifice of Taharqa*, p. 35, translates the phrase by "naming his members". See also n. 55 on that page for a further name added to this invocation.

21. See Hornung, *Das Buch der Anbetung des Re*, Part 2, p. 120, n. 194.

22. This is the name of the temple at Heliopolis housing the pyramidal Benben-stone.

23. Piankoff, *The Litany of Re*, p. 28, n. 70, interpreting this phrase, says: "i.e., makes time continuous". In line with what I have already said about Egyptian conceptions of time, this probably means he put together "eternity" and "everlastingness" or past and future time to produce eternal time.

Document II.6: Introduction

The *Book of the Divine Cow*

I have mentioned on more than one occasion in Chapter Two the work that is known to modern readers as the *Book of the Divine Cow* (or often, the *Book of the Celestial Cow*). It has considerable interest for the student of ancient Egyptian cosmogony. The first part of this text (columns 1-35) has been known in modern times as the *Destruction of Man* since the subject of that part is man's plotting against the aging Re and the consequences of that act. Of that first part I shall include only the beginning in Document II.6 below.

As with Documents II.4 and II.5, its earliest extant copies appear in the royal tombs of the New Kingdom: those of Tutankhamen (only fragments), Seti I, Ramesses II, Ramesses III, and Ramesses VI. Its opening sentence contains a standard reference to Re as "the god who created himself". But as I said earlier and point out in the notes below, a later assertion of Nun speaks of Re as being greater than the god that made him (by which Nun apparently means himself) and older than the gods that created him (the Ogdoad in the Abyss?). If indeed these gods that created him are the Ogdoad of the Abyss, it is probably they whom Re addresses along with Nun when he mentions the plot of mankind against him. Whoever they are, he calls them the primeval or ancestor gods. It may also be these same gods that were intended when Re called to his side "the fathers and mothers who were with me when I was in the Abyss" along with his Eye, the first generations of the Ennead after him (Shu and Tefenet, Geb and Nut), as well as Nun. Hence it is obvious that the text does not confine itself to the simple Heliopolitan doctrine of autogenesis.

It is also clear that the *Book of the Divine Cow* introduces a distinct variation in the Heliopolitan plan, for it describes in the beginning an earlier time in the evolution of creation when Re ruled the gods and men together on the earth, a time when no heaven existed, no support gods, and no stars. At this simple stage of creation a complication was introduced by the fact that Re had aged and his human subjects began to plot against him. The result told here is that mankind was partially extinguished by Re's Eye in the form of Hathor, and, though Re spared the rest of mankind, he made firm his decision to leave the earth, to return to the Abyss, or at least to a place in the sky. The consequences of that decision are briefly given here: the transformation of Nut into the sky (in the form of the celestial cow on whose back Re himself had climbed in order to be removed from the earth), the assignment to Shu of the task of lifting and holding Nut up with the help of newly created Heh-gods (support gods), and the creation of the stars for gods and deceased spirits and their location in the sky.

Further, we are told that Geb was to look out for the snakes in the earth and their magic. Geb was to have the help of Thoth's skill in writing. But we see that Thoth himself was to be not only Re's scribe but his vicar in the Netherworld, Re having removed himself from the earth and the underworld and being content to be located in the sky. The preoccupation with snakes, those early cosmogonic figures, is of interest. It is not by accident that so much of the last part of the work is concerned with magical spells that would be effective protection against the enemies of Re. It is accordingly of great interest that in the last section we find, in the so-called "theology of the bas", that Re's ba is Magic. I note also that the bas listed for the gods seem to be the personifications of their most important powers or attributes, sometimes in the form of another god (see note 26 of Document II.6).

Hornung draws the conclusion that man's rebellion and Re's departure from earth meant a rupture of the initial union of man and gods, a union to be replaced by their hoped-for, future association as celestial beings or spirits, an association by no means elaborated on in this work.[1] In this intriguing leap Hornung sees the main features of the Egyptian cosmogonic and funerary doctrines accepted in the New Kingdom. He regards as an important aspect of the new form of creation described in this work the attention given to the supporting of the sky, and indeed not only does Shu play an important role in the support of Nut, along with the support given by the newly created Heh-gods, but the Pharaoh too is conceived as having a role similar to that of Shu (see Fig. II.51), being like Shu a son of Re.

Furthermore, Eternity and Everlastingness, those two components of eternal time, are personified, and, as seen in Fig. II.52, play their role in supporting the sky. The importance of the creation of the sky and its support is shown by the centrality of the celestial cow herself. The description of the picture (Fig. II.2a) appears at the center of the work and in itself embraces the main results of the celestial creation: the sky as the belly of the cow, the stars of an extended Ennead on the belly, the solar bark moving across the belly, the depicted Heh-gods, the "Millions" (of stars), all consequences of Re's decision to retreat from his earthly rule. On the whole, Hornung gives more coherence than earlier students to the ideas found in the *Book of the Divine Cow*.

Texts and Studies of the *Book of the Divine Cow*

The earlier texts and studies are conveniently listed in E. Hornung, *Der ägyptische Mythos von der Himmelskuh: Ein Ätiologie des Unvollkommenen* (Göttingen, 1982), p. xi. Hornung's work displaces most of the earlier studies. It not only contains the versions of all the copies, but it also provides a new and very revealing translation, with perceptive notes and an overall discussion of the work.

We should mention that the first edition of all the versions is found in C. Maystre, *Le livre de la vache du ciel dans les tombeaux de la Valle des Rois (Bulletin de l'Institut Français d'Archéologie Orientale,* Vol. 40) (Cairo, 1940).

Note also the complete German translation in G. Roeder, *Urkunden zur Religion des alten Ägypten* (Jena, 1923), pp. 142-49, and the complete English translation of A. Piankoff, *The Shrines of Tut-ankh-Amon (Bollingen Series,* Vol. 40.2) (New York, 1955), pp. 17-34.

There are several partial texts and translations listed by Hornung. One of the most important of these is the translation of E. Brunner-Traut, *Altägyptische Märchen* (Düsseldorf and Cologne, 1963), pp. 69-72, 266-67.

The English Translation

My translation is heavily dependent on the German translation of Hornung. The reader is advised to examine Hornung's notes in which he attempts to justify his many departures from the more common interpretations. Most of these departures are convincing, but in a few cases I have seen fit to substitute my own renderings. The numbers employed in the translation are (approximately) the column numbers of the version in Seti I's tomb.

Note to the Introduction of Document II.6

1. Hornung, *Der ägyptische Mythos*, pp. 74-105, discusses all aspects of the work.

Document II.6

The *Book of the Divine Cow*

/1/ It happened that Re arose [or shone], the god who created himself, after he took over the kingship, when men and gods were [still] united [on earth]. Then men devised /2/ plots against Re, for his majesty, L.P.H., had grown old, and his bones became like silver, his flesh like gold, his hair like real lapis lazuli.[1] His majesty had learned /3/ about the plots that were being devised by men against him. Then his majesty, L.P.H., said to those [gods] who were in his Following: "Call to me my Eye,[2] Shu, /4/ Tefenet, Geb and Nut, together with the fathers and the mothers who were with me when I was in the Abyss (the four pairs of the Ogdoad?), as well as Nun (the god of the Abyss). Let him (Nun) bring his attendants /5/ with him. Bring them secretly so that men will not see [them] and their hearts will not flee.[3] Come with them to the palace and let them tell their private plans /6/ so that I may go [back] into the Abyss to the place in which I came into being."[4]

These gods were brought and they were lined up on both sides of him with their heads bent to the earth /7/ before his majesty, that he might speak his words before the father of the eldest ones (*or* before the eldest father), the maker of men, the king of people (*rhyt*). Then they said to his majesty: "Speak /8/ to us that we might hear." Then Re said to Nun: "O Eldest

God in whom I came into being[5] and you primeval gods, behold, men, who came into being /9/ from my Eye,[6] devise plots against me. Tell me what you would do about it. Behold, I do not seek to kill them before having heard what you will say /10/ on the matter."

Then spoke the majesty of Nun: "My son Re, a god greater than he who made him and older than those who created him,[7] stay on thy throne, /11/ for great is the fear of thee when thine Eye is on those who scheme against thee"[8] ..... [They advise him to send his Eye out against rebellious mankind, and he does so, she taking the form of Hathor. She slays many men the first day and reports back to Re, who relents and decides to thwart a second day of slaughter. This is accomplished by the mixing of red ochre with beer and spreading it "3 palms (deep)" over the place where the destruction is to take place. Then when the goddess shows up the next morning, she drinks the beer, becomes drunk, and does not recognize man. The latter is thus spared. The other worthwhile consequence of Re's action, we are told, is the custom of the preparation of intoxicating drinks by servant girls on the Feast of Hathor!]

[In the second part of the work, Re decides to withdraw from his earthly rule and go back to the Abyss, assigning other gods their places and creating still others by authoritative commands.]

/27/.... Then the majesty of Re said: "As truly as I live, my heart is very weary of being with them (mankind). I shall kill them, without exception, and the /28/ reach of my hand (i.e. power) will not be narrow."[9] The gods who were in his Following said: "Do not escape into your weariness, for thou hast power over the things thou likest (or what thou wishest)." Then said the majesty /29/ of this god to the majesty of Nun: "My limbs are (or body is) weak as in (or for) the first time. I will not return so that another will attack me."

Then the majesty of Nun said: "My son Shu, [keep] thine eye /30/ upon [thy] father as his protection. My daughter Nut, put him [on thy back]." Then said Nut: "What meanest thou, my father Nun?" Then Nut said [further]: /31/ "In ... Nun." Nut transformed herself [into a cow] and the majesty of Re [climbed] on her back. These men /32/ [returned from the place to which they had withdrawn] and then they saw him on the back of the cow.[10] Then these men spoke /33/ to him: ".... [Come?] to us, for we shall overthrow thine enemies, who have devised plots against him who created them." [But] his majesty proceeded /34/ to [his] palace [on the back of] this cow. [Nothing?] came with them. And so the land was in darkness. When the land had become light [again] at dawn, these men /35/ came out carrying their bows[11] .... [A battle ensues, from which, the Great God remarks, human battles ("massacre among men") arose. Then follows a series of creations by authoritative command.] /36/ .... Then this god spoke to Nut: "I have set myself upon thy (*miswritten* my) back in order to lift myself up." /37/ "What does this mean?" asked Nut. And she became [thereupon] the Dweller in Both Skies. The majesty of this god said: "Be far from them and near to me in order to see /38/ me." So she was transformed into the sky. Then the majesty of this god looked inside her and she said: "Would that I were provided with /39/ a multitude of beings." Thereupon the ... [the stars or some epithet of Nut?][12] came into being. Then said his majesty, L.P.H.: "Peaceful (*ḥtp*) is the field here." Thereupon came into being the Field of Offering

(Htp).[13]   "I will cause green plants to grow /40/ in it"--and thus came into being the Field of Rushes.[14]   "I will provide it (the sky) with all kinds of things," which are the *iḥiḥw*-stars (i.e. the ever-shining stars).[15]   Then Nut /41/ began to shake because of the height, and then the majesty of Re spoke: "Would that I had *ḥḥ*-gods to support her"; and thus the Heh-gods were caused to come into being.[16]   Thereupon the majesty of Re spoke: /42/ "My son Shu, place thyself under my daughter Nut, and the Heh-gods shall guard for me the Millions [of bas] there,[17] that they might live [again] in the twilight. Take /43/ her upon thy head to stabilize her." And [so] it came about that a nursemaid is given to a son or daughter and it came about that a son is placed by a father on his /44/ head.

This spell is to be spoken over [a picture of] a cow [see Fig. II.2a][18] .... /56/ Then the majesty of this god said to Thoth:[19] "Pray call to me the majesty of Geb in these words: 'Come immediately.'" And the majesty of Geb came there. Thereupon spoke the majesty of this god: "Watch out /57/ for thy snakes which are in thee (i.e. in the earth). Behold I am afraid of them myself when I am there. But thou knowest their [magical] power. Go then to the place in which my father Nun is and say to him: 'Watch out /58/ for snakes which are in the land and the water and for every place in which thy snakes are put in writing the following: *Guard against playing with anything*'. They know that I am here. /59/ Behold I rise for them [also], and as for their need it will be [given to them] in the land forever. Beware also of those magicians /60/ who know their spells, for the god Heka (Magic) is there himself .... I do not myself need to guard the great god who came into being /61/ before me, for I have assigned them (the snakes) to Osiris, who guards their children, and I have caused the hearts of their elders to forget. Give their [spiritual] power, which has been made according to /62/ their desires, to the whole land as their magical words which are in their bodies."

Thereupon spoke the majesty of this god: "Now call Thoth to me." And he was brought immediately. Then spoke the majesty /63/ of this god to Thoth: "Behold, I am here in the sky /64/ in my place. Since I /65/ shall produce light and brilliance /66/ in the Netherworld and in the Island of the Double Soul,[20] /67/ thou shalt write there and thou shalt make impotent those who are in /68/ them which we created and who [later] produced /69/ rebellions .... /71/ Thou shalt be in my place, my representative. Men shall address thee as 'Thoth the Representative of Re', and I shall cause thee to send out [emissaries] who are greater than thou." So came into being the ibis of Thoth.[21] "I shall /72/ cause thy hand to be stretched out in the presence of the primeval gods, who are greater than thou, and it will be good for me when thou dost [it]." So came into being the ibis-bird of Thoth.[22] "I shall cause thee to /73/ encompass both skies with thy beauty and with thy light." So came into being the moon of Thoth.[23] "I shall cause thee to turn back the Haunebu." So came into being the baboon of Thoth.[24] ... [Then follows a description of spells and their magical effects, the last of which is effective "a million times".]

/84/ .... Nun was embraced by the Eldest God himself (i.e. Re), who said to the gods who came forth from the East of the Sky [with him]: "Give praise to the [truly] Eldest God (i.e. Nun), [for] I have come into being /85/ from him. I have made the sky and stabilized [it] in order that the bas of the gods might be set in it. I am with them an eternity *(nḥḥ)*, which the years have formed *(ms)*.[25] Magic (Heka) is my ba and accordingly it (Magic?) is older than it (i.e. the eternity?)."

The ba of Shu is Khnum (*var.* the air).[26] /86/ The ba of Eternity is the rain. The ba of Darkness is the night. The ba of Nun is Re (*var.* water). The ba of Osiris is the Ram of Mendes. The bas of Sobek are the crocodiles. The ba /87/ of every god and every goddess is in the snakes. The ba of Apep is in the Eastern Mountain, while the ba of Re is throughout the whole earth (*var.* is in Magic) .... [The remainder of the work continues with magical spells, of which I note only that one spell contains the following:] /89/ I am his (Re's) ba, Heka. O Lord of Eternity *(nḥḥ)*, who has created Everlastingness *(dt)*, who causes the years of the gods to pass (*lit.* to be swallowed up), and in whom Re has descended, lord of his own god[liness] and ruler of him who has created him ....

Notes to Document II.6

1.   This appears to be an almost conventional way of referring to the aged Re, as I noted above in Chapter Two (see also Document II.11, H,3-4). The god is being depicted in terms applicable to a statue made of precious metals and stones.

2.   The Eye of Re had long since assumed its separate existence as a divinity, and when taking the form of the Uraeus on his forehead she was the Great God's protector, as she was later the protector of his son, the king.

3.   For an explanation of this expression, see Hornung, *Der ägyptische Mythos*, p. 53, n. 14, referring to its use in medicine for a fainting spell. But this still does not make clear what it means in this context. What reaction in man was secrecy supposed to forestall?

4.   Here he declares the conventional Heliopolitan view of Re's coming into being in the Abyss.

5.   The Abyss has become personified as the father of Re, but a father inferior to his son.

6.   Here we have a reference to the view that mankind arose in the tears of the creator.

7.   The creation of Re is even more confused in this passage since the first reference is apparently to his creation by Nun, only to be followed by an allusion to "those who created him". It could perhaps reflect the doctrine of the Ogdoad, which, as we have seen, presents the negative characteristics of the Abyss as four pairs of divinities.

8.   See Hornung, *Der ägyptische Mythos*, p. 54, n.

30.

9. *Ibid.*, p. 59, n. 75.

10. For the reconstruction of this sentence, see *ibid.*, nn. 82-85.

11. This obscure account of a war between factions of mankind (those for Re and those against him) is given as the origin of human massacres. The passage is however not at all clear, at least so far as the identification of the contestants is concerned.

12. For these two alternative interpretations, see Hornung, *Der ägyptische Mythos*, p. 61, n. 102. I lean toward the second of the alternatives, for the actual creation of the stars is mentioned later.

13. Here begins the series of word-plays by which creation of different places and divinities takes place. For this example I have given the Egyptian words expressing this word-play in parentheses. In the succeeding examples I have cited the relevant notes of Hornung.

14. These two celestial fields, which I have discussed in Chapter Two above, have been successively created. The punning involved in the second case, that is, in this sentence, is quite contrived. See Hornung, *Der ägyptische Mythos*, p. 61, n. 105.

15. These stars are usually called *'ḥ'ḥw*-stars rather than *iḥiḥw*-stars. For the punning involved, see Hornung, *ibid.*, n. 106.

16. We have already mentioned the Heh-gods, which appeared so often in the *Coffin Texts*. They are shown supporting the legs of the celestial cow in Fig. II.2a.

17. The same hieroglyph is used for the Heh-gods and for the Millions [of stars]. Hence once more word play is crucial to the creation. These

Millions are, of course, the bas of the gods and the deceased that populate the sky.

18.   There follows in the text a detailed description of the picture of the celestial cow. For a comparison between the description and the surviving pictures, see Hornung, *ibid.*, pp. 62-64, nn. 113-35; pp. 81-85.

19.   It is evident that here Thoth is performing his secretarial duties as the Scribe of Re. After Geb has been called, he is charged with guarding against the snakes that are in the earth and the water. Nun in the Abyss is also to be brought into the control of the snakes, presumably those in the waters that constitute the Chaos. Osiris was to have control over the snakes in the Underworld. All of the commands issued by Re in regard to the snakes are a part of his divesting himself of control in the lower regions since he has removed himself therefrom into the sky.

20.   Here Re declares that he is in his place in the sky, but, since he will give light to the Netherworld and to the so-called Island of the Double Soul (i.e. the United One, consisting of Osiris and Re?), he commands that Thoth will become his representative in those regions, and Thoth is to exercise the power of his pen to "write there" and render impotent those whom Re created (i.e. men) and who later rebelled against him.

21.   The play on words for the creation process here is the similarity between "send out" *(h'b)* and "ibis" *(hby)*.

22.   For the word play, see Hornung, *Der ägyptische Mythos*, p. 67, n. 164.

23.   For the word play, again see Hornung, *ibid.*, n. 166.

24.   *Ibid.*, n. 168.

25. My translation differs from that of Hornung. My emphasis is that the years [of the past] have already formed an eternity (*nḥḥ*). Hence Re's ba, which is Magic, must share the timelessness of Re and therefore be older than the Eternity which the years since the first creation have brought into being.

26. This excursus on bas is of considerable interest. The bas all seem to involve word play, while revealing important attributes of the gods named. See the various notes to this passage given by Hornung, *ibid.*, pp. 69-70, nn. 191-96.

Document II.7: Introduction

Hymns

All but the first of the various hymns I have presented here as Document II.7 touch rather directly on the subject of Chapter Two, cosmogony and cosmology. The first, the Great Hymn to Osiris, has been included because it contains the most extensive references to the Osiris legend in Egyptian sources. It does however in its second stanza emphasize the power that Osiris possessed in the already created world, as I have pointed out in Chapter Two above.

The cosmogonic doctrines found in the remaining hymns (those to Amon-Re in Documents II.7b[1-4], to Aten in II.7c, to Ptah in II.7d, and to Khnum and Neith in II.7e) have been summarized in Chapter Two and this summary should serve as an adequate introduction to this collection of hymns celebrating and describing the creation as assigned to each of these gods. Before presenting the documents themselves I give the usual bibliographical references to texts and translations. In doing so I depart from the usual style by including the remarks on my English translations within these bibliographic sections. I further note that the translations of II.7a and II.7c are those done by Miriam Lichtheim, while all the others are my own (in which I have made free use of renderings from previous translations where they seem to me to be felicitous).

ANCIENT EGYPTIAN SCIENCE

Texts and Studies of the Hymns

(*Doc. II.7a*). The text of the stela (Louvre C 86) on which the Great Hymn to Osiris is inscribed was published by A. Moret, "La légende d'Osiris à l'époque thébaine d'après l'hymne à Osiris du Louvre," *Bulletin de l'Institut Français d'Archéologie Orientale* Vol. 30 (1931), pp. 725-50, and plates I-III. It was republished by A. de Buck, *Egyptian Reading Book*, Vol. 1 (Leiden, 1948), pp. 110-13. A German translation was given by G. Roeder, *Urkunden zur Religion des alten Ägypten* (Jena, 1923), pp. 22-26, and more recently by J. Assmann, *Ägyptische Hymnen und Gebete* (Zurich and Munich, 1975), pp. 443-48, 625-26. An English version appeared in the English translation by A. M. Blackman of A. Erman, *The Literature of the Ancient Egyptians* (London, 1927), pp. 140-45 (later reprinted as *The Ancient Egyptians. A Sourcebook of their Writings*, with an introduction by W. K. Simpson [Harper Torchbook, New York, 1966]). The best translation seems to me to be that of M. Lichtheim, *Ancient Egyptian Literature. A Book of Readings*, Vol. 2 (Berkeley/Los Angeles/London, 1976), pp. 81-86, and indeed I have given that translation here, adding the line numbers from the Moret text. I have occasionally given Egyptian words and topographical identifications within parentheses.

(*Doc. II.7b[1]*) The initial publication of the text of this hymn to Amon-Re (or Min-Amon) from Papyrus Boulaq 17 in the Cairo Museum was done by A. Mariette, *Les papyrus égyptiens du Musée de Boulaq*, Vol. 2 (Paris, 1872), p. 6 and plates 11-13. This papyrus was from the eighteenth dynasty. It was studied and edited by E. Grébaut, *Hymne à Ammon-Ra* (Paris, 1874), who also gives a French translation. This hymn also appeared in incomplete fashion on statue no. 40950 of the British Museum, dating from the Middle Kingdom, as noted and studied by S. Hassan, *Hymnes religieux du moyen empire* (Cairo, 1928), pp. 157-93. Hassan gives the text of the parts on the statue alongside the text given by Grébaut (with French translation), and he adds a running commentary. Parts of the text have been translated often: Roeder, *Urkunden*, pp. 4-8; Erman, *The Literature*, pp. 282-88; John Wilson, in J. Pritchard, ed., *Ancient Near Eastern Texts Relating to the Old Testament* (Princeton, 1950, 2nd ed., 1955), pp. 365-67; Assmann, *Ägyptische Hymnen*, pp. 199-207. I have employed Assman's divsions A-G in my translation, with the column and line numbers of the papyrus in parentheses.

(*Doc. II.7b[2]*) This hymn appears in two stelas from the Theban tomb of the Overseers of the Works of Amun under Amenhotep III: Seth and Horus. The stelas are no. 34051 of the Museum at Cairo and no. 826 of the British Museum. The stela from the Cairo Museum is in wretched condition; therefore my translation is made from that in the British Museum and I employ its line numbers in my translation. For the text, see I.E.S. Edwards, *British Museum, Hieroglyphic Texts from Egyptian Stelae*, Part VIII (1939), pp. 22-25, plate XXI; A. de Buck, *Egyptian Reading Book*, pp. 113-15; A. Vareille, "L'hymne au soleil des architectes d'Amenophis III Souti et Hor," *BIFAO*, Vol. 41 (1942), pp. 25-30 (with two plates), including a French translation; and W. Helck, *Urkunden der 18. Dynastie* (=*Urkunden* IV), Heft 21 (Berlin, 1958), pp. 1943-46. There are many translations, the earlier ones being listed in the article by Vareille. Of the more recent ones I found the English translation by Wilson in Pritchard, *Ancient Near Eastern Texts*, pp. 367-68, and the German translation of Assmann, *Ägyptische Hymnen*, pp. 209-12, 555-57, the most useful.

(*Doc. II.7b[3].*) The initial text and English translation were made by A. H. Gardiner, "Hymns to Amon from a Leiden Papyrus," *ZÄS*, Vol. 42 (1905), pp. 12-42. See also the text of J. Zandee in his study of this papyrus: *Hymnen aan Amon van Papyrus Leiden I 350* (Leiden, 1948), Bijlage 1, Hieroglyphische Tekst, plates I-VI. Wilson gives extracts in Pritchard, ed., *Ancient Near Eastern Texts*, pp. 368-69. Consult the extracts in German translation by Assmann, *Ägyptische Hymnen*, pp. 312-21, 586-89. My translation owes much to Gardiner's. Note that the chapter numbers were artificially selected for the purpose of punning and do not represent truly successive ordinals, the first nine chapters being numbered by units, the next nine by tens, and the final eight by hundreds.

(*Doc. II.7b[4].*) I have used the text published by W. Golénisheff, ed., *Papyrus hiératiques (Catalogue général des antiquités du Musée du Caire*, Vol. 83, Nos. 58001-58036) (Cairo, 1927), pp. 171-77. The bracketed numbers given in my translation are the line numbers given by Golénisheff. See also the German translation of E. Meyer, "Gottesstaat, Militärherrschaft und Ständewesen in Ägypten," *Sitzungsberichte der Preussischen Akademie der Wissenschaften, Jahrgang 1928, Phil.-hist. Klasse* (Berlin, 1928), pp. 503-08 (whole study, pp. 495-532). It was later translated into English by A. Piankoff, *Mythological Papyri, Edited with a*

Chapter on the Symbolism of the Papyri by N. Rambova, (*Bollingen Series*, Vol. 40.3) (New York, 1957), pp. 13-14. Piankoff failed to translate the text after "heart" in line 39. The most accurate of the modern translations is that in German by Assmann, *Ägyptische Hymnen und Gebete*, pp. 308-12, 584-85.

(*Doc. II.7c.*) For the text of the Great Hymn to the Aten, see N. de G. Davies, *The Rock Tombs of El Amarna*, Part 6 (London, 1908), pp. 29-31, plates. xxviii and xli, and M. Sandman, *Texts from the Time of Akhenaten (Bibliotheca Aegyptiaca VIII)* (Brussels, 1938), pp. 93-96. There are a great many translations in various languages. I have given that of M. Lichtheim, *Ancient Egyptian Literature*, Vol. 2, pp. 96-100, because it is particularly appealing to me. She notes three other English translations in her introduction to this hymn. The numbers included in the translation given here I have taken from the text itself.

(*Doc. II.7d.*) The hieroglyphic transcription of the Hymn to Ptah found in Berlin Papyrus 3048 (dating to the time of Ramesses VI), was given by W. Wolf, "Der Berliner Ptah-Hymnus (P 3048, II-XII)," *ZÄS*, Vol. 64 (1929), pp. 17-44. A German translation with notes is also included. See as well the more recent translation of Assmann, *Ägyptische Hymnen*, pp. 322-33, 589-92. Many passages from this hymn are quoted and discussed by M. Sandman Holmberg, *The God Ptah* (Lund, 1946), see Index (under "Papyrus"). The letters used for the main divisions are those given by Wolf.

(*Doc. II.7e.*) The pertinent texts from Esna for the hymns of Neith and Khnum are found in S.

Sauneron, *Esna*, Vol. 3: *Le temple d'Esna* (Cairo, 1968): Texts No. 206, pp. 28-34; No. 319, pp. 243-46; No. 378, pp. 349-53; and No. 394, pp. 375-76. For a French translation with commentary on the pertinent parts of the hymns, see *Esna*, Vol. 5: *Les fêtes religieuses d'Esna aux derniers siècles du Paganisme* (Cairo, 1962): No. 206, pp. 253-71; No. 319, pp. 238-42; No. 378, pp. 212-19; and No. 394, pp. 206-07.

After the completion of this multiple document, I acquired the very useful work of A. Barucq and F. Daumas, *Hymnes et prières de l'Égypte ancienne* (Paris, 1980). It contains good general and special introductions, with a helpful bibliography. It has complete French translations of the following hymns I have included in Document II.7: Doc. II.7a, pp. 91-97; II.7b[1], 191-201; II.7b[2], 187-91; II.7b[3], 206-33; II.7b[4], 255-61; II.7d, 389-407. In addition, it contains many more hymns.

Document II.7a

The Great Hymn to Osiris

(1) Adoration of Osiris by the overseer of the cattle of [Amun], [Amen]mose, and the lady Nefertari. He says:

Hail to you, Osiris,
Lord of eternity *(nḥḥ)*, king of gods,
Of many names, of holy forms,
Of secret rites in temples!
Noble of ka he presides in Djedu (Busiris),
He is rich in sustenance (2) in Sekhem (Letopolis),
Lord of acclaim in Andjty (the ninth nome of Lower Egypt),
Foremost in offerings in On (Heliopolis).
Lord of remembrance in the Hall of Justice (*lit.* the Two Truths),
Secret ba of the lord of the cavern,
Holy in White-Wall (Memphis and its nome),
Ba of Re, his very body.
Who reposes in (3) Hnes (Heracleopolis Magna),
Who is worshipped in the *narat*-tree,
That grew up to bear his ba.
Lord of the palace in Khmun (Hermopolis),
Much revered in Shashotep (Hypselis),
Eternal Lord who presides in Abydos,
Who dwells distant in the graveyard,
Whose name endures in peoples' (4) mouths.

ANCIENT EGYPTIAN SCIENCE

Oldest in the joined Two Lands,
Nourisher before the Nine Gods,
Potent spirit among the spirits.
Nun has given him his waters,
Northwind journeys south to him,
Sky makes wind before his nose,
That his heart be satisfied.
(5) Plants sprout by his wish,
Earth grows its food for him,
Sky and its stars obey him.
The great portals open for him.
Lord of acclaim in the southern sky,
Sanctified in the northern sky,
(6) The imperishable stars are under his rule,
The unwearying stars are his abode.
One offers to him by Geb's command,
The Nine Gods adore him,
Those in *dat* (the Netherworld) kiss the ground,
Those on high bow down.
The ancestors rejoice (7) to see him,
Those yonder are in awe of him.

.... (9) ....
The eldest of the Nine Gods,
Who set Maat through the Two Shores,
Placed the son on his father's seat.
.... (10) ....
Who vanquishes the evil-plotters,
Whose heart is firm when he crushes the rebels.

Geb's heir (in) the kingship of the Two Lands,
Seeing his worth he gave (it) to him,
To lead (11) the land to good fortune.
He placed the land into his hand,

Its water, its wind,
Its plants, all its cattle.
All that flies, all that alights,
Its reptiles and its desert game,
Were given to the (12) son of Nut,
And the Two Lands are content with it.
.... (13) ....
He is the leader of all the gods
Effective in the Word of Command,
....
His sister was his guard,
She who drives off the foes,
(14) Who stops the deeds of the disturber
By the power of her utterance.
The clever-tongued whose speech fails not,
Effective in the word of command,
Mighty Isis who protected her brother,
Who sought him without wearying,
(15) Who roamed the land lamenting,
Not resting till she found him,
Who made a shade with her plumage,
Created breath with her wings.
Who jubilated, joined her brother,
(16) Raised the weary one's inertness,
Received the seed, bore the heir,
Raised the child in solitude,
His abode unknown.
Who brought him when his arm was strong
Into the broad hall of (17) Geb.

The Ennead was jubilant:
"Welcome, Son of Osiris,
Horus, firm-hearted, justified,
Son of Isis, heir of Osiris!"

The Council of Maat assembled for him
The Ennead, the All-Lord himself,
The Lords of Maat, united in her.
(18) Who eschew wrongdoing,
They were seated in the hall of Geb,
To give the office to its lord,
The kingship to its rightful owner.
Horus was found justified,
His father's rank was given him,
He came out crowned (19) by Geb's command,
Received the rule of the two shores.
....

Document II.7b

Hymns to Amon-Re

Document II.7b[1]

A

(I,1) Adoration to Amon-Re, the bull who resides in On (Heliopolis), chief of all gods, the good god, the beloved one who gives life to everything warm and to all good cattle.

Hail to you, Amon-Re,
Lord of the thrones of the Two Lands, presiding over Thebes,
Kamatef (Bull of his Mother), presiding over his fields,
Far-strider, presiding over Upper Egypt,
Lord of the Madjoi (Nubia) and ruler of Punt,
Eldest of heaven, first-born of earth,
Lord of what is (i.e. of all that exists),
(I,5) Enduring in all things, enduring in all things.

Unique in his nature among the gods,
Beautiful bull of the Ennead, chief of all gods,
Lord of maat (truth *or* order), father of the gods,
Who made mankind and created beasts,
Lord of what is, who created the fruit tree (*or* the plants of life),
Who made herbage and caused cattle to live.

Divine Power whom Ptah made,
Beautiful youth, (II,l) beloved one,
To whom the gods give praise,
Who made the things below and the things above,
Who illuminates the Two Lands,
Who traverses the heaven in peace.
....
The chief one who made the entire earth,
More distinctive as to plans than any god,
In whose beauty the gods rejoice,
To whom is given jubilation in Per-wer,
And [ceremonial] appearances in Per-nezer,
Whose fragrance the gods love when he comes from Punt,
Rich in perfume (II,5) when he comes from Madjoi,
The one beautiful of face who comes [from] God's land,
....
Jubilation to you who made the gods,
Raised the heaven and laid down the ground.

B

(III,l) He who awakes in health, Min-Amon,
Lord of eternity *(nḥḥ)*, who made everlastingness *(ḏt)*,
Lord of praise, presiding over the Ennead
.... (IV,l) ....

C

Hail to you, O Re, lord of maat,
Whose shrine is hidden, lord of the gods,
Khepri in the midst of his bark,
Who gave commands (*lit.* decreed words) and the gods
came into being.
Atum, who made the people,
Distinguished their natures, made their life,
And separated colors (skin pigments?), one from
another,
.... (IV,5) ....
Lord of Understanding *(si¹)*, with Command *(ḥw)* in his
mouth,
....
In whose beauty the gods rejoice;
(V,1) Their hearts live when they see him.

D

O Re, adored in Karnak,
Great in appearances in the Benben-house,
The pillared-one (i.e. the Heliopolitan), lord of the New
Moon Feast,
For whom the Feasts of Sixth Day and
Third-Quarter-Month Day are held.
.... (VI,1) ....

E

You are the sole one who made (VI,3) everything that
is,
Solitary sole [one], who made that which is,
From whose eyes mankind came forth,
And on whose mouth the gods came into being,
He who made herbage that cattle might live,

And the fruit trees (*or* plants of life) for the sun-people
(i.e. man),
Who made that [on which] the fish (VI,5) of the river
may live,
And the birds which belong to the sky,
He who gives breath to that which is in the egg,
He who causes to live the offspring of the serpent,
And makes that on which gnats may live,
And worms and flies in like manner;
He who makes provision for the mice in their holes,
And gives life to flying things (*or* beetles?) in every
tree (*or* wood).
Hail to you who did all of these things,
The solitary, sole one, with his many arms;
(VII,1) ....
The gods bow (VII,5) down to your majesty
And exalt the might of him who produced them,
Rejoicing at the approach of him who begot them.
They say to you: "Come in peace (i.e. Welcome),
Father of the fathers of all the gods (i.e. the Ogdoad or
first gods),
Who raised the heavens and laid down the ground,
Who made what is and created what exists."
.... (VIII,1) ....

F

Hail to you (VIII,2) who made all that is,
Lord of maat, father of the gods,
Who made man and created beasts,
Lord of the grain,
Who made nourishment for the beasts of the desert.
....
(VIII,5) The chief of the Great Ennead,
The solitary, sole one, without his peer

....
And living on maat every day.
.... (IX,1) ....

G

Sole king (IX,3) among the gods,
Rich in names, the number of which nobody knows,
Rising in the eastern horizon,
And setting in the western horizon
....

Document II.7b[2]

(1) Praising Amun when he rises as Harakhti, by the overseer of the works of Amun, Seth, and by the overseer of the works of Amun, Horus. They say:

Hail to you, beautiful Re, every day, who rises (2) at dawn without ceasing, Khepri who wearies himself with work. Your rays are in the face [of every one] without [his] knowing it. Electrum is not like (i.e. not so bright as) your radiance. (3) You are a Ptah, you who have fashioned your body, the shaper who is not [himself] shaped, the unique one who runs through eternity on high, [through] ways by the millions [which] carry his image; (4) just as is your radiance, so is the radiance of heaven, [but] your color is more sparkling than its surface.

When you cross the sky all faces see you, [but] when you depart you are hidden from their (5) faces .... One brief day and you have raced a course of millions and hundreds of thousands of leagues (itrw). (6) Every day under you is [but] an instant, and you set when it passes .... (8) ....Hail to you sun disk of the daytime, creator of all and maker of their living, great falcon (9) with bright plumage, who came into being in order to lift himself up and to create himself, one who was not born, Horus, the first-born in the midst of the sky-goddess, for whom they jubilate (10) when he rises, as well as when he sets, the fashioner of that which the earth produces, the Khnum and Amun of man, ... beneficent mother (11) of the gods and man, a patient craftsman, who becomes weary with [the toil of] making them without number ....

Document II.7b[3]

Ninth Chapter. The Ennead, which came forth from Nun. They unite at (II,3) the sight of you, great of majesty, Lord of lords, fashioning himself, the Lord of the Two Ladies .... (II,4) .... His eyes gleam, his ears are opened. All bodies are clothed [in light] when (II,5) his brightness comes. The sky is gold, Nun is lapis lazuli, the earth is overspread with emerald, when he arises in it. The gods see (II,6), their temples are opened, men come to see, seeing by means of him .... (II,10). There is nothing made without him, great god, life of the Ennead.

Tenth Chapter .... [Here Thebes is praised as the place where creation was enacted in the first time, i.e. where water and land first existed; see above, Chapter Two, note 3.]

Twentieth Chapter. How you sail (II,16), Harakhti, doing your customary routine of yesterday in the course of each day, maker of years, one who forms the months. Days, nights, and (II,17) hours are according to his march. You are created (i.e. renewed) this day beyond yesterday ....

Fortieth Chapter. (II,26) Crafting himself, no one knows his shapes, a fair color, becoming holy, building his [own] images, creating himself, beautiful Power,

(II,27) making beautiful his heart, uniting his seed with his body in order to bring into being his egg within his secret self, becoming a form, (II,28) image of birth, completely himself ....

Fiftieth Chapter .... (III,l) Adoration to your majesty. The Disk of the sky shines for your face. The Nile flows from its cavern for your primordial gods. Earth is provided for your statue. (III,2) Yours alone is what Geb causes to grow, your name is strong, your might is weighty .... [Amun's might is further described in terms of a hawk, a bull, and a lion.]

Sixtieth Chapter .... (III,7) .... his boundary was strong when he was on the earth throughout the whole land and up to the sky. (III,8) The gods begged their sustenance from him. He gave them food from his possessions, lord of fields, and *idbw-* and *nḥb-*lands .... (III,10) Sated (*or* satisfied, i.e. satisfactorily established?) was the royal cubit, which measures blocks of stone. Stretching the cord over the ... of the ground, establishing the Two Lands upon its foundation, and shrines and temples ....

Eightieth Chapter. The Ogdoad (III,23) were your first form, until you completed them, and you were one. Secret is your body among the great ones, concealing yourself as (III,24) Amun at the head of the gods. You made your transformation as Tatenen in order to cause the primordial gods to come into being. Exalting (III,25) your beauty (i.e. becoming?) as Kemphis, you removed yourself as one who dwells in the sky, being established as Re .... (III,26) .... You came into being when there was no other being. No land was without you on the first occasion. All gods came into being after you ....

[Nintieth Chapter.] (IV,l) The Ennead was still joined with your members [in the beginning]. As for your form, all gods were still joined to your body. You emerged first. You began before [all], Amun hiding (IV,2) his name from the gods, great aged one, older than these, Tatenen who shaped himself as Ptah. The toes (or nails) (IV,3) of his limbs were the Ogdoad. Rising as Re from Nun, he became young again. Spitting .... Shu and Tefenet, (IV,4) joined with his might, rising on his throne according to the granting of his heart (i.e. according to the desire of his heart). He ruled everything that existed by his power (?). He arranged the kingdom of eternity (nḥḥ) (IV,5) down to everlastingness (ḏt), established as sole lord. His form shone at the first time. Everything was silent on account of his majesty. [Then] he cackled, (IV,6) being the Great Cackler, in the place where he was created, he alone. He began to speak in the midst of silence. He opened all eyes (IV,7) and made them see. He commenced to cry when the earth was inert. His cry spread about when there was no one else in existence but him. (IV,8) He brought forth all things which exist. He caused them to live. He made all men understand the way to go and their hearts came alive when they saw him ....

(IV,9) Hundredth Chapter. "Coming into being" (i.e Creation) began in the first time and Amun came into being in the [very] beginning. [Hence] his form was unknown. No god came into being before him; there was no other (IV,10) god with him [to whom] he might tell his form. He had no mother to produce his name and had no father who begot him [and thus] could say "This is I" (or "He is mine"). Shaping (IV,11) his own egg, mysterious force of births who created his [own] beauty, divine god who created himself. All [other] gods came into being after he began [to make] himself.

(IV,12) Two-hundredth Chapter. Secret of forms, gleaming of shapes, marvelous god, multiple of forms, all other gods boast (IV,13) of him in order to magnify themselves through his beauty, according as he is divine. Re himself is united with his body. He is the great one who is in On (i.e. Heliopolis). He is called (IV,14) Tatenen and Amun who came forth from Nun. He leads the people. Another of his forms is the Ogdoad. The begetter *(p' wtt)* of the primordial (IV,15) gods *(p'wtyw)* who brought Re to birth. He completed *(tm-f)* himself as Atum *(tm)*, a single body with him. He is the All-Lord, the beginning of that which is (i.e. of everything that exists). His soul, they say, is that (IV,16) which is in heaven, ... his body is in the West. His statue is in Hermonthis, (IV,17) exalting his appearances.

One is Amun, hiding himself from them, concealing himself from the gods, so that his [very] color (*or* complexion) is not known. He is far (IV,18) from heaven, he is absent (?) from the Netherworld, [so that] no gods know his true form. His image is not displayed (?) in writings ....

Three-hundredth Chapter. All gods are three: Amun, Re, and Ptah, and there is none like them (i.e. equal to them). His name is "hidden" *(imn)* as (IV,22) Amun, he is Re in face (i.e. appearance), and his body is Ptah. Their cities are on the earth: Thebes, Heliopolis, and Memphis, enduring for an eternity ....

[Six-hundredth Chapter.] Understanding *(si')* is his heart, Command *(ḥw)* is his lips, (V,17) his ka, every existing thing which is in his mouth. When he enters the two caverns (V,18) which are under his feet, the Nile comes forth from the grotto under his sandals. (V,19) His ba is Shu, his heart is Tefnut (?). He is

Harakhti who dwells in (V,20) heaven; his right eye is day and his left eye is night. He [thus] leads (V,21) people to every way. His body is Nun, and he who is in it is the Nile, giving (V,22) birth to whatever is, and causing to live whatever exists. His warm air is breath for every nostril (V,23). Fate and fortune are with him for everybody (i.e., he has within himself the fate and fortune of everyone). His wife is the fertile field; he impregnates her; his seed is the fruit tree and his fluid is the grain ....

Document II.7b[4]

[1] This august god, Lord of all Gods, Amon-Re, Lord of the Thrones of the Two Lands, He who is at the fore of Karnak.

[2] August Soul *(b)*, who came into being at the beginning, the Great God who lives on Truth, the first primeval one who fashioned the [3] primeval gods, he out of whom every other god came into being.

The Unique one, who created what existed, who at the first time created the earth.

[4] Mysterious of births, numerous of appearances, whose secret image is not known.

August power, who calls forth love, [5] Majestic one, rich in his appearances, Lord of magnificence, with mighty forms, out of whose form [6] came every form, he who first came into being when nothing existed but him.

He who gave light to the earth in the first time, the Noble sundisk [7] with light-producing rays, Radiating one [such that] when he appears every countenance comes alive.

When he sails the sky, he is not weary.

[8] Morning upon morning his routine is fixed.

Ancient One who in the morning arises as a youth who brings forth (i.e. reaches to) [9] the limits of eternity *(nḥḥ)*, who circles the sky, passing through the Netherworld *(d't)* in order to give light to the Two Lands for that which he has created.

[10] Divine god who formed himself, who made heaven and earth in his heart.

Ruler of rulers, Greatest [11] of the great, the Noble one who is older than the gods.

Young Bull with sharp horns, [12] before whose mighty name the Two Lands tremble.

Eternity comes under his might, he who reaches the end of everlastingness.

[13] Great God who created being, who seized the Two Lands with his might.

Ram-headed one with lofty form, [14] more gracious than all the gods, a lion of terrible aspect when he lifts up both *wadjat*-eyes,

[15] Lord of flames against his enemies, the great Watery Abyss *(nw)*, who manifests himself at his hour to [16] make live what comes forth from his potter's wheel.

He who wanders about the sky, who circles in the Netherworld, giving light to the earth as he gave it [17] yesterday.

Lord of Might, of unapproachable majesty, his rays keep his body hidden.

His right [18] eye and his left eye are the sun-disk and the Moon, the sky and the earth are full of the beauty of his light.   [19] Benevolent king, who is never weary, [but has] a strong heart at rising and setting.

Mankind came forth from [20] his two divine eyes, the gods from the speech of his mouth, he who makes food, who creates nourishment, who shapes [21] all that is.

Everlasting one, who wanders through the years without end to his lifetime.

Ancient one [22] who becomes young, who traverses eternity.

When old he makes himself young.

With numerous pairs of eyes and [23] with many pairs of ears, he who guides millions when he shines.

Lord of Life, who gives to whom he wishes.

The circumference of [24] the earth is under his charge, who gives a command and it happens without opposition, whose work will never perish.

[25] He whose name is pleasing, whose love is sweet.

In the morning all faces pray to him.

[26] He is great of terror, he whom every god fears.

The young bull who repulses [27] the adversary, strong of arm, he who strikes down his enemy.

This god who created the earth according to his plans.

Soul (*bʒ*) [28] who shines with both *wadjat*-eyes.

He who manifests himself while coming into being, he who becomes the Holy One, without being known.

[29] He is the King who creates kings, who ties together lands with the order he has made.

The gods and the goddesses [30] bow down before his might because [their] reverence of him is so great.

He who came in the beginning accomplishes the end.

He has created [31] lands according to his plans.

He whose forms are mysterious is he whom nobody knows, who has hidden himself from all the gods, [32] who drew himself into the disk, the Unknown One, who conceals himself from those who came out of him.

Radiant torch, [33] Great of Light, one sees because he is seen.

He is seen the whole day [34] without being understood therein.

When he dawns, all faces adore him, [35] the one brilliant in appearances in the midst of the Ennead.

His form is that of every god.

The flood comes, [36] the Northwind proceeds upstream in this mysterious god, he who makes decrees for millions of millions.

[37] His decisions do not waver, his word is stable, his decrees are efficacious and do not ever fail.

[38] He grants existence, he doubles the years of him who is in his favor, he is a good protector of him whom [39] he has placed in his heart, he is the supporter for ever and ever; King of Upper and Lower Egypt, Amon-Re, King of the Gods, Lord of the Heaven and the Earth, of the Waters and the Mountains, [40] who created the land by means of his forms, he is greater and more elevated (*lit.* more White-crown-like) than all of the primeval gods of the first time ....

Document II.7c

The Great Hymn to the Aten

(1) Adoration of *Re-Harakhti who rejoices in*

Lightland In his name Shu who is Aten, living forever; the great living Aten who is in jubilee, the lord of all that the Disk encircles, lord of sky, lord of earth, lord of the house-of-Aten in Akhet-Aten; (and of) the King of Upper and Lower Egypt ... *Neferkheprure, Sole one of Re*, the Son of Re who lives by Maat .... and his beloved great Queen, the Lady of the Two Lands, *Nefer-nefru-Aten Nefertiti* ...; (2) he says:

Splendid you rise in heaven's lightland.
O living Aten, creator of life!
When you have dawned in eastern lightland,
You fill every land with your beauty.
....
Your rays embrace the lands,
To the limit of all that you made.
.... (4) ....
Earth brightens when you dawn in lightland,
When you shine as Aten of daytime;
As you dispel the dark,
As you cast your rays,
The Two Lands are in festivity.
Awake they stand on their feet.
You have roused them;
.... (5) ....
All beasts browse on their herbs;
Trees, herbs are sprouting,
Birds fly from their nests,
Their wings greeting your ka.
All flocks frisk on their feet,
All that fly up and alight,
(6) They live when you dawn for them.
....

ANCIENT EGYPTIAN SCIENCE

Who makes seed grow in women.
Who creates people from sperm;
Who feeds the son in his mother's womb,
Who soothes him to still his tears.
Nurse in the (7) womb,
Giver of breath,
To nourish all that he made.

....

How many are your deeds,
Though hidden from sight,
(8) O Sole God beside whom there is none!
You made the earth as you wished, you alone,
All people, herds, and flocks;
All upon earth that walk on legs,
All on high that fly on wings,
The lands of Khor and Kush,
The land of Egypt.
You set every man in his place;
Everyone has his food,
His lifetime is counted.
Their tongues differ in speech,
Their characters (9) likewise;
Their skins are distinct,
For you distinguished the peoples.

You made Hapy (the Nile god) in *dat* (the Netherworld),
You bring him [as flood waters] when you will,
To nourish the people;
....
You made a heavenly Hapy (i.e. rain) descend for them;
(10) He makes waves on the mountains like the sea
....

Your rays nourish all fields,
When you shine they live, they grow for you;
You made the seasons to foster all that you made,
(11) Winter to cool them, heat that they taste you.
You made the far sky to shine therein,
To behold all that you made;
You alone, shining in your form of living Aten,
Risen, radiant, distant, near.
You made millions of forms *(ḫprw)* from yourself alone,
Towns, villages, fields, the river's course;
....

(12)....
⟨Those on⟩ earth come from your hand as you made them,
When you have dawned they live,
When you set they die;
You yourself are lifetime, one lives by you.
....

Document II.7d

Hymn to Ptah

A
(1) Adoration of Ptah, the father of the gods, Tatenen,
The eldest of the primordial gods, in the morning.

B
(1) Say: Hail to you, Ptah, the father of the gods, Tatenen,
The eldest of the primordial gods,

....
(10) Lightgiver who caused the gods to live,

....
(15) Who travels the sky and passes through the Netherworld,

....
(19) He who begot himself when no being had [yet] come into being,
Who crafted the earth from the plans of his heart.
Whose forms came into being [from himself].
You are he who has fashioned every thing that is, who has produced and formed that which exists.

C

(1) Hail to you, Ptah-Tatenen,
Great god whose image is hidden,

....
(4) Father of the father of all the gods,

....
(9) May you awaken in peace, you who carries Nut and lifts up Geb,

....
(12) May you awaken in peace, you [as] Khnum and Mut who bore the gods,
Who begot and made all eyes (i.e. men) and made their provisions of life.
....

D

....
(3) Hail to you in front of your primordial gods,
Whom you have made after you came into being as a god,
(5) Body who has modeled his own body,

When the sky had not [yet] come into being and the earth had not [yet] come into being,
When no flood had arisen.
You have knotted the land,
You have assembled your body and have counted your members,
(10) You have found yourself to be the Sole One, who has made his [own] place,
God who has fashioned the Two Lands,
You have no father who engendered you when you came into being,
You have no mother who bore you,
You who are your own Khnum,
(15) The provider from whom [every] provision has come forth.
You stood on the land when it was inert,
[You] by whom it was later put together,
You in your form of Tatenen,
In your form of Uniter of the Two Lands,
(20) That which your mouth has engendered and your hands have fashioned,
You have drawn it forth from Nun
By means of your two hands in imitation of your beauty;
Your son (i.e. the sun), ancient in his forms,
Has chased away the darkness and the shadows,
(25) By means of the rays of his two eyes.
....
(30) You have set him right along his secret ways.
His two barks (day and night) travel along the sky
By means of the wind which comes forth from his (! your?) mouth.
Your feet are on the ground and your head is in the sky

In your form of "He who is in the Netherworld".

....

(42) Your power lifts up water into the sky.
The exhalations of your mouth are the clouds.

....

(63) Your right eye is the sun disk, your left eye the moon.
Your guides are the unwearying stars (i.e. the non-circumpolar ones).

....

E

(1) Hail, let us praise him,
The god who has lifted high the sky, who causes his sun disk to sail forth into the belly of Naunet
And leads him into the belly of Nut,
In his name of Re.

(5) Hail, let us praise him,
Who fashioned the gods, the people, and the animals (?),
He who made all the lands and the ocean shore,
In his name as Crafter (or Builder) of the Earth.

Hail, let us praise him,
(10) Who brings the Nile out of its cavern,
Who makes green the fruit trees (or plants of life)
And makes provision for those who came forth from him,
In his name of Nun.

Hail, let us praise him,
(15) Who sets in motion (or lifts up) Nun (i.e. water) into the sky

And makes water come forth upon the mountains,
In order to bring other people to life,
In his name of Maker of Life.

Hail, let us praise him,
(20) Who made the Netherworld in its plan,
Who overcomes the heat of the bas in their caverns,
In his name of King of the Two Lands.

Hail, let us praise him,
King of eternity *(nḥḥ)* and everlastingness *(ḏt)*,
(25) Lord of life in the Lake of Knives, chief of the
Desert of the Land of the Dead,
In his name of Chief of the Netherworld.
....

F

(1) O, You who have opened the Way.
....

(3) O, You who arise as Re,
O, You who come forth as Khepri, who lives among the
dwellers in the horizon.
You have awakened them,
That they may point out to you the correct way of
Nut,
In your coming forth as the Great Form in order to
make the plans of eternity *(nḥḥ)* forever *(ḏt)*.

G

(1) Come let us celebrate him.
Let us praise his august image
In all his beautiful names,
Child who is born every day.

(5) Come let us celebrate him,
Let us praise his august image
In all his beautiful names,
Old one who dwells at the limits of eternity *(nḥḥ)*.

Come let us celebrate him,
(10) Let us praise his august image
In all of his beautiful names,
Elder one who will continue to travel everlastingly *(ḏt)*.

....

(32) Ba of the Lord of years, who gives life to whom
he wishes.

Come let us jubilate him,
Let us praise his august image
In all his beautiful names,
You who commands and it will be done without
opposition.

H

(1) Hail to you, the ways are opened to you,
The paths of everlastingness are opened up for you,
Founded for you are the sky and the earth, the
Netherworld and Nun.
You have made plans for those who are in them (i.e.
these areas)
(5) You give life, you set the years (i.e. life times) for
men and for gods.
....
(9) You come and go, [in] both heavens ... as living ba.
....
(13) Your eldest son, he worships you
In your form "More beautiful than the [other] gods",

In your form "He is beautiful as to form",
As your son has said to you:
"Mighty is my father, from whom I have come forth,
The Lord of mankind, who created me in Nun,
He who has lifted high the heaven for me and
supported the earth for me,
(20) Who has caused me to travel along the body of
Nut,
And has led me on the secret ways."
....

I

....
(5) Hail to you, Ptah,
Hail to your gods, who come into being from your
body,
How great are you in respect to (i.e. compared to) your
primordial gods.
....

(13) They have been with him for an eternity *(nḥḥ)*,
they will be with him time everlasting *(ḏt)*.
...

K

....
(9) Heka (magic) has power over the gods;
His fame *(or* power) is great under the Ennead.
....

(23) And their [majesty] caused his (i.e. Ptah's) son Re
to ascend
To protect men and gods
By the skill of his power.

L

(1) Hail, Come, and protect the King of Upper and Lower Egypt, Ramesses IX, L. P. H.

As you protect the gods who came into being in this land,

Whose king you were....

(7) Eternity *(nḥḥ)* belongs to you; it is your possession.

Document II.7e

Cosmogonies at the Temple of Esna

Text 206: Neith

(1) Father of the fathers, mother of the mothers, the divinity who began to come into being in the beginning was in the midst of the Abyss. She appeared out of herself while the land was [still] in the shadows and no land had [yet] appeared and no plant had sprouted .... She turned herself into a cow so that no divinity wherever he would be could recognize her. Then she changed herself into (*lit.* renewed her appearance as) a lates-fish (2) and started off. She made luminescent the glances of her eyes, and light came into being. Then she said: "Let this place (where I am) become for me a platform of land in the midst of the Abyss in order that I might stand on it." And this place became a platform of land in the midst of the Abyss, just as she said. And [thus] came into being "the land of the waters" (=Esna), which is also Sais ....

Everything which her heart conceived came into being immediately. (3) Thus she felt happy about this emergence [of the land] and so Egypt came into being in this happiness.

She created thirty gods by pronouncing their names, one by one, and she became happy when she saw them. They said: "Hail to you, Mistress of divinities, our mother, who has brought us into being. You have made our names before we knew them (i.e. yet had cognizance of them) ... you have made [for us] the land upon which we can stand, you have separated [for us] the night from the day .... How very beneficial is everything which comes from your heart, O Sole One, created in the beginning. Eternity (*nḥḥ*) and everlastingness (*dt*) pass before your face .... [(4) Then Neith establishes the gods on the emergent land, and they ask (5) what is going to be created.]

Neith then said: "I shall cause you to know what is coming into being. Let us count the four spirits (*ʾḥw*). Let us give form to what is in our bodies (i.e. in our hearts?) and then let us pronounce our forms. So, we shall recognize everything the same day." Everything she said took place, and the eighth hour (i.e. the culminating time) occurred in the space of a moment.

The Ahet-cow (*or* Ihet-cow; *here* Neith) began to think about what she was going to create. She said: "An august god will come into being today. When he opens his eyes, light will come into being; when he closes them, (6) darkness will come into being. People will come into being from the tears of his eye, gods from the spittle of his lips. I will strengthen him by my strength, I will make him effective by my efficacy, I will make him vigorous by my vigor. His children will rebel against him, but they will be beaten on his behalf and struck down on his behalf, for he is my son issued from my body, and he will be king of this land forever (*dt*). I will protect him with my arms .... (7) I am going to tell you his name: it will be Khepri in the morning and Atum in the evening; and he will be the radiating god in his rising forever, in his name of Re, every day."

Then these gods said: "We are ignorant *(hm-n)* of the things we have heard." So the "Eight" *(hmnw)* became the name of these gods (i.e. the Ogdoad) and also the name of this city (i.e. "Eighttown", i.e. Hermopolis or Ashmunein).

So this god was born from the excretions that came forth from the body of Neith and which she placed in the body of this [primordial] egg .... (8) .... When it broke the shell, it was Re who was hidden in the midst of the Abyss in his name of Amun the Elder and who fashioned the gods and the goddesses with his rays in his name of Khnum.

His mother, the cow goddess, called out loudly: "Come, come, you whom I have created. Come, come, you whom I have conceived. Come, come, (9) you whom I have caused to come into being .... I am your mother, the cow goddess." This god then came forth, his mouth open, his arms opened toward this goddess .... And this day [of the sun's birth] then became the beautiful day of the beginning of the year *(tp rnpt)*.

Then he cried in the Abyss when he did not see his mother, the cow goddess, and mankind came into being from the tears of his eye; and he salivated when he saw her again, and the gods came into being from the saliva of his lips.

(10) These primordial gods [now] rest in their shrines; they have been pronounced *(dm)* [by creative word] just as this goddess conceived them in her heart ....

They (the ancestor gods) thrust aside (11) a wad

of spittle from her mouth which she had produced in the Abyss, and it was transformed into a serpent of 120 cubits, which was named Apep (Apophis). Its heart conceived the revolt against Re, its cohorts coming from its eyes.

Thoth emerged from his (i.e. Re's) heart in a moment of bitterness *(dḥr)*, which accounts for his name of Thoth *(dḥwty)*. He speaks with his father, who sent him against the revolt, in his name of Lord of the Word of God. And this is how Thoth, Lord of Khmun, came into being, in this place, as well as that of the Eight-gods of the first company of gods.

.... [Then Neith goes to her city of Esna (i.e. Sais) with her son to establish his name there. She will suckle him until he is strong enough to massacre those plotting against him. Then we are told that the seven propositions that she declared in the course of creation became seven divinities] .... (13) And so came into being the Seven Proposition-Goddesses of Methyer ....

Text 319: To Khnum
(13) Praise, praise to your good countenance,
Khnum, our Lord of All (i.e. to the Limits),
Sacred god who created (14) all his aspects in just account.
....
See, you are the divine god who organizes the country by his work.

Your are my ruler and I am your servant.
....
[My] nose lives from the breath of your mouth.
.... (16) ....

You are the [All-]Powerful, rich in aspects, great in
favor, who acts according to your perfection.
You have modeled men on the potter's wheel,
You have made the gods,
You have modeled large and small cattle,
You have formed everything upon your (17) wheel,
each day,
[in] your name of Khnum the Potter.

You have distinguished the king in the womb from
among those you have created
In order that he administer the universe to the limits
[of time].
.... (20) ....

How great is your might among the gods and how great
is the awe of you among the Ennead ....

Text 378
Another Hymn to Khnum
(9) You are the Lord of Esna,
The god of the [potter's] wheel,
Who turned the gods [on it],
Who modeled men as well as animals [on it];
....
No god can equal him, the best-beloved ram,
Who made that which is and that which is not,
(10) Who bore the gods and engendered the goddesses.

You are the august god who came into being in the
beginning,
The one whom one goddess accompanies, the uraeus
serpent,
The mysterious one whose form no one knows,

The secret god whose appearance no one knows,
The one who came forth from the Abyss,
Who arose in the horizon as a flame.
The One under whose feet the flood rises from two caverns.
The One from whom wafts the sweet North Wind for the nostrils of gods and men.
....

(12) You are the infant who spreads his light at dawn,
....
You illuminate the Two Lands with your two eyes.
His right eye is the sun and his left is the moon;
And he has created creatures in [the heaven] and on earth, in the inferior world and in the Abyss.

(13) You are Tatenen, more eminent than the gods,
.... (14) ....
All of this he has made in totality,
And there is no god who could do what he has done,
He, this august god,
Khnum-Re, Lord of Esna.

You are the god who lifted the sky,
Heh who supported Nut,
Shu, the eldest son of Atum, (15) the solitary one,
Who has produced everything which exists,
Who has created the light by his two eyes,
To give light to the Two Lands,
....
Who has made life (16) for all those he has created.
....

Text 394

(23) Another hymn to Khnum. To be said:

Hail to you, Khnum-Re, Lord of Esna,
Ptah-Tatenen, who has fashioned the primordial gods,
Great god, who came into being at the very beginning,
Magnificent ram, at the first time.
He lifted the earth and supported the sky,
And he diffused the light there in the form of radiance.
He installed the soul of the spirits in the midst of the waters (?).
....
He acted the god (25) when he began to come into being,
....
The most significant of the significant, who is greater than all the gods,
Mysterious of aspect, who is eminent (26) above the gods,
Modeler of the modelers,
Eldest of the primordial gods,
Father of the fathers, mother of the mothers,
Who made the superior beings and created the inferior beings,
The august ram, who made the rams,
Khnum who made the Khnums,
....
Having made (27) the cities and separated the countries,
Having created the Two Lands and made firm the mountains,
He made man on the [potter's] wheel,
He bore the gods, in order to fill up (i.e. populate) the land and the circle of the great ocean,
Coming in time to bring to life all those who came forth from his wheel,

Making herbage to provide for cattle and the tree of
life for men,
He comes forth at the right time without (28) cease.
....

Document II.8: Introduction

The Destruction of Apep

This work from which our document has been taken is the third and longest text to be found in Papyrus British Museum No. 10188, a papyrus, which, at the latest, dates from the fourth century B.C.; it is written in Middle Egyptian but shows evidence of having been artificially composed in that idiom, since it includes some tell-tale Late-Egyptian expressions.[1] Faulkner, the modern editor of this papyrus, suggests that it was a collection of religious texts originally written for a temple library.[2]

The interest of this work for us lies primarily in its fifth section entitled "The Book of Knowing the Creations (*ḫprw*) of Re and the Felling of Apep". The section on creation is cast as a monologue of Re. It actually appears in two not very different forms in separate parts of the work (26,21-27,5 and 28,20-29,6). I have given only the first version, but I have also included before it some ritualistic preparations for using the book against the enemies of Re and of the pharaoh (see 26,2-7).

In the part on creation we first see Re's autogenesis in the Abyss as the self-generator, Khepri, at the time when there was no other being, no sky, no earth, no ground, and no reptiles. He planned a multitude of beings in his heart and they came forth from his mouth, an apparent reference to the doctrine of the creative word. With his own genesis the very nature of coming into being itself came into being. The subsequent creation of land on which he might stand was followed in turn by the creation of Shu and Tefenet, the old views of their creation by expectoration and masturbation being put forth. Other cosmogonic acts follow: the separation of these first children from their father in the Abyss and their recovery by Re's Eye, the creation of man from the tears of his Eye, the placing of the Eye on his face (i.e. forehead) in the form of the uraeus serpent, the successive procreation of Geb and Nut by Shu and Tefenet, and of Osiris, Horus Mekhantenirti, Seth, Isis, and Nephthys by Geb and Nut, and finally the begetting of their multitudes in the land.

Text and Study of *The Destruction of Apep*

The most important edition of this work is that included in R.O. Faulkner's edition of the papyrus: *The Papyrus Bremner-Rhind (British Museum No. 10188): Bibliotheca Aegyptiaca* III (Brussels, 1933), from column 22,1 to 32,12, pp. 42-88. Furthermore, Faulkner prepared a translation with commentary: "The Bremner-Rhind Papyrus-III [and -IV]," *JEA*, Vol. 23 (1937), pp. 166-85; Vol. 24 (1938), pp. 41-53. I have employed this translation with very few changes; e.g., I have, for the most part, stripped it of its biblical form of language and I have used "Abyss" for the non-personified form of *nw(n)* but have retained "Nun" when the reference is to the Deep as a god, the father of Re. Faulkner used "Nun" throughout. I have used the simple transcriptions of the names Re, Khepri, Tefenet, and Apep, which I have used elsewhere in the volume.

Notes to the Introduction of Document II.8

1. Faulkner, *The Papyrus Bremner-Rhind*, p. IX.
2. *Ibid.*, p. VI.

Document II.8

The Destruction of Apep

/**26**/ .... (2) .... To be recited by a man who is pure and clean. You shall depict (?) every foe of Re and every foe of pharaoh, whether dead or alive, and every accused one whom he has in mind, [also] the names of their fathers, their mothers (3), and their children ..., they having been drawn in green ink on a new sheet of papyrus, their names written on their breasts, [these] having been made of wax and also bound with bonds (?) of black thread; they are to be spat upon, (4) and [they are] to be trampled with the left foot, felled with the spear and knife and cast on the fire in the melting furnace of the coppersmiths. Afterwards the name of Apep is to be burnt in a fire of bryony when Re manifests himself, when Re is at noontide, (5) and when Re sets in the West; in the first hour of the day and of the night and in the second hour of the night down to the third hour of the night; at dawn, and likewise every hour of the night and every hour of the day; at the festival of the New Moon, at the sixth-day festival, (6) at the fifteenth-day festival, and likewise at the monthly festival .... This book is to be employed in this manner which is in writing .... (7) .... It will go well with the man who makes conjurations for himself [from] this book in the presence of the august god, a true matter, [tested] a million times .... (21) ....

The Book of Knowing the Creations of Re and of
Felling Apep. Recite: Thus spoke the Lord of All after
he had come into being: "It was I who came into being
as Khepri. When I came into being, 'Being' (22) came
into being, and all beings came into being after I came
into being; manifold were the beings which came forth
from my mouth ere the sky had come into being, ere
the earth had come into being, ere the ground and
reptiles had been created in this place. I created [some
of] them in the Abyss (23) as Inert Ones when I could
as yet find no place where I could stand. I considered
(?) in my heart, I surveyed with my sight, and I alone
made every shape ere I had spat out Shu, ere I had
expectorated Tefenet, ere there had come into being
any other (24) who could act with me. I planned with
my own heart and there came into being a multitude of
forms of living creatures, namely the forms of children,
and the forms of their children. I indeed made
excitation with my /27/ fist, I copulated (1) with my
hand, I spat with my own mouth; I spat out Shu, I
expectorated Tefenet, and my father Nun brought them
up, my Eye following after them since the aeons when
they were far from me. After I had come into being as
sole god, (2) there were three gods in addition to
myself. I came into being in this land and Shu and
Tefenet rejoiced in the Abyss, in which they were.
They brought back to me my Eye with them after I had
united my members; I wept over them, and that is how
men came into being (3) from the tears which came
forth from my Eye, for it was angry with me when it
returned and found I had set another in its place,
having replaced it with Glorious [Eye]. So I promoted it
in my face, and when it exercised governance over (4)
this entire land, its wrath died away (??), for I had

replaced what had been (?) taken from it. I came forth from the roots, I created all reptiles and all that exists among them. Shu and Tefenet begot (5) Geb and Nut, and Geb and Nut begot Osiris, Horus Mekhantenirti, Seth, Isis, and Nepththys from the womb, one after another, and they begot their multitudes in this land ...."

Document II.9: Introduction

The *Memphite Theology*

The so-called *Memphite Theology* (this is a modern title coined by Erman) exists in only one copy, which is inscribed on a black granite stone (British Museum No. 408). The stone was prepared by the order of King Shabaka (712-698 B.C.) of the twenty-fifth dynasty, i.e. the Ethiopic dynasty, purportedly copied from an earlier "worm-eaten" copy (see line 2 of Document II.9 below). The stone has inscribed on it two horizontal lines at the top and sixty-two vertical columns, which begin on the left side of the stone (see Fig. II.53). The middle part of the text is almost completely effaced, since the stone was reused as a millstone and so a rectangular hole was cut in the middle. Grooves radiate out from the hole and the upper millstone revolved in the middle of the inscribed surface. The first person to understand the significance of the cosmogonic part of this document was Breasted, who called it the philosophy of a Memphite priest of Ptah.[1] Since his time it has been widely studied, as I note in the bibliographic paragraph below. The older view was that the text was made from an original composition that was written either in Archaic times or at least no later than the Old Kingdom. But recently F. Junge cast doubt on the advanced age of this document and instead concluded that it was composed by those who inscribed it on the stone in the twenty-fifth dynasty and that they employed archaizing language in order to enhance its and their importance.[2] An even later study suggests a nineteenth-dynasty origin.[3] Be that as it may, it seems clear that the concept of creation by word, which is its most important doctrinal feature for us in our cosmogonic studies, reflects an old doctrine,[4] though perhaps one never elaborated to such a degree. Again and again I have mentioned in Chapter Two and the preceding documents that the earlier traces of this doctrine seem to have solid roots in the Heliopolitan conceptions of Sia (Understanding or Intelligence) and Hu (Authoritative Utterance), and so I need not repeat this earlier history here. Nor indeed need I summarize once more the main features of the creative activity of Ptah that is outlined in the document, for they too have been described in some detail in Chapter Two. In presenting as Document II.9 those extracts of the Shabaka Stone which are designated the *Memphite Theology*, I have given only the first two lines on the stone and the cosmogonic section itself. The rest of the stone, or at least that which is readable, is available in the texts and translations noted below in the next section of this introduction.

Studies and Texts of the *Memphite Theology*

The first accurate text and useful study of this work was that of J. H. Breasted, "The Philosophy of a Memphite Priest," *ZÄS*, Vol. 39 (1901), pp. 39-54. Breasted has partially reconstructed the text on the stone in accompanying plates (see Fig. II.53). The most complete study of the Shabaka Stone, which includes text, translation, and commentary, is that of K. Sethe,

Dramatische Texte zu altägyptischen Mysterienspielen (Untersuchungen zur Geschichte und Altertumskunde Ägyptens, Vol. 10) (Leipzig, 1928; repr. Hildesheim, 1964). For the cosmogonic part, see H. Junker, *Die Götterlehre von Memphis (Schabaka-Inschrift) (Aus den Abhandlungen der Preussischen Akademie der Wissenschaften, Jahrgang, 1939 Phil.-hist. Klasse*, Nr. 23) (Berlin, 1940). Among other studies, excerpts, and translations, we can note A. Erman, "Ein Denkmal memphitischer Theologie," *Sitzungsberichte der Königlichen Preussischen Akademie der Wissenschaften, Jahrgang 1911* (Berlin, 1911), pp. 916-50; J. H. Breasted, *The Dawn of Conscience* (New York, 1933), pp. 29-42; J. A. Wilson, "The Nature of the Universe," in H. and H. A. Frankfort, J. A. Wilson, T. Jacobsen, and W. A. Irwin, *The Intellectual Adventure of Ancient Man* (Chicago and London, 1946; Phoenix ed. 1977), pp. 55-60; M. Sandman Holmberg, *The God Ptah* (Lund and Copenhagen, 1946), *passim*, and especially pp. 19-23, 42-45; and Wilson, "Egyptian Myths, Tales, and Mortuary Texts," in J. B. Pritchard, ed., *Ancient Near Eastern Texts Relating to the Old Testament* (Princeton, 1950; 2nd ed. 1955), pp. 4-6; M. Lichtheim, *Ancient Egyptian Literature*, Vol. 1 (Berkeley/Los Angeles/London, 1975), pp. 51-57; F. Junge, "Zur Fehldatierung der sog. Denkmals memphitischer Theologie oder Der Beitrag der ägyptischen Theologie zur Geistesgeschichte der Spätzeit," *MDAIK*, Vol. 29 (1973), pp. 195-204; and H. A. Schlögel, *Der Gott Tatenen* (Freiburg, Switz., and Göttingen, 1980), pp. 110-17.

ANCIENT EGYPTIAN SCIENCE

The English Translation

In my English translation I have paid close
attention to Sethe's and Junker's commentaries as well
as to the English translations of Wilson and Lichtheim.
The numbers in parentheses are those of the columns
on the stone (except for the first two numbers which
refer to the horizontal lines at the top of the stone). I
have indicated in footnote 2 of the document the
alternate translations of columns 53 and 54 proposed by
Lichtheim on the basis of Sethe's and Junker's
renderings.

Notes to the Introduction of Document II.9

1. See the article of Breasted, "The Philosophy of
a Memphite Priest," given in the bibliographical section
of this Introduction.
2. See the article of Junge, "Zur Fehldatierung...,"
given in the bibliographical section of this Introduction.
3. See H. A. Schlögel, *Der Gott Tatenen*, pp.
110-17.
4. J. Zandee, "Das Schöpferwort im alten
Ägypten," *Verbum: Essays on Some Aspects of the
Religious Function of Words dedicated to Dr. H. W.
Obbink (Rheno-Traiectina*, Vol. 6, 1964), pp. 33-66.

Document II.9

The *Memphite Theology*

(1) The living Horus: he who makes prosper the Two Lands; [the one of] the Two Ladies: he who makes prosper the Two Lands; the Golden Horus: he who makes prosper the Two Lands; the King of Upper and Lower Egypt: Neferkare; Son of Re: Sha[baka], beloved of Ptah South of his Wall, who lives like Re forever.[1] (2) His majesty copied this book anew in the House of his father Ptah South of his Wall. His majesty had found it as a work of his ancestors, which was worm-eaten, so that it could not be known from the beginning to the end. Then his majesty copied it anew so that it was in a better state than it had been in before. [He did this] in order that his name might endure and in order that his monument be made to last in the House of his father Ptah South of his Wall through the length of everlastingness (*dt*), as a work done by the Son of Re [Shabaka] for his father Ptah-Tatenen so that he might be given life forever ....

(48) The gods who came into being in Ptah:

(49a) Ptah-upon-the-Great-Throne ...;

(50a) Ptah-Nun, the father who made (?) Atum;

(51a) Ptah-Naunet, the mother who gave birth to Atum;

(52a) Ptah-the-Great, who is the heart and the tongue of the Ennead;

(49b) [Ptah] ... who bore the gods;

(50b) [Ptah] ....

(51b) [Ptah] ....

(52b) [Ptah] ..., Nefertem at the nose of Re every day.

(53) There came into being in the heart [of Ptah] and there came into being on the tongue [of Ptah] something in the shape (i.e. form) of Atum, for Ptah is the very great one who transmitted [life] to all the gods and to their kas by means of the heart in which Horus has taken shape and by means of the tongue in which Thoth has taken shape, ... [each] (54) as [an agent or form of] Ptah.[2]

[Thus] it happened that heart and tongue gained mastery over [every] member [of the body] according to the teaching that he (Ptah) is in every body [as heart] and in every mouth [as tongue]: [i.e. in the bodies and mouths] of all gods, all men, all cattle, all creeping things, and of everything which lives. Accordingly [as heart] he thinks out (i.e. conceives) and [as tongue] he commands what he wishes [to exist].

(55) His (Ptah's) Ennead is before him as teeth and lips, [they are] as the semen and hands of Atum, for [it is said that] the Ennead of Atum came into being by means of his semen and his fingers. But the Ennead [of Ptah] is the teeth and lips in this mouth which pronounced the name of everything, from which Shu and Tefnut came forth, (56) and which gave birth to the Ennead.

The sight of the eyes, the hearing of the ears, the breathing of the nose, they report to the heart, and it (the heart) causes every understanding (i.e. completed concept) to come forth, and it is the tongue which repeats what the heart has thought out (i.e. devised). Thus all gods were born and his (Ptah's) Ennead was completed. For every word of the god came into being by means of what the heart thought out and the tongue commanded.

(57) So were made all the kas, and the hemsut were determined, [i.e.] those [spirits or faculties and qualities] which make all foods and provisions, by means of this word [which the heart thought out and which came forth on the tongue]. [Thus justice is done] to him who does what is liked [and punishment to] him who does what is disliked. Thus life is given to him who has peace (i.e. is peaceful and law-abiding) and death to him who has sin (i.e. who is a criminal). Thus were made all works and all crafts, the action of the hands and the movement of the legs, (58) and the activity of every member, in accordance with the command which is thought out by the heart and comes forth on the tongue and creates the performance of everything.[3]

And so is said of Ptah [this epithet]: "He who made all and brought the gods into being." He is Tatenen, who gave birth to the gods and from whom every thing came forth: food, provisions, offerings for the gods, and all good things. So it was discovered and understood that he is the mightiest of the gods. And so Ptah was satisfied after he had made everything and all the divine words.

He gave birth to the gods,

He made the towns,

He founded the nomes,

He put the gods into their (60) shrines.

He settled their offerings,

He founded their shrines,

He made their bodies (i.e. their statues) as they desired [them].

So the gods entered into their bodies,
Of every wood, every stone, every clay,
And of every thing which grows upon him [as Tatenen]
(61) In which their forms resided.
Thus all gods were gathered in him, and also their kas,
Content and united with the Lord of the Two Lands.

Notes to Document II.9

1.   This first line contains the full five-fold titulary of King Shabaka. The special relationship with Ptah that is stressed in the second line is immediately suggested after the titulary in line 1. In both lines Ptah is given the very old epithet South of his Wall. In the second line he is also called Ptah-Tatenen, which name is probably of New-Kingdom origin.

2.   Following the suggestions of Sethe and Junker, Miss Lichtheim, *Ancient Egyptian Literature*, Vol. 1, p. 54, gives the following alternative translations of this paragraph: "(53) There took shape in the heart, there took shape on the tongue the form of Atum. For the very great one is Ptah, who gave [life] to all the gods and their *kas* through this heart and through this tongue, (54) in which Horus had taken shape as Ptah, in which Thoth had taken shape as Ptah. [Alternative reading: (53) Heart took shape in the form of Atum, Tongue took shape in the form of Atum. It is Ptah, the very great, who has given [[life]] to all the gods and their *kas* through this heart and through this tongue, (54) from which Horus had come forth as Ptah, from which Thoth had come forth as Ptah."]

3.   For the translation of this last clause, see Lichtheim, *Ancient Egyptian Literature*, Vol. 1, pp. 55 and 57, n. 13.

Document II.10: Introduction

A Dream-Book

This work appears on the recto of a papyrus of the nineteenth dynasty (British Museum, Papyrus Chester Beatty III=10683) and is written in Middle Egyptian "with hardly a trace of the Late-Egyptian Idiom, either in vocabulary or grammar."[1] Gardiner believes that[2]

> it is, indeed, the earliest Dream-book in existence and may well date back to the Twelfth Dynasty ..., though this cannot be proved with certainty. The core of the work consists of a long enumeration of dreams in clear tabular form, accompanied by their interpretations. A vertical column of large hieratic signs occupies the entire height of the page, and yields the sense *'If a man see himself in a dream ....'*. This clause has to be read before each separate horizontal line to the left of the column, only one horizontal line being allowed for each dream. The descriptions of the dreams are necessarily very terse, and are divided off from the interpreter's equally terse judgment upon them by a small space. The

general scheme is thus as follows: '*If a man see himself in a dream [doing-so-and-so], good* (or *bad*); it means such-and-such a thing will happen.' .... From twenty-five to twenty-eight dreams go to the page, after which the scribe starts afresh from the top with a new vertical column. The good dreams form a solid block, like the less numerous bad dreams that follow. The word '*bad*', being a word of ill omen, is written in red [represented by caps in Doc. II.10 below], the colour of blood .... It seems legitimate to suppose that the work opened with some such general title as '*The book of the interpretation of dreams*', to which may even have been appended the name of the reputed author .... Such a beginning of the book [in which good Horian dreams are distinguished from bad Sethian ones], if ever it existed, is completely lost, and in the first preserved page we find ourselves in the midst of the good Horian dreams.

I have already suggested in Chapter Two above that the gods were often considered to be the source of dreams, but that a god did not directly communicate with ordinary people as he occasionally did with the king. It is perhaps because of the divine origin of dreams that the message of the dream is only indirectly indicated. Indeed the interpretation of the dream rests on using words that are either identical or similar to those in which the action of the dream is expressed. In the latter case it is clear that the similiarity is in the sounds of the words and so rests on punning. The reader will recall how popular punning was with the creator god (whoever he was) when he created by asserting the names of the entities he had conceived in his heart. Something like this process was evident when the god introduced the dream into the dreamer, and I have followed Gardiner in noting in parentheses some of the identities or similarities in the key words of dream and interpretation. But other relationships between dream and interpretation are also evident in this dream-book, as Gardiner points out:[3]

> Between dream and interpretation there was necessarily a certain correspondence of idea, and it was this correspondence which enabled the interpreter to declare the former portended the latter. Often the resemblance was restricted simply to the occurrence of the same word in the verbal descriptions .... Elsewhere the correspondence is more symbolical. For example, to see one's face in a mirror is to discover a second self, which second self must naturally have a wife; hence the interpretation *it means another wife* (7,11). Again to dream you are bringing in the cattle means that your people will be assembled to you ..., or if you see a large cat this signifies that your harvest will be a

big one (4,3) .... A rarer form of correspondence is that manifesting itself in contraries. Thus to see oneself dead is to have long life in prospect (4,13) .... Beyond the dichotomy of good and bad dreams, no attempt at classification is visible .... A point of special interest is the frequent reference to a man's dependence upon 'his god', which probably refers, not to the good god (Horus) or the bad god (Seth) represented by the man himself, but rather to the 'city-god'... so often mentioned in the texts of the Middle Kingdom.

No separate bibliographical section is necessary for this document, since I am wholly dependent on A. H. Gardiner, *Hieratic Papyri in the British Museum. Third Series. Chester Beatty Gift*, 2 vols. (London, 1935); see Vol. 1, pp. 9-23, and Vol. 2, plates 5-8. I have simply used the Gardiner translation in Vol. 1 and referred to the text in the plates of Vol. 2. I have changed Gardiner's formatting somewhat in order to accommodate it to my computerized printing program.

Notes to the Introduction of Document II.10

1.   Gardiner, *Hieratic Papyri .... Third Series. Chester Beatty Gift*, Vol. 1, p. 10.
2. *Ibid.*, pp. 9-10.
3. *Ibid.*, pp. 21-22.

Document II.10

A Dream-Book

[IF A MAN SEE HIMSELF IN A DREAM]
..... ---- [good]
..... ---- [good]
     (The number of dreams lost is unknown)
(1,13) .... ---- good; it means putting .... in his hand.

     (I skip the rest of the lines on this page, the dreams being lost, though the interpretations are for the most part complete.)

(2,1) with his mouth split open *(sd)*, ---- good; it means something he was afraid of will be opened up *(sd)* by the god.

     (skipping one line)

(2,3) .... a crane *(d̲ʾt)*, ---- good; it means prosperity *(wd̲ʾ)*.

     (skipping two lines)

(2,6) munching lotus leaves (?), ---- [good]; it means something he will enjoy.

(2,7) shooting at a mark, ---- [good]; it means something good will happen to him.

     (skipping several lines)

(2,11) his penis becoming large, ---- [good]; it means his possessions will multiply.

     (skipping two lines)

(2,14) seeing the god who is above, ---- good; it means much food.

     (skipping a number of lines)

(2,23) up a growing tree *(nht)*, ---- good; it means his loss *(nhy)* of ....

(2,24) looking out a window, ---- good; it means the hearing of his cry by his god.

(2,25) rushes being given to him, ---- good; it means the hearing of his cry.

(2,26) seeing himself on a roof, ---- good; it means finding something.

(skipping two lines)

(3,2) seeing himself [in] mourning, ---- good; the increase of his possessions.

(skipping a line)

(3,4) white *(ḥḏ)* bread being given to him, ---- good; it means something given to him, at which his face will brighten up *(ḥḏ)*.

(3,5) drinking wine, ---- good; it means living in righteousness.

(skipping a line)

(3,7) copulating with his mother..., ---- good; [his?] clansmen will cleave fast to him.

(3,8) copulating with his sister, ---- good; it means the bequeathing of something to him.

(skipping a number of lines)

(4,3) seeing a large cat *(my ꜥꜣ)*, ---- good; it means a large harvest *(šmw ꜥꜣ)* will come to [him].

(skipping three lines)

(4,7) seeing himself upon a tree, ---- good; the destruction of all his ills.

(4,8) killing an ox, ---- good; killing his enemies.

(skipping two lines)

(4,11) climbing up a mast, ---- good; his being suspended aloft by his god.

(4,12) destroying his clothes, ---- good; his release from all ills.

(4,13) seeing himself dead, ----good; a long life [in] front of him.

 (skipping a number of lines)

(4,22) being given victuals *(ᶜnḫw)* belonging to a temple, ---- good; the bestowing of life *(ᶜnḫ)* upon him by his god.

(4,23) sailing in a boat, ----good; it means sitting among his [distant] townsfolk.

 (skipping a number of lines)

(5,9) reading aloud from a papyrus, ---- good; the establishment of a man in his house.

 (skipping a number of lines)

(5,22) [seeing] the moon shining, ---- good; the pardoning of him by his god.

 (skipping a number of lines)

(6,21) praising [Re?], ---- good; his being found innocent before his god.

 (skipping two lines)

(6,24) seeing Bedouins, ---- good; the love of his father when he dies will come into his presence.

 (skipping three lines)

(7,3) copulating with a female jerboa, ---- BAD; the passing of a judgment against him.

(7,4) drinking warm beer, ---- BAD; it means suffering will come upon him.

(7,5) eating ox-flesh, ---- good; it means something will accrue to him.

 (skipping four lines)

(7,10) removing one of his legs, ---- BAD; judgment upon him(?) by those yonder [i.e. the dead].

(7,11) seeing his face in a mirror, ---- BAD; it means another wife.

(7,12) the god making his tears for him ---- BAD; it means fighting.

(skipping one line)

(7,14) eating hot meat, ---- BAD; it means his not being found innocent.

(skipping two lines)

(7,17) copulating with a woman,---- BAD; it means mourning.

(7,18) being bitten by a dog, ---- BAD; a cleaving fast to him of magic.

(7,19) being bitten by a snake, ---- BAD; it means the arising of words with him.

(7,20) measuring barley, ---- BAD; it means the arising of words with him.

(7,21) writing on a papyrus, ---- BAD; the reckoning up of his misdeeds by his god.

(skipping one line)

(7,23) having a spell put upon his mouth by another (?), ---- BAD; it means mourning.

(skipping a number of lines)

(8,2) seeing his penis stiff *(nḫtw)*, ---- BAD; victory to *(nḫtw)* his enemies.

(skipping two lines)

(8,5) looking into a deep well, ---- BAD; his being put in prison.

(skipping three lines)

(8,9) folding wings around himself(?), ---- BAD; he will not be found innocent with his god.

(8,10) copulating with a kite, ---- BAD; it means robbing him of something.

(skipping two lines)

(8,13) seeing a dwarf, --- BAD; the taking away of half of his life.

(8,14) falling a prey to (?) the council ---- BAD; his being driven from his office.

(skipping two lines)

(8,17) shaving his lower parts,---- BAD; it means mourning.

(skipping a number of lines)

(9,2) being made into an official, ---- BAD; death is close at hand.

(9,3) an Asiatic garment upon him, ---- BAD; his removal from his office.

(9,4) seeing people afar off, ---- BAD; his death is at hand.

(skipping three lines)

(9,8) seeing the heavens raining, --- BAD; words have come up against him.

(skipping one line)

(9,10) uncovering his own backside *(pḥwy)*, ----BAD; he will be an orphan later *(ḥr pḥwy)*.

(9,11) eating figs and grapes, ----BAD; it means illness.

(skipping two lines)

(9,14) putting his face to the ground, ---- BAD; the requirement of something from him by those yonder (i.e. the dead).

(9,15) seeing a burning fire, ---- BAD; the removal of his son or his brother.

(9,16) copulating with a pig, ---- BAD; being deprived of his possessions.

(skipping one line)

(9,18) drinking blood, ---- BAD; a fight awaits him.

(skipping three lines)

(9,22) copulating with his wife in daylight, ---- BAD; the seeing of his misdeeds by his god.

(skipping three lines)

(9,26) seizing wood belonging to the god in his hand, ---- BAD; finding misdeeds in him by his god.

(9,27) looking after monkeys, ---- BAD; a change awaits him.

(9,28) fetching mice from the field, ---- BAD; a sore heart.

(10,1) sailing downstream, ---- BAD; violent words.

    (skipping six lines)

(10,8) putting beer into a vessel, ---- BAD; the removal of something from his house.

(10,9) breaking a vessel with his feet, ---- BAD; it means fighting.

[Incantation to Isis][1]

(10,10) To be recited by a man when he wakes in his [own] place. 'Come to me, come to me, my mother Isis. Behold I am seeing what is (?) far from me in my (?) city.' 'Here am I, my son Horus, come out with what thou hast seen, in order that my afflictions(?) throughout thy dreams may vanish, and fire go forth against him that frighteneth thee. Behold, I am come that I may see thee and drive forth thy ills and extirpate all that is filthy.' 'Hail to thee, thou (?) good dream which art seen [by] night (10,15) or by day. Driven forth are all evil filthy things which Seth, the son of Nut, has made. [Even as] Re is vindicated against his enemies, [so] I am vindicated against my enemies.'

This spell is to be spoken by a man when he wakes in his [own] place, there having been given to him *pesen*-bread in [his] presence and some fresh herbs moistened with beer and myrrh. A man's face is to be rubbed therewith, and all evil dreams that [he] has seen are driven away.

Note to Document II.10

1.   Gardiner, *Hieratic Papyri ... Third Series.*

Chester Beatty Gift, Vol. 1, p. 19: "At this point the catalogue of dreams gives place to an incantation for the protection of the dreamer, and here for the first time we receive a hint that Horus was regarded as the prototype of the normal Egyptian man whose nocturnal visions were interpreted in the first half of the book. The text of the spells is unhappily rather corrupt, though not to the extent of rendering it wholly unintelligible. The form of a dialogue is adopted, Horus calling upon his mother Isis to shield him from the baneful consequences portended by his dreams. These consequences are of course ascribed to the machinations of Seth."

Document II.11: Introduction

The Harris Magical Papyrus

The papyrus which is known as the Harris Magical Papyrus is Harris Papyrus 501 (=British Museum Papyrus 10042). It was purchased by the British Consul-General Mr. A.C. Harris in Alexandria in 1855. At that time it was complete and was almost nine feet in length.[1] This was its condition when tracings were made from it by Chabas for his edition of it in 1860 (see the bibliographical paragraph below).[2] But by the time it was purchased from Mr. Harris's daughter by the British Museum in 1872, it was in a seriously mutilated state as the result of a powder explosion in Alexandria, and the two remaining portions together are but five feet in length. Hence the tracings of Chabas played an important role in the best hieroglyphic text and study of the papyrus produced by H.O. Lange in 1927. The recto of the papyrus is in a book-hand of the nineteenth or twentieth dynasty, i.e., it is a Ramesside document. The verso is in a careless hand not much later than that of the recto.[3]

The varied content of hymns and magical spells (primarily against the crocodile) has been neatly described by Lange,[4] and I confine myself to noting that the parts of interest for cosmogony (which I have extracted as Document II.11) are primarily those which are hymns. Pieces B-D are three hymns to Shu (or three parts of a single hymn). Shu is called the heir of Re and his power is like that of Re himself as the Lord of Transformations. I have already mentioned that by the time of the *Coffin Texts* Shu was described as having an important auxiliary role in creation. Re has supplied his heir with magic (B,14-15). Shu's kas (his spiritual attributes) rest in Re's, the latter nourishing the former. In normal Egyptian fashion a will was drawn up in Hermopolis by Reharakhti's scribe, Thoth, who is called Lord of Hermopolis, the transfer of power being thereby "established, confirmed, and perpetuated in writing" (B,22). Shu is further described as "begotten of Atum himself, who has created himself, without there being a mother" (C,2-3). Shu is said to have brought the Eye of his father to him, rather reversing the usual story that Shu and Tefenet were lost in the Abyss and Atum accordingly sent his eye to fetch them. The hymn in D which celebrates the various attributes of Shu epitomized in his various names resembles the *Litany of Re* and shows once again what an important role Shu has assumed in this account.

In hymn E, Sepu, an epithet of Re as personifying the "First Time" of creation (?), is addressed as the one who made his own body, the One Lord who came out of the Abyss. He is called Hu, the self-creator who made the food in him, the one who created both his father and mother.

In hymn F the creator is also called "you Five Great Gods who came out of the City of Hermopolis," a title I have discussed in Chapter Two, and it is asserted that his own creation preceded his being in the heaven and on the earth and was before the sun first shone, a poetic description of the first stage of creation.

Hymn G to Amon-Re was quoted in Chapter Two above and its cosmogonic aspects need only brief mention here: the Hermopolitan influence which is evident in the insertion of the Ogdoad into the creation process (G,3), creation by the word (G,7), Amun's transformation of himself into "millions" (G,8), his limitless size (G,9), his early form as a serpent (G,11), and his richness of magic and secret forms (G,12).

The creator also takes on the name of Tatenen "who lifted himself up" (G,20), thus showing that Amun borrows from the traditions of Ptah at Memphis as well as from those of Re at Heliopolis and the Ogdoad at Hermopolis.

The influence of the Hermopolitan doctrines is also apparent in pieces H and K. In H the Ogdoad worship the god who is in their midst (H,2), the great god being described in the common fashion as having bones of silver, flesh of gold, and hair of real lapis lazuli (H,3-4).[5] The Ogdoad speak of Amun's hiding himself in his pupil, i.e. in his sun disk (H,6-7), as "wondrous of forms" and as the "sacred one whom nobody knows" (H,8). In K the god is called in Hermopolitan terms the Egg of the water, the Seed of the earth, and the Essence (?) of the Ogdoad of Hermopolis (K,5-6). There is also reference to the Lake of the Two Knives, apparently located in Hermopolis. The hymn in K shows the not uncommon union of Amun and Min (K,11-12). The theme of the primordial egg is continued in the final assertion that the spell is "to be recited over the egg of the navel" held in his hand by a person in the bow of a ship (K,13-14).

My final extract from the Harris Magical Papyrus celebrates the creator as the "Image of millions of millions" and the "single Khnum of his son" (M,2-3). This is a poetic reference to Khnum's creating activity on his potter's wheel and thus means simply that by himself the creator god has fashioned or crafted his son. A final reference to the doctrine of the Great Cackler, honking in the primordial night, is worth noting (M,9).

Texts and Studies of the Harris Magical Papyrus

As I have indicated above, the first text of the papyrus was produced through tracings made by F. Chabas, *Le Papyrus Magique Harris* (Chalons-sur-Saône, 1860). This work included a translation and commentary and was a remarkable work considering the knowledge of hieratic writing at this time. Chabas made another translation later: Chabas, *Mélanges égyptologiques*, 3rd Series, Vol. 2 (Chalons-sur-Saône and Paris, 1873), pp. 242-78.

E.A.W. Budge published facsimiles of the text (along with Chabas's tracings of the then lost pages): *Facsimiles of Egyptian Hieratic Papyri in the British Museum, With Descriptions, Translations, etc.* (London, 1910), plates 20-30. This work also included the first hieroglyphic transcription (separate pag., pp. 34-40) and a now greatly out-of-date translation (separate pag., pp. 23-27).

The next step (with no great progress in understanding the text) was taken by E. Akmar, who published a hieroglyphic transcription of the text along with a French translation: *Le Papyrus Magique Harris transcrit et publié* (Upsala, 1916).

French extracts were published by F. Lexa, *La magie dans l'Égypte antique de l'Ancien Empire jusqu'à l'Époque Copte*, Vol. 2 (Paris, 1925), pp. 35-44.

By far the most important work on this papyrus was that of H.O. Lange, *Der magische Papyrus Harris* (Copenhagen, 1927), which includes a new hieroglyphic transcription, a fine German translation, and an extensive commentary. It is this work on which I have depended almost exclusively in making my translation. I have used Lange's letters to distinguish the various pieces that make up the text.

Notes to the Introduction of Document II.11

1. Budge, *Facsimiles of Hieratic Papyri*, p. xv.
2. Lange, *Der magische Papyrus Harris*, p. 6.
3. *Ibid.*, p. 7.
4. *Ibid.*, pp. 8-11.
5. See above, Document II.6, (2), note 1.

Document II.11

The Harris Magical Papyrus

A.
The Beautiful Spells for Singing
Which Drive Away the Swimming One (i.e. the
Crocodile)

B.

Hail to you, Heir of Re,
Eldest son, who comes forth from his body,
The one whom he selected before all of his children,
Whose power is like that of the Lord of
Transformations,
(5) Who kills the enemies every day.
When the ship [of the sun] has a good wind, your heart
is pleased;
The morning bark is in jubilee,
When they [in the bark] see Shu, the son of Re,
vindicated,
When he has put his spear into the serpent villain.
(10) When Re travels the sky each morning,
Then Tefenet rests upon his head
And hurls her flame against his enemies,
So that they (*corr. from* he) are made as ones
nonexistent.
You whom Re has amply provided with a divinity of
magic,
(15) As an heir upon the throne of his father,

Whose kas rest in the kas of Re
As food for the nourishment of that which is near him,
[You] for whom he has made a will,
Written out by the Lord of Hermopolis [i.e. Thoth],
(20) The ... scribe of Reharakhti,
In the palace of the Great Double House of Heliopolis;
It is established, confirmed, and perpetuated in writing,
Under the feet of Reharakhti,
And he shall transmit it to the son of his son for ever
and ever.

C.

Hail to you, you son of Re,
Begotten of Atum himself,
Who has created himself, without there being a mother,
The True One, the Lord of the Double Truth,
(5)You Powerful One, who has power over the gods,
Who brings the Eye of his father Re,
To whom they bring gifts in their own hands,
....

D.

You are mysterious, you are greater than the [other]
gods
In your name of Shu, son of Re.
....
(5) You are greater and richer than the [other] gods
In your name of The Very Great Divine One.
.... [and there follows a series of attributes of Shu
reflected in special names.]

E.

O Sepu (i.e. One of the First Time?), who made his own
body.

O One Lord, who came out of the Abyss.
O Hu, who created himself.
O One who made this food which is in him.
(5) O One who has created his father and borne his mother.

F.

Hail to you, you Five Great Gods
Who came out of the City of Eight (Hermopolis),
You who were not yet in heaven (i.e., who existed before the heaven).
You who were not yet upon the earth (i.e., who existed before the earth),
(5) You who were not yet illumined by the sun (i.e., who existed before the sun shone).
....

G.

Adoration of Amon-Reharakhti, who came into being by himself,
Who founded the land when he began [to create],
Made by the Ogdoad of Hermopolis (or Who made the Ogdoad of Hermopolis) in the first primordial time.
....
(5) Amun the primordial god of the Two Lands
When he arose from Nun and Nunet.
What was said [came into being] on the water and on the land.
Hail to you, the One who made himself into millions,
Whose length and breadth are without limit,
(10) Ready Power who bore himself,
The primordial serpent who is powerful of flame,
The One rich in magic with secret forms,
....

(20) Tatenen who lifted himself up (*or* extolled himself)
over the gods,
The Elder one who rejuvenates himself and lives
through eternity *(nḥḥ)*,
Amun who is established in every thing,
This god who brought forth the earth by means of his
plans,
Come to me, O Lord of the Gods, L.P.H.
....

H.

To be said by the Ogdoad of the primordial company of
gods,
Great ones, who worship the god who is in their midst,
His bones are of silver and his flesh of gold,
His hair of real lapis lazuli.
(5) The Ogdoad says:
O Amun, who hides himself in the pupil [of] his [eye],
Soul who shines in his sound eye *(wḏ't)*,
One wondrous of forms, sacred one whom nobody
knows,
Who radiates his forms to make himself seen by means
of his brilliance,
(10) One mysterious of mysteries, whose secrets are not
known,
Praise to you up to the body of Nut (i.e. to heaven)!
Righteous are your children, the gods;
Maat is joined with your secret chapel,
....

K.

....
(5) O you Egg of the water, Seed of the earth,
... (Essence?) of the Ogdoad of Hermopolis,

Great One in the Heaven and Great One in the Netherworld,
Residing in the nest in front of the Lake of the Two Knives,
I have emerged with you from the water,
(10) I have left with you from your nest,
I am Min of Gebtu (Koptos=Qift),
I am Min, Lord of the Land of Gebtu,
This spell will be recited over the egg of the navel,
Which is placed in the hand of a person in the bow of a ship.

....

M.

Another spell:
Come to me, Come to me, O Image of millions of millions,
The single Khnum of his son (i.e. unique parent of his son),
The One who was conceived yesterday and born today,
(5) The One whose name I knew,
The One who has seventy-seven [pairs of] eyes,
And seventy-seven [pairs of] ears,
Come to me that you may hear my voice,
As the voice of the Great Cackler was heard in the Night.
(10) I am the Great Flooded-one, the Great Flooded-one.
To be recited four times.

....

Section Three

Appendixes

Chronology

In preparing the following chronology I have primarily used J. Baines and J. Málek, *Atlas of Ancient Egypt* (Oxford, 1980), pp. 36-37, and O. Neugebauer and R. A. Parker, *Egyptian Astronomical Texts*, Vol. 1 (Providence, Rhode Island, and London, 1960), p. 129. The latter I have consulted for the general dates of the periods and the dynasties through the First Intermediate Period, i.e. during the third millennium. The dates given by Baines and Málek for the periods and the individual monarchs of that millennium are reported in the footnotes. The names of monarchs and their order for the first five dynasties are those which I adopted in Chapter One above. Use has also been made of Parker's detailed chronology published in *The Encyclopedia Americana*, Vol. 10, article "Egypt". From the Middle Kingdom through the Greco-Roman Period, the main source of the dates is the work of Baines and Málek. For Dynasties 21-26 I have also referred to K. A. Kitchen, *The Third Intermediate Period in Egypt (1100-650 B.C.)* (Warminster, 1973, with Supplement, 1986). Except where indicated by asterisks, there is still considerable uncertainty about precise dates, though in most cases the divergencies of dating from the Middle Kingdom onward are small. Some of the problems arising in efforts to achieve precise dating will be discussed in Chapter Three in the next volume. The names of monarchs are given in the forms often used in this volume, i.e. with vowels supplied but with only common abbreviated phonetic renderings. Overlapping dates indicate joint or collateral regencies.

The reader may wish to consult the detailed chronology developed by A. H. Gardiner, *Egypt of the Pharaohs* (Oxford, 1961), pp. 429-33, since it correlates the names of the kings found in Manetho, the king-lists, and the monuments.

EARLY DYNASTIC PERIOD: 3110-2665 B.C.[1]

Dynasty 1: Menes (Narmer?), Aha (sometimes thought to be Menes, as I have noted in Chapter One above), Djer, Djet or Wadji, Den, Adjib, Semerkhet, and Ka^{c}a.[2]

Dynasty 2: Hetepsekhemuy, Nebre, Ninetjer, Sekhemib, Peribsen (these last two kings are sometimes thought to be the same king; sometimes a King Weneg and King Sened are placed before Peribsen), Sened, Neferka, Neferkaseker, Khasekhem, and Khasekhemuy (the last two names may designate one king who changed his name; see Doc. I.1, n. 64).[3]

OLD KINGDOM: 2664-2155.[4]

Dynasty 3: 2664-2615: Zanakht (=Nebka?), Djoser (Netjeri-khet), Sekhemkhet (Djoser-tety), Khaba, and Huny.[5]

Dynasty 4: 2614-2502: Sneferu, Khufu (Cheops), Djedefre or Redjedef, Rekhaef (Chephren) (sometimes a king Baufre is added here), Menkaure (Mycerinus), and Shepseskaf.[6]

Dynasty 5: 2501-2342: Weserkaf, Sahure, Neferirkare Kakai, Shepseskare Ini, Reneferef, Niuserre Izi, Menkauhor, Djedkare Izezi or Issy, and Wenis or Unis.[7]

Dynasty 6: 2341-2181: Teti, Pepy I (Meryre), Merenre Nemtyemzaf, and Pepy II (Neferkare).[8]

Dynasties 7 and 8: Interregnum and 2174-2155: As Baines and Málek note, there are numerous ephemeral kings assigned to these dynasties, and they date these dynasties 2150-2134.

FIRST INTERMEDIATE PERIOD: 2154-2052

(Neugebauer and Parker) or 2134-2040 (Baines and Málek). From this point on I quote in the main chronology the dates of the dynasties and of their kings given by Baines and Málek, preserving however the spelling of the kings' names that I have adopted throughout this and the succeeding volumes.

Dynasties 9 and 10: These are the Heracleopolitan dynasties with several kings called Khety; Merykare; and Ity. In Thebes during this same period several princes ruled locally, the last of whom, Nebhepetre (Mentuhotep), conquered the Heracleopolitans and assumed the kingship as the first Theban king of all Egypt in the eleventh dynasty.

MIDDLE KINGDOM: 2040-1640

Dynasty 11: 2040-1991: Nebhepetre Mentuhotep 2061-2010, Sankhare Mentuhotep 2010-1998, Nebtawyre Mentuhotep 1998-1991.

Dynasty 12: *1991-1783: Sehetepibre Amenemhet I *1991-1962, Kheperkare Senwosret or Senusert (Sesostris) I *1971-1926, Nubkaure Amenemhet II *1929-1892, Khakeperre Senwosret II *1897-1878, Khakaure Senwosret III *1878-1841?, Nimaatre Amenemhet III 1844-1797, Maakkerure Amenemhet IV 1799-1787, Sebekkaure Nefrusobk 1787-1783.

Dynasty 13: 1783-after 1640. About 70 kings. Some of the more important monarchs are given by Baines and Málek. See also Gardiner, *Egypt of the*

Pharaohs, pp. 440-41. The fourteenth dynasty (at Xois?) included a number of minor kings probably contemporary with the kings of Dynasty 13 or Dynasty 15.

SECOND INTERMEDIATE PERIOD: 1640-1532

Dynasty 15: This was a Hyksos dynasty at Avaris and included Kings Salitis, Seshi, Khian, Apohis c. 1585-1542, and Khamudi c. 1542-1532.

Dynasty 16: Minor Hyksos kings or vassals contemporary with those of Dynasty 15.

Dynasty 17: 1640-1550: A number of Theban kings, the last two being Seqenenre Tao (or Djehutio) and Wadjikheperre Kamose c. 1555-1550.

NEW KINGDOM: 1550-1070

Dynasty 18: 1550-1307: Nebpehtire Ahmose 1550-1525, Djeserkare Amenhotep (Amenophis) I 1525-1504, Akheperkare Tuthmosis I 1504-1492, Akheperenre Tuthmosis II 1492-1479, Menkheperre Tuthmosis III 1479-1425, Maatkare Hatshepsut (the pharaoh queen) 1473-1458 who ruled with Tuthmosis III in the background until he took over the sole rule, Akheprure Amenhotep II 1427-1401, Menkheprure Tuthmosis IV 1401-1391, Nebmaatre Amenhotep III 1391-1353, Neferkheprure waenre Amenhotep IV (=Akhenaten) 1353-1335, Ankhkeprure Smenkhkare 1335-1333, Nebkheprure Tutankhamun 1333-1323, Kheperkheprure Ay (or Aya) 1323-1319, Djeserkheprure Haremhab 1319-1307.

Dynasty 19: 1307-1196: Menpehtire Ramesses I 1307-1306, Menmaatre Seti or Sety (Sethos) I 1306-1290, Usermaatre setepenre Ramesses II 1290-1224, Baenre hotephirmaat Merneptah or Merenptah 1224-1214,

Userkheprure setepenre Seti II 1214-1204, Menmire Amenmesse, a usurper in the reign of Seti II, Akhenre setepenre Siptah 1204-1198, Sitre meritamun Twosre or Tausert, a queen, 1198-1196.

Dynasty 20: 1196-1070: Userkhaure meryamun Sethnakhte 1196-1194, Usermaatre meryamun Ramesses III 1194-1163, Heqamaatre setepenamun Ramesses IV 1163-1156, Usermaatre sekheperenre Ramesses V 1156-1151, Nebmaatre meryamun Ramesses VI 1151-1143, Usermaatre setepenre Ramesses VII 1143-1136, Usermaatre akhenamun Ramesses VIII 1136-1141, Neferkare setepenre Ramesses IX 1131-1112, Khepermaatre setepenre Ramesses X 1112-1100, Menmaatre setepenptah Ramesses XI 1100-1070.

THIRD INTERMEDIATE PERIOD:[9] 1070-712

Dynasty 21: 1070-945: Hedjkheperre setepenre Smendes 1070-1044,[10] Neferkare Amenemnisu 1044-1040,[11] Akheperre setepenamun Pausennes I 1040-992,[12] Usermaatre setepenamun Amenemope 993-984, Akheperre setepenre Osorkon[13] I 984-978, Netjerkheperre setepenamun Siamun 978-959, Titkheprure setepenre Pausennes II 959-945.

Dynasty 22: 945-712: Hedjkheperre setepenre Shoshenq I 945-924, Sekhemkheperre setepenre Osorkon II 924-909, Usermaatre setepenamum Takelot I 909-, Heqakheperre setepenre Shoshenq II -883, Usermaatre setepenamun Osorkon III 883-855, Hedjkheperre setepenre Takelot II 860-835, Usermaatre setepenre Shoshenq III 835-781, Usermaatre setepenre/amun Pami 783-773, Akheperre Shoshenq V 773-735, Akheperre setepenamun Osorkon V 735-712.[14]

Dynasty 23: c. 828-712: Baines and Málek remark concerning this dynasty: "Various contemporary lines of kings recognized in Thebes, Hermopolis, Herakleopolis, Leontopolis and Tanis; precise arrangement and order are still disputed." Pedubaste I 828-803, Osorkon IV 777-749, Neferkare Peftjauawybast 740-725.[15]

Dynasty 24: 724-712 at Sais: (Shepsesre? Tefnakhte 724-717), Wahkare Bocchoris 717-712.[16]

Dynasty 25: 770-712 (Nubia and Theban area): Nimaatre Kashta 770-750, Usermaatre (and other names) Piye 750-712.

LATE PERIOD: 712-332

Dynasty 25: 712-657 (Nubia and all Egypt): Neferkare Shabaka 712-698, Djedkaure Shebitku 698-690, Khure nefertem Taharqa 690-664, Bakare Tantamani 664-657 (possibly later in Nubia).[17]

Dynasty 26 (Saite): "664-525: (Necho I "672-664), Wahibre Psammetichus I "664-610, Wehemibre Necho II "610-595, Neferibre Psammetichus II "595-589, Haaibre Apries "589-570, Khnemibre Amasis "570-526, Ankhkaenre Psammetichus III "526-525.

Dynasty 27 (Persian): "525-404: Cambyses "525-522, Darius I "521-486, Xerxes I "486-466, Artaxerxes I "465-424, Darius II "424-404.

Dynasty 28: Amyrtaios "404-399.

Dynasty 29: "399-380 Baenre merynetjeru Nepherites I "399-393, Userre setepenptah Psammuthis "393, Khnemmaatre Hakoris "393-380, Nepherites II "380.

Dynasty 30: "380-343: Kheperkare Nectanebo I "380-362, Irmaatenre Teos "365-360, Senedjemibre setepenanhur Nectanebo II "360-343.

2nd Persian Period (sometimes called Dynasty 31): "343-332: Artaxerxes III Ochus "343-338, Arses "338-336, Darius III Codoman "335-332. Period interrupted by a native ruler Senetanen setepenptah Khababash.

GRECO-ROMAN PERIOD: *332 B.C.-395 A.D.

Macedonian Dynasty: *332-304: Alexander III the Great *332-323, Philip Arrhidaeus *323-316. Alexander IV *316-304.

Ptolemaic Dynasty: *304-30: Ptolemy I *304-284, Ptolemy II *285-246, Ptolemy III *246-221, Ptolemy IV *221-205, Ptolemy V *205-180, Ptolemy VI *180-164, *163-145, Ptolemy VIII *170-163, *145-116, Ptolemy VII *145, Cleopatra III and Ptolemy IX *116-107, Cleopatra III and Ptolemy X *107-88, Ptolemy IX *88-81, Cleopatra Berenice *81-80, Ptolemy XI *80, Ptolemy XII *80-58, *55-51, Berenice IV *58-55, Cleopatra VII *51-30, Ptolemy XIII *51-47, Ptolemy XIV *47-44, Ptolemy XV Caesarion *44-30.

Roman Emperors down to the last use of hieroglyphics: *30 B.C.-395 A.D.: I mention only the first two, Augustus *30 B.C.-14 A.D. and Tiberius *14-37, and the twelfth, Trajan *98-117.

Notes to the Chronology

1. Baines and Málek have a preceding period entitled Late Predynastic, in which they include Kings Zekhen and Narmer. This period they date as c. 3000. Presumably the Neugebauer-Parker chronology of the Early Dynastic Period would include Narmer as the first dynastic king, since Parker in his earlier chronology in the *Encyclopedia Americana* has the same dates as in his later chronology and the former does include Narmer as the first king. Also note that Baines and Málek insert the third dynasty as a part of the Early Dynastic Period, and they date that period as 2920-2575.

2. Parker in his chronology in the *Encyclopedia Americana* dates the various kings of this dynasty as follows: Menes (Narmer), Ity (Hor-Aha), dated together as 3110-3056; Iteti (Djer) 3055-3009; Interregnum 3008; Iti? (Djet), Zemti (Udimu) (Den), the last two kings dated together 3007-2975; Merpabia (Adjib) 2974-2917; Iryneter (Semerkhet) 2917-2909); and Qaa (Qaa Sen) 2908-2884.

3. Parker in his detailed chronology gives the following monarchs and their dates: Hetep (Hetepsekhemuy), Nubnefer (Reneb), these two kings are dated together 2883-2811; Nineter 2810-2766; Weneg 2766-2747; Sened 2747-2733; Peribsen 2733-2718; Neferkasokar 2718-2711; Khasekhem 2711-2691; and Khasekhemuy 2691-2665.

4. Since Baines and Málek include the third dynasty with the Early Period, they date the Old Kingdom with the beginning of the fourth dynasty,

suggesting the span 2575-2134.

5. Baines and Málek date the third dynasty between 2649 and 2575, with the individual reigns as follows: Zanakht 2649-2630, Djoser 2630-2611, Sekhemet 2611-2603, Khaba 2603-2599, and Huny 2599-2575. Parker in his detailed chronology gives the following kings and dates: Nebka (Sanakht) 2664, Djoser 2663-2645, Djoser Teti 2644-2639, and Huny 2638-2615.

6. Baines and Málek date the fourth dynasty as 2575-2465, with the individual monarchs dated as follows: Sneferu (written as Snofru) 2575-2551, Khufu 2551-2528, Redjedef 2528-2520, Chephren 2520-2494, Menkaure 2490-2472, and Shepseskaf 2472-2467.

7. Baines and Málek date the fifth dynasty 2465-2323, with the individual monarchs dated as follows: Weserkaf (written Userkaf) 2465-2458, Sahure 2458-2446, Neferirkare 2446-2426, Shepseskare 2426-2419, Reneferef 2419-2416, Niusserre (written Neussere) 2416-2392, Menkauhor 2396-2388, Djedkare Izezi 2388-2356, and Wenis 2356-2323.

8. Baines and Málek date the sixth dynasty 2323-2150, with the individual monarchs dated as follows: Teti 2323-2291, Pepy I 2289-2255, Merenre 2255-2246, and Pepy II 2246-2152.

9. Parker calls the period from Dynasty 21 through Dynasty 25 the Late Period, while Baines and Málek use that title for the time from Dynasty 25 through the second Persian period (=Dynasty 31). K. A. Kitchen uses the designation Third Intermediate Period as the title of his recent authoritative work, which I mentioned above in the introduction to this Chronology.

10. Kitchen has 1069-1043. Kitchen also gives alternate dates for this and the following reigns (see p.

406 of his work).

11. Kitchen gives 1043-1039.

12. Kitchen gives 1039-991.

13. Parker writes Osokhor and Kitchen Osochor.

14. Parker has the following monarchs and dates for Dynasty 22: Sheshonk (Sheshonq) 940-919, Osorkon I 919-883, Takelot I 883-860, Osorkon II 860-833, Sheshonk II 837, Takelot II 837-823, Shesonk III 823-772, Pami 772-767, Sheshonk IV 767-730. Kitchen (p. 588) has the following monarchs and dates (I follow his renditions of the names of the kings): Shoshenq I 945-924, Osorkon I 924-889, Shoshenq II c. 890, Takeloth I 889-874, Osorkon II 874-850, Harsiese c. 870-860, Takeloth II 850-825, Shoshenq III 825-773, Pimay 773-767, Shoshenq V 767-730, Osorkon IV 730-715?/713?.

15. Parker has the following: Pedibast c. 761-738, Sheshonk V, Osorkon III, Takelot III, Amenrud, Osorkon IV, these last five dated c. 738-715. Kitchen (p. 588) has: Pedubast I 818-793, Input I 804-803?, Shoshenq IV 793-787, Osorkon III 787-759, Takeloth III 764-757, Rudamun 757-754, Input II 754-720 (or 715), (Shoshenq VI 720-715, existence doubtful).

16. Parker has: Tefnakht 725-715, Bocchoris 715-710. Kitchen (p. 589) has: Tefnakht I 727-720 (or less likely 727-719), Bakenranef 720-715 (or less likely 719-713).

17. Parker has no split of the dynasty but simply gives: (Ethiopian) Piankhi [=Piye] 736-?710, Shabaka (?) 710-696, Shabataka 698-685, Taharka 690-664, Tanutamon 664-657. Kitchen (p. 589) first gives a Proto-Saite Dynasty: Ammeris, Nubian governor?, 715-695, Stephinates, Tefnakht II?, 695-688, Nekauba 688-672, Menkheperre Necho I 672-664. He then

follows this dynasty with the twenty-fifth (Nubian) dynasty: Alara c. 780-760, Kashta c. 760-747, Piankhy 747-716, Shabako 716-702, Shebitku 702-690, Taharqa 690-664, Tantamani 664-656. Kitchen also presents a less likely set of dates for this dynasty.

Abbreviations Used in Text and Bibliography

ASAE = *Annales du Service des Antiquités de l'Égypte.*

BIFAO = *Bulletin de l'Institut Français d'Archéologie Orientale.*

JEA = *The Journal of Egyptian Archaeology.*

JNES = *Journal of Near Eastern Studies.*

MDAIK = *Mitteilungen des Deutschen Archäologischen Instituts, Abteilung Kairo.*

PSBA = *Proceedings of the Society of Biblical Archaeology.*

Urkunden = G. Steindorff, ed., *Urkunden des ägyptischen Altertums* (For *Urkunden* I and IV, see the Bibliography, Sethe, K., *Urkunden des Alten Reichs* and *Urkunden der 18. Dynastie.* For *Urkunden* V, see Grapow, H., *Religiöse Urkunden*).

Wb, see the Bibliography, Erman, A., and H. Grapow, *Wörterbuch.*

ZÄS = *Zeitschrift für ägyptische Sprache und Altertumskunde.*

Bibliography

Akmar, E., *Le Papyrus Magique Harris transcrit et publié* (Upsala, 1916).

Aldred, C., *Egypt to the End of the Old Kingdom* (London, 1965).

Aldred, C., *Egyptian Art in the Days of the Pharaohs* (New York and Toronto, 1980).

Allen, T.G., *Horus in the Pyramid Texts* (Chicago, 1916).

Allen, T.G., *The Book of the Dead or Going Forth by Day. Ideas of the Ancient Egyptians Concerning the Hereafter as Expressed in Their Own Terms* (The Oriental Institute of the University of Chicago Studies in Ancient Oriental Civilization, No. 37) (Chicago, 1974).

Allen, T.G., *The Egyptian Book of the Dead. Documents in the Oriental Institute Museum at the University of Chicago* (The University of Chicago Oriental Institute Publications, Vol. 82) (Chicago, 1969).

Altenmüller, H., "Hu," in W. Helck and W. Westendorf, eds. *Lexikon der Ägyptologie*, Vol. 3 (Wiesbaden, 1980), cc. 65-68.

Altenmüller, H., "Toten-Literatur. 22. Jenseitsbücher, Jenseitsführer," *Handbuch der Orientalistik*, Abt. 1, Vol. 1, Part 2 (Leiden, 1970), pp. 69-81.

Andrae, W., see Schäfer, H. and.

Andrews, C., see Faulkner, R.O., *The Ancient Egyptian Book of the Dead*.

Anthes, R., "Egyptian Theology in the Third Millennium B.C.," *JNES*, Vol. 18 (1959), pp. 169-212.

Anthes, R., "Mythology in Ancient Egypt," in S.N. Kramer, ed., *Mythologies of the Ancient World* (New York, 1961), pp. 1-92.

Anthes, R., "The Original Meaning of m^{3c} hrw," *JNES*, Vol. 13 (1954), pp. 21-51.

Assmann, J., *Ägyptische Hymnen und Gebete* (Zurich and Munich, 1975).

Assmann, J., *Zeit und Ewigkeit im alten Ägypten* (Heidelberg, 1975).

Baedeker, K., *Egypt and the Sûdân* (Leipzig, 1929).

Baines, J. and J. Málek, *Atlas of Ancient Egypt*

(Oxford, 1980).

Barta, W., "Das Jahr in Datumsangaben und seine Bezeichnungen," *Festschrift Elmar Edel: 12 März 1979* (Bamberg, 1980), pp. 35-42.

Barta, W., "Die Chronologie der 1. bis 5. Dynastie nach den Angaben des rekonstruierten Annalenstein," *ZÄS*, Vol. 108 (1981), pp. 11-23.

Barucq, A., and F. Daumas, *Hymnes et prières de l'Égypte ancienne* (Paris, 1980).

Baumgartel, E.J., "Scorpion and Rosette and the Fragment of the Large Hierakonpolis Mace Head," *ZÄS*, Vol. 93 (1966), pp. 9-13.

Baumgartel, E.J., *The Cultures of Prehistoric Europe*, 2 vols. (London/New York/Toronto, 1955-60).

Bayoumi, A., *Autour du champ des souchets et du champ des offrandes* (Cairo, 1941).

Beckerath, J. von, "Die Lesung von 𓍹𓇳 'Regierungsjahr': Ein Vorschlag," *ZÄS*, Vol. 95, 2 (1969), pp. 88-91.

Beckerath, J. von, "*šmsj-Ḥrw* in der ägyptischen Vor- und Frühzeit," *MDAIK*, Vol. 14 (1956), pp. 1-10.

Beckerath, J. von, *Handbuch der ägyptischen Königsnamen* (Munich and Berlin, 1984), pp. 1-42.

Bell, B., "The Oldest Records of the Nile Floods," *The Geographical Journal*, Vol. 136 (1970), pp. 569-73.

Bissing, F.W. von, ed., *Das Re-Heiligtum des Königs Ne-Woser-Re*, Vol. 3: H. Kees, *Die grosse Festdarstellung* (Leipzig, 1928).

Bleeker, C.J., *Egyptian Festivals. Enactments of Religious Renewal* (Leiden, 1967).

Bonnet, H., *Reallexikon der ägyptischen Religionsgeschichte* (Berlin, 1952).

Borchardt, L., "Nilmesser und Nilstandsmarken," *Abhandlungen der Königlichen Preussischen Akademie der Wissenschaften*, 1906 (*Abhandlungen nicht zur Akademie gehöriger Gelehrter, Phil.-hist. Abh.*, I, pp. 1-55).

Borchardt, L., *Die Annalen und die zeitliche Festlegung des Alten Reiches der ägyptischen Geschichte* (Berlin, 1917).

Borchardt, L., *Denkmäler des alten Reiches (Ausser den Statuen) (Catalogue général des antiquités égyptiennes du Musée du Caire: Nos. 1295-1808)*, Part I (Berlin, 1937), Part II (Cairo, 1964).

Borchardt, L., *Das Grabdenkmal des Königs Saʾḥu-Reᶜ*, Vol. 2 (Leipzig, 1913).

Borghouts, J.F., *Ancient Magical Texts* (Leiden, 1978).

Boylan, P., *Thoth: the Hermes of Egypt* (Oxford, 1922).

Breasted, J.H., "The Philosophy of a Memphite Priest," *ZÄS*, Vol. 39 (1901), pp. 39-54.

Breasted, J.H., "The Predynastic Union of Egypt," *BIFAO*, Vol. 30 (1931), pp. 709-14.

Breasted, J.H., *Ancient Records of Egypt*, 5 vols. (Chicago, 1906).

Breasted, J.H., *The Dawn of Conscience* (New York, 1933).

Breasted, J.H., *Development of Religion and Thought in Ancient Egypt* (New York, 1912; Harper Torchbook, 1959).

Breasted, J.H., *A History of Egypt*, 2nd ed. (New York, 1912).

Brugsch, H., "Bau und Maasse des Tempels von Edfu," *ZÄS*, Vol. 9 (1871), pp. 32-45.

Brugsch, H., *Religion und Mythologie der alten Ägypter* (Leipzig, 1891).

Brugsch, H., *Thesaurus inscriptionum*

aegyptiacarum (Leipzig, 1883-84; reprint, Graz, 1968).

Brunner, H., *Die Lehre des Cheti, Sohnes des Duauf (Ägyptologische Forschungen*, 13) (Glückstadt and Hamburg, 1944).

Brunner, H., *An Outline of Middle Egyptian Grammar for Use in Academic Instruction* (Graz, 1970).

Brunner-Traut, E., *Altägyptische Märchen* (Düsseldorf and Cologne, 1963).

Buck, A. de, *The Egyptian Coffin Texts*, 7 vols. (Chicago, 1935-61).

Buck, A. de, *Egyptian Reading Book*, Vol. 1 (Leiden, 1948).

Buck, A. de, *De Egyptische Voorstellingen betreffende den Oerheuvel* (Leiden, 1922).

Budge, E.A.W., *The Book of the Dead. An English Translation of the Chapters, Hymns, etc. of the Theban Recension*, 2nd ed., revised and enlarged, 3 vols. (London, 1909).

Budge, E.A.W., *The Chapters of Coming Forth by Day or the Theban Version of the Book of the Dead. The Egyptian Hieroglyphic Text Edited from Numerous Papyri*, 3 vols. (London, 1910).

Budge, E.A.W., *An Egyptian Hieroglyphic Dictionary*, 2 vols. (London, 1920).

Budge, E.A.W., *Facsimiles of Egyptian Hieratic Papyri in the British Museum, With Descriptions, Translations, etc.* (London, 1910).

Budge, E.A.W., *The Mummy* (New York, 1972), Collier Books ed.

Butzer, K.W., "Die Naturlandschaft Ägyptens während der Vorgeschichte und den dynastischen Zeitalter," *Akademie der Wissenschaften und der Literatur, Mainz, Math.-Naturwiss. Kl., Abhandlung* No. 2 (1959).

Caminos, R.A., *Late-Egyptian Miscellanies* (London, 1954).

Capart, J., and M. Werbrouck, *Memphis à l'ombre des pyramides* (Brussels, 1930).

Carter, H., *The Tomb of Tut-ankh-Amen*, Vol. 2 (London, 1927).

Cenival, J. L. de, "Un nouveau fragment de la Pierre de Palerme," *Bulletin de la Société Française d'Égyptologie*, Vol. 44 (1965), pp. 13-17.

Černý, J., *Ancient Egyptian Religion* (London, 1952, repr. 1957).

Chabas, F., *Le Papyrus Magique Harris* (Chalons-sur-Saône, 1860).

Chabas, F., *Mélanges égyptologiques*, 3rd Series, Vol. 2 (Chalons-sur-Saône and Paris, 1873), pp. 242-78.

Chace, A.B. *The Rhind Mathematical Papyrus*, 2 vols. (Oberlin, Ohio), 1927-29.

Champollion, J.F., *Lettres écrits d'Égypte et de Nubie en 1828 et 1829*, new ed. (Paris, 1868).

Champollion, J.F., *Monuments de l'Égypte et de la Nubie*, Vol. 1 (Paris, 1835).

Chassinat, E., *Le temple d'Edfou*, Vols. 2 and 3 (Cairo, 1918, 1928).

Clark, R.T.R., *Myth and Symbol in Ancient Egypt* (London, 1959; paperback ed., 1978).

Daressy, G., "La Pierre de Palerme et la chronologie de l'Ancien Empire," *BIFAO*, Vol. 12 (1916), pp. 161-214.

Daumas, F., see A. Barucq and.

David, R., *A Guide to Religious Ritual at Abydos* (Warminster, 1981).

Davies, N. de G., see Gardiner, A.H., and.

Davies, N. de G., *The Rock Tombs of El Amarna*, Part 6 (London, 1908).

BIBLIOGRAPHY

Davies, N. de G., *The Temple of Hibis in El-Kargeh Oasis*, Part III, *The Decoration* (New York, 1952).

Derchain, P., *Le Papyrus Salt 825 (B.M. 10051), rituel pour la conservation de la vie en Égypte, Académie Royale de Belgique, Classe des Lettres, Mémoires*, Vol. 38 (Brussels, 1965).

Desroches-Nobelcourt, C., *Life and Death of a Pharaoh: Tutankhamen* (New York, 1963).

Drioton, E., "La religion égyptienne," in M. Brillant and R. Aigrain, eds., *Histoire des religions* (Paris, 1955), pp. 13-141.

Dümichen, J., *Altägyptische Tempelinschriften in den Jahren 1863-65*, Vol. 1 (Leipzig, 1867).

Edel, E., "Zur Lesung von ⌅ 'Regierungsjahr'," *JNES*, Vol. 8 (1949), pp. 35-39.

Edgerton, W.F., "Critical Note," *The American Journal of Semitic Languages and Literature*, Vol. 53 (1937), pp. 187-97.

Edwards, I.E.S., *British Museum, Hieroglyphic Texts from Egyptian Stelae*, Part VIII (London, 1939).

Emery, W.B., *Archaic Egypt* (Harmondsworth, England, 1961).

Emery, W.B., *Hor-aha* (Cairo, 1963)

Erman, A., "Ein Denkmal memphitischer Theologie," *Sitzungsberichte der Königlichen Akademie der Wissenschaften, Jahrgang 1911* (Berlin, 1911), pp. 916-50.

Erman, A., *The Ancient Egyptians: A Sourcebook of Their Writings*, tr. by A.M. Blackman; intro. by W.K. Simpson, Torchbook ed. (New York, 1966). Original title: *The Literature of the Ancient Egyptians* (London, 1927).

Erman, A., *A Handbook of Egyptian Religion* (London, 1907).

Erman, A., *Life in Ancient Egypt* (London, 1894)

Erman, A., *Die Märchen des Papyrus Westcar (Mittheilungen aus den orientalischen Sammlungen*, Hefte 5-6) (Berlin, 1890).

Erman, A., and H. Grapow, *Wörterbuch der ägyptischen Sprache*, 7 vols. (Leipzig, 1926-53); *Die Belegstellen*, 5 vols. (Leipzig, 1953). The whole work was reprinted in Berlin in 1971. It is abbreviated as *Wb*.

Faulkner, R.O., "The Bremner-Rhind Papyrus--III (and--IV)," *JEA*, Vol. 23 (1937), pp. 166-85; Vol. 24 (1938), pp. 41-53.

Faulkner, R.O., "The King and the Star-Religion in the Pyramid Texts," *JNES*, Vol. 25 (1966), pp. 153-61.

Faulkner, R.O., *The Ancient Egyptian Book of the Dead, Translated [into English]*, Revised edition by C. Andrews (London, 1985). Published earlier by the Limited Editions Club under the title *The Book of the Dead. A Collection of Spells* (New York, 1972).

Faulkner, R.O., *The Ancient Egyptian Coffin Texts*, 3 vols. (Warminster, 1973-78).

Faulkner, R.O., *The Ancient Egyptian Pyramid Texts: Translated into English* (Oxford, 1969), with a *Supplement of Hieroglyphic Texts* (Oxford, 1969).

Faulkner, R.O., *A Concise Dictionary of Middle Egyptian* (Oxford, 1962).

Faulkner, R.O., *The Papyrus Bremner-Rhind (British Museum No. 10188): Bibliotheca Aegyptiaca* III (Brussels, 1933).

Fazzini, A., *Images for Eternity. Egyptian Art from Berkeley and Brooklyn* (New York, 1975).

Fischer, C.S., see Reisner, G.A., and.

Fischer, H.G., see Terrace, E.L.B., and.

Frankfort, H., *Ancient Egyptian Religion: An Interpretation* (New York, 1948).

Frankfort, H., *Kingship and the Gods* (Chicago, 1948).

Gardiner, A.H., "Horus the Beḥdetite," *JEA*, Vol. 30 (1944), pp. 23-60.

Gardiner, A.H., "The House of Life," *JEA*, Vol. 24 (1938), pp. 157-79.

Gardiner, A.H., "Hymns to Amon from a Leiden Papyrus," *ZÄS*, Vol. 42 (1905), pp. 12-42.

Gardiner, A.H., "Magic (Egyptian)," in J. Hastings, ed., *Encyclopedia of Religion and Ethics*, Vol. 8 (1915), pp. 262-69.

Gardiner, A.H., "The Mansion of Life and the Master of the King's Largesse," *JEA*, Vol. 24 (1938), pp. 83-91.

Gardiner, A.H., "Professional Magicians in Ancient Egypt," *PSBA*, Vol. 39 (1917), pp. 31-44, 138-40.

Gardiner, A.H., "Regnal Years and Civil Calendar in Pharaonic Egypt," *JEA*, Vol. 31 (1945), pp. 11-28.

Gardiner, A.H., "Some Personifications. I. *ḤỊKE*, The God of Magic," *PSBA*, Vol. 37 (1915), pp. 253-62.

Gardiner, A.H., "Some Personifications. II. *ḤU*, 'Authoritative Utterance.' *SIA*', 'Understanding,' " *PSBA*, Vol. 38 (1916), pp. 43-54, 83-94.

Gardiner, A.H., *Ancient Egyptian Onomastica*, 3 vols. (Oxford, 1947).

Gardiner, A.H., *Egypt of the Pharaohs* (Oxford, 1961).

Gardiner, A.H., *Egyptian Grammar*, 3rd ed. (London, 1957).

Gardiner, A.H., *Egyptian Hieratic Texts: Series I. Literary Texts of the New Kingdom. Part I. The Papyrus Anastasi and the Papyrus Koller* (Leipzig, 1911).

Gardiner, A.H., *Hieratic Papyri in the British Museum. Third Series. Chester Beatty Gift*, 2 vols.

(London, 1935).

Gardiner, A.H., *Late-Egyptian Miscellanies* (Brussels, 1937).

Gardiner, A.H., *The Royal Canon of Turin* (Oxford, 1959).

Gardiner, A.H., and N. de G. Davies, *The Tomb of Amenemhet* (London, 1915).

Garstang, J., *Tombs of the Third Egyptian Dynasty* (Westminster, 1904).

Gauthier, H., "Quatre nouveaux fragments de la Pierre de Palerme," *Le Musée Égyptien*, Vol. 3 (1915), pp. 29-53.

Gauthier, H., *Le Livre des rois d'Égypte*, 5 vols. (Cairo, 1907-17).

Ghalioungui, P., see Habachi, L, and.

Ghalioungui, P., *The House of Life: Per Ankh. Magic and Medical Science in Ancient Egypt* (Amsterdam, 1963; 2nd ed., 1973).

Ghalioungui, P., *The Physicians of Pharaonic Egypt* (Cairo, 1983).

Giustolisi, V., "La 'Pietra di Palermo' e la cronologia dell' Antico Regno," *Sicilia archeologica*, Anno I, Nr. 4, Dec., 1968, pp. 5-14; Nr. 5, March, 1969, pp. 38-55; Anno II, Nr. 6, June, 1969, pp. 21-38.

Gödecken, K.B., *Eine Betrachtung der Inschriften des Meten im Rahmen der sozialen und rechtlichen Stellung von Privatleuten im ägyptischen Alten Reich* (Wiesbaden, 1976).

Godron, G., "Quel est le lieu de provenance de la 'Pierre de Palerme'?" *Chronique d'Égypte*, Vol. 27 (1952), pp. 17-22.

Goedicke, H., "Diplomatic Studies in the Old Kingdom," *Journal of the American Research Center in Egypt*, Vol. 3 (1964), pp. 31-41.

BIBLIOGRAPHY

Goedicke, H., "Die Laufbahn des Mṭn," *MDAIK*, Vol. 21 (1966), pp. 1-71.

Golénisheff, W., ed., *Papyrus hiératiques (Catalogue général des antiquités du Musée du Caire: Nos. 58001-58036)*, Vol. 83 (Cairo, 1927).

Goyon, J.-C., see Parker, R.A.

Grapow, H., "Die Welt vor der Schöpfung," *ZÄS*, Vol. 67 (1931), pp. 34-38.

Grapow, H., *Grundriss der Medizin der alten Ägypter*, Vol. 1: *Anatomie und Physiologie* (Berlin, 1934).

Grapow, H., *Religiöse Urkunden (Urkunden des ägyptischen Altertums*, Abt. V [=*Urkunden V*]), 3 Heften (Leipzig, 1915-17).

Grapow, H., and W. Westendorf, "37. Wörterbücher, Repertorien, Schülerhandschriften," *Handbuch der Orientalistik*, Abt. 1, Vol. 1, Part 2 (Leiden, 1970), p. 221.

Grapow, H., see Erman A. and.

Grdseloff, B., "Notes sur deux monuments inédits de l'ancien empire," *ASAE*, Vol. 42 (1943), pp. 107-25.

Grébaut, E., *Hymne à Ammon-Ra* (Paris, 1874).

Griffith, F.Ll., "Notes on Egyptian Weights and Measures," *PSBA*, Vol. 14 (1891-92), pp. 403-50.

Griffith, F.Ll., and W.M.F. Petrie, *Two Hieroglyphic Papyri from Tanis* (London, 1889).

Griffith, F. Ll., and H. Thompson, *The Demotic Magical Papyrus of London and Leiden* (London, 1904).

Griffiths, J.G., see Plutarch.

Griffiths, J.G., *The Origins of Osiris and His Cult*, 2nd ed. (Leiden, 1980).

Habachi, L., "Khatâᶜna-Qantir: Importance," *ASAE*, Vol. 52 (1954), p. 450 et seq.

Habachi, L., and P. Ghalioungui, "Notes on Nine

Physicians of Pharaonic Egypt," *Bulletin de l'Institut d'Égypte*, Vol. 51 (1969), pp. 15-24.

Habachi, L., and P. Ghalioungui, "The 'House of Life' of Bubastis," *Chronique d'Égypte*, Vol. 46 (1971), pp. 59-71.

Hall, H.R., Review of A. de Buck's *De Egyptische Voorstellingen*, in *JEA*, Vol. 10 (1924), pp. 185-87.

Harris, J.R., *Lexicographical Studies in Ancient Egyptian Minerals* (Berlin, 1961).

Harris, J.R., see Lucas, A., and.

Hassan, S., *Hymnes religieux du moyen empire* (Cairo, 1928).

Hayes, W.C., *The Scepter of Egypt: A Background for the Study of the Egyptian Antiquities of the Metropolitan Museum of Art*, 2 vols. (New York, 1953-59).

Helck, W., "Bemerkungen zum Annalenstein," *MDAIK*, Vol. 30 (1974), pp. 31-35.

Helck, W., "Nilhöhe und Jubiläumsfest," *ZÄS*, Vol. 93 (1966), pp. 74-79.

Helck, W., *Die Lehre des Dwȝ-Ḥtjj* (Wiesbaden, 1970).

Helck, W., *Untersuchungen zu den Beamtentiteln des ägyptischen Alten Reiches* (Glückstadt / Hamburg / New York, 1954).

Helck, W., *Untersuchungen zu Manetho und den ägyptischen Königslisten* (Berlin, 1956).

Helck, W., see Sethe, K., *Urkunden der 18. Dynastie*.

Holmberg, M.S., *The God Ptah* (Lund and Copenhagen, 1946).

Hornung, E., "Chaotische Bereiche in der geordneten Welt," *ZÄS*, Vol. 81 (1956), pp. 28-32.

Hornung, E., *Der ägyptische Mythos von der*

Himmelskuh: Ein Ätiologie des Unvollkommenen (Göttingen, 1982).

Hornung, E., *Ägyptische Unterweltsbücher* (Munich, 1972, 2nd ed., 1984).

Hornung, E., *Das Amduat: Die Schrift des Verborgenen Raumes* (*Ägyptologische Abhandlungen*, Vols. 7, 13), 3 parts (Wiesbaden, 1963, 1967).

Hornung, E., *Das Buch der Anbetung des Re im Westen (Sonnenlitanei)* (*Aegyptiaca Helvetica*, Vol. 3), 2 parts (Geneva, 1975-76).

Hornung, E., *Conceptions of God in Ancient Egypt. The One and the Many* (Ithaca, 1982).

Hornung, E., with the assistance of A. Brodbeck and E. Staehelin, *Das Buch von den Pforten des Jenseits* (*Aegyptiaca Helvetica*, Vol. 7), 2 parts (Basel and Geneva, 1979-84).

Hornung, E. and C. Seeber, *Studien zum Sedfest* (Geneva, 1974).

Hurry, J.B., *Imhotep, the Vizier and Physician of King Zoser and Afterwards the Egyptian God of Medicine* (Oxford, 1926).

Iversen, E., *Papyrus Carlsberg Nr. VII: Fragments of a Hieroglyphic Dictionary* (Copenhagen, 1958).

James, T.G.H., *Pharaoh's People. Scenes from Life in Imperial Egypt* (Chicago and London, 1984).

Jenkins, N., *The Boat beneath the Pyramid: King Cheops' Royal Ship* (New York, 1980).

Jéquier, G., *Considérations sur les religions égyptiennes* (Neuchâtel, 1946).

Jéquier, G., *Le monument funéraire de Pepi II* (Cairo, 1936-).

Jéquier, G., *La pyramide d'Oudjebten; Les pyramides des reines Neit et Apouit; La pyramide d'Aba* (Cairo, 1928-35).

Jonckheere, F., *Les médecins de l'Égypte pharaonique. Essai de prosopographie* (Brussels, 1958).

Junge, F., "Zur Fehldatierung des sog. Denkmals memphitischer Theologie oder Der Beitrag der ägyptischen Theologie zur Geistesgeschichte der Spätzeit," *MDAIK*, Vol. 29 (1973), pp. 195-204.

Junker, H., *Die Götterlehre von Memphis (Schabaka-Inschrift) (Aus den Abhandlungen der Preussischen Akademie der Wissenschaften, Jahrgang 1939, Phil.-hist. Klasse*, Nr. 23) (Berlin, 1940).

Kaiser, W., "Einige Bemerkungen zur ägyptischen Frühzeit. II. Zur Frage einer über Menes hinausreichenden ägyptischen Geschichtsüberlieferung," *ZÄS*, Vol. 86 (1961), pp. 39-61.

Kaplony, P., "Gottespalast und Götterfestungen in der ägyptischen Frühzeit," *ZÄS*, Vol. 88 (1962), pp. 5-16.

Kaplony, P., *Die Inschriften der ägyptischen Frühzeit*, Vols. 1-3 (Wiesbaden, 1963); *Supplementband* (Wiesbaden, 1964).

Kaplony, P., *Kleine Beiträge zu den Inschriften der ägyptischen Frühzeit* (Wiesbaden, 1966).

Kees, H., "Archaisches [hieroglyphs] [tt-ʾtt] 'Erzieher'?," *ZÄS*, Vol. 82 (1957), pp. 58-62.

Kees, H., "Die Feuerinsel in Sargtexten und im Totenbuch," *ZÄS*, Vol. 78 (1942), pp. 41-63.

Kees, H., "Toten-Literatur. 20. Pyramidentexte," *Handbuch der Orientalistik*, Abt. 1, Vol. 1, Part 2 (Leiden, 1970), pp. 52-60; "21. Sargtexte und Totenbuch," *Ibid.*, pp. 61-69.

Kees, H., *Ancient Egypt* (Chicago and London, 1961; Phoenix ed., 1977).

Kees, H., *Der Götterglaube im alten Ägypten*, 2nd ed. (Berlin, 1956).

Kees, H., *Totenglauben und Jenseitsvorstellungen*

der alten Ägypter (Leipzig, 1956; 4th print., 1980).

Kees, H., see Bissing, F.W. von.

Kitchen, K.A., *Pharaoh Triumphant, The Life and Times of Ramesses II* (Warminster, England, 1982).

Kitchen, K.A., *The Third Intermediate Period in Egypt (1100-650 B.C.)* (Warminster, 1973)

Köhler, U., *Das Imiut*, 2 vols. (Wiesbaden, 1975).

Lacau, P., "Textes religieux égyptiens," *Receuil de travaux relatifs à la philologie et à l'archéologie égyptiennes et assyriennes*, Vols. 26-37 (1904-15), partially reprinted in *Textes religieux égyptiens* (Paris, 1910).

Lacau, P., and J. P. Lauer, *Fouilles à Saqqara; la pyramide à degrés*, Vol. 4: *Inscriptions gravées sur les vases* (Cairo, 1959-61).

Lange, H.O, *Der magische Papyrus Harris* (Copenhagen, 1927).

Lanzone, R.V., *Dizionario di mitologia egizia*, 3 vols. (Turin, 1881-86; reprint, Amsterdam, 1974, with a fourth previously unpublished volume, 1975).

Lauer, J.P., *Saqqara: The Royal Cemetery of Memphis. Excavations and Discoveries since 1850* (London, 1976).

Leclant, J., "Recherches récentes sur les Textes des Pyramides et les pyramides à textes de Saqqarah," *Académie Royale de Belgiques: Bulletin de la Classe des Lettres et des Sciences Morales et Politiques*, 5e série, Tome LXXI (1985, 10-11), pp. 292-305

Leclant, J., see Parker, R.A.

Lefebvre, G., "L'oeuf divin d'Hermopolis," *ASAE*, Vol. 23 (1923), pp. 65-67.

Lefebvre, G., *Tableau des parties du corps humain mentionnées par les égyptiens (Supplément aux Annales du Service des Antiquités, 17)* (Cairo, 1952).

Lefebvre, G., *Le tombeau de Petosiris* (Cairo, 1923-24).

Lepsius, C.R., *Denkmäler aus Ägypten und Äthiopien*, Abt. I-VI (Berlin 1849-58; photographic reprint, Geneva, 1972); Text, 5 vols., ed. E. Naville et al. (Leipzig, 1897-1913; phot. repr., Geneva, 1975).

Lepsius, C.R., *Das Todtenbuch der Ägypter nach dem hieroglyphischen Papyrus in Turin* (Turin and Leipzig, 1842; repr. Osnabrück, 1969).

Lepsius, C.R., *Über die Götter der vier Elemente bei den Ägyptern* (Berlin, 1856).

Lesko, L.H., *The Ancient Egyptian Book of Two Ways* (Berkeley/Los Angeles/London, 1972).

Lexa, F., *La magie dans l'Egypte antique de l'Ancien Empire jusqu'à l'Époque Copte*, Vol. 2 (Paris, 1925).

Lichtheim, M., *Ancient Egyptian Literature*, 3 vols. (Berkeley/Los Angeles/London, 1975-80).

Lucas, A., and J.R. Harris, *Ancient Egyptian Materials and Industries* (London, 1962).

Lythgoe, A.M., "Excavations at the South Pyramid at Lisht in 1914," *Ancient Egypt*, 1915, pp. 145-53.

Málek, J., see Baines, J. and.

Manetho, see Wadell, W.G.

Mariette, A., *Dendérah*, 5 vols. (Paris, 1870-80).

Mariette, A., *Les mastabas de l'Ancien Empire*, ed. by G. Maspero (Paris, 1882-89).

Mariette, A., *Les papyrus égyptiens du Musée de Boulaq*, Vol. 2 (Paris, 1872).

Maspero, G., "La carrière administrative de deux hauts fonctionnaires égyptiens," *Études égyptiennes*, Vol. 2 (1890), pp. 113-272.

Maspero, G., "De quelques termes d'architecture

égyptienne," *PSBA*, Vol. 11 (1889), pp. 304-17.

Maystre, C., *Le livre de la vache du ciel dans les tombeaux de la Valle des Rois, BIFAO*, Vol. 40 (Cairo, 1940).

Maystre, C., and A. Piankoff, *Le livre des portes*, 3 vols. (Cairo, 1939-62).

Mercer, S.A.B., *Horus, Royal God of Egypt* (London, 1942).

Meyer, E., "Gottesstaat, Militärherrschaft und Ständewesen in Ägypten," *Sitzungsberichte der Preussischen Akademie der Wissenschaften. Jahrgang 1928. Phil.-hist. Klasse* (Berlin 1928), pp. 495-532.

Meyer, E., *Ägyptische Chronologie* (Berlin, 1904).

Montet, P., *Everyday Life in the Days of Ramesses the Great* (Philadelphia, 1981).

Morenz, S., *Egyptian Religion* (Ithaca, 1973).

Moret, A., "La Légende d'Osiris à l'époque thébaine d'après l'hymne à Osiris du Louvre," *BIFAO*, Vol. 30 (1931), pp. 725-50, and plates I-III.

Moret, A., *Du charactère religieux de la royauté pharaonique (Annales du Musée Guimet: Bibliothèque d'études*, Vol. 15) (Paris, 1902).

Moret, A., *Le rituel du cult divin journalier en Égypte* (Paris, 1902).

Moss, R.L.B., see Porter, B., and.

Müller, H., *Die formale Entwicklung der Titulatur der ägyptischen Könige* (Glückstadt/Hamburg/New York, 1938).

Murnane, W.J., *United with Eternity. A Concise Guide to the Monuments of Medinet Habu* (Chicago and Cairo, 1980).

Murray, M.A., *Index of Names and Titles of the Old Kingdom* (London, 1908).

Naville, E., "The Litany of Ra," in S. Birch, ed.,

Records of the Past: Being English Translations of the Assyrian and Egyptian Monuments, Vol. 8 (London, 1876), pp. 103-28.

Naville, E., "La Pierre de Palerme," *Recueil de travaux relatifs à la philologie et à l'archéologie égyptiennes et assyriennes*, 25me année (1903), pp. 64-81.

Naville, E., *Das ägyptische Todtenbuch der XVII. bis XX. Dynastie aus verschiedenen Urkunden zusammengestellt und herausgegeben*, 3 vols. (Berlin, 1886).

Naville, E., *La Litanie du Soleil: Inscriptions recueillies dans les tombeaux des rois à Thèbes* (Leipzig, 1875).

Neugebauer, O., and R.A. Parker, *Egyptian Astronomical Texts*, 3 vols. (Providence, Rhode Island, and London, 1960-69).

Newberry, P.E., "The Set Rebellion of the IInd Dynasty," *Ancient Egypt*, 1922, pp. 40-46.

Newberry, P.E., and G.A. Wainwright, "King Udy-mu (Den) and the Palermo Stone," *Ancient Egypt*, 1914, pp. 148-55.

Nims, C.F., *Thebes of the Pharaohs, Pattern for Every City* (London, 1965).

O'Mara, P.F., *The Chronology of the Palermo and Turin Canons* (La Canada, Calif., 1980).

O'Mara, P.F., *The Palermo Stone and the Archaic Kings of Egypt* (La Canada, Calif., 1979).

Otto, E., *Das ägyptische Mundöffnungsritual*, 2 vols. (Wiesbaden, 1960).

Otto, E., *Egyptian Art and the Cults of Osiris and Amon* (London, 1968).

Parker, R.A.. "Egyptian Chronology," in "Egypt," *Encyclopedia Americana*, Vol. 10, pp. 32-33.

Parker, R.A., J. Leclant, and J.-C. Goyon, *The Edifice of Taharqa by the Sacred Lake of Karnak (Brown Egyptological Studies,* VII) (Providence, Rhode Island, and London, 1979).

Parker, R.A., see Neugebauer, O., and.

Patrick, R., *All Color Book of Egyptian Mythology* (London, 1972).

Peet, T.E., Review of Borchardt's *Die Annalen* in *JEA*, Vol. 6 (1920), pp. 149-54.

Pellegrini, A., "Nota sopra un'iscrizione egizia del Museo di Palermo," *Archivio storico siciliano*, n. s., Anno xx (1895), pp. 297-316.

Petrie, W.M.F., "The Earliest Inscriptions," *Ancient Egypt*, 1914, pp. 61-77.

Petrie, W.M.F., "New Portions of the Annals," *Ancient Egypt*, 1916, pp. 114-20.

Petrie, W.M.F., *History of Egypt*, Vol. 1 (XIth ed. rev., London, 1924).

Petrie, W.M.F., *The Making of Egypt* (London and New York, 1939).

Petrie, W.M.F., *The Royal Tombs of the Earliest Dynasties*, Part II (London, 1901).

Petrie, W.M.F., *The Royal Tombs of the First Dynasty*, Part I (London, 1900).

Petrie, W.M.F., see Griffith, F.Ll., and.

Piankoff, A., *The Litany of Re (Bollingen Series*, Vol. 40.4) (New York, 1964).

Piankoff, A., *Le livre des quererts* (Cairo, 1946); published earlier in *BIFAO*, Vols. 41-45 (1942-47).

Piankoff, A., *Mythological Papyri, Edited with a Chapter on the Symbolism of the Papyri by N. Rambova (Bollingen Series*, Vol. 40.3) (New York, 1957).

Piankoff, A., *The Pyramid of Unas (Bollingen Series*, Vol. 40.5) (Princeton, 1968).

Piankoff, A., *The Shrines of Tut-Ankh-Amon* (*Bollingen Series*, Vol. 40.2) (Princeton, 1955).

Piankoff, A., *The Tomb of Ramesses VI*, ed. by N. Rambova (*Bollingen Series*, Vol. 40.1) (New York, 1954).

Piankoff, A., see Maystre, C., and.

Pirenne, J., *Histoire de la civilisation de l'Égypte ancienne* (Neuchâtel, 1961).

Plutarch, *De Iside et Osiride*, ed., transl., and comm. by J. G. Griffiths (Cardiff, 1970).

Porter, B., and R.L.B. Moss, *Topographical Bibliography of Ancient Egyptian Hieroglyphic Texts, Reliefs, and Paintings*, Vols. 1-3, 2nd ed. (Oxford, 1960-78).

Posener, G., *De la divinité du pharaon* (Paris, 1960).

Posener, G., *Ostraca hiératiques littéraires de Deir el Medineh*, Vol. 2 (Cairo, 1951).

Posener, G., et al., *Dictionnaire de la civilisation égyptienne*, 2nd ed. (Paris, 1970).

Pritchard, J.B., ed., *Ancient Near Eastern Texts Relating to the Old Testament* (Princeton, 1950), with the translations of many Egyptian texts by J.A. Wilson.

Quibell, J.E., *Hierakonpolis*, Part 1, with Notes by W.M.F. Petrie (London, 1900).

Quibell, J.E., and F.W. Green, *Hierakonpolis*, Part 2 (London, 1902).

Ranke, H., *Die ägyptischen Personennamen*, Vol. 2 (Glückstadt/Hamburg/New York, 1952).

Read, F.W., "Nouvelles remarques sur la Pierre de Palerme," *BIFAO*, Vol. 12 (1916), pp. 215-22.

Redford, D.B., *Pharaonic King-lists, Annals and Day-Books: A Contribution to the Study of the Egyptian Sense of History* (Mississauga, Ontario, Canada, 1986).

Reeves, C.N., "A Fragment of Fifth Dynasty Annals at University College London," *Göttinger Miszellen. Beiträge zur ägyptologischen Diskussion*, Heft 32 (1979), pp. 47-50.

Reisner, G.A., "A Scribe's Tablet Found by the Hearst Expedition at Giza," *ZÄS*, Vol. 48 (1911), pp. 113-14.

Reisner, G.A., *A History of Giza Necropolis*, Vol. 1 (Cambridge, Mass., and London, 1942).

Reisner, G.A., and C.S. Fischer, "Preliminary Report on the Work of the Harvard-Boston Expedition in 1911-13," *ASAE*, Vol. 13 (1914), pp. 227-52.

Roeder, G., "Die Kosmogonie von Hermopolis," *Egyptian Religion*, Vol. 1 (1933), pp. 1-37.

Roeder, G., "Zwei hieroglyphische Inschriften aus Hermopolis (Ober-Ägypten)," *ASAE*, Vol. 52 (1954), pp. 315-74.

Roeder, G., *Urkunden zur Religion des alten Ägypten* (Jena, 1923).

Sandman, M., *Texts from the Time of Akhenaten (Bibliotheca Aegyptiaca* VIII) (Brussels, 1938).

Sandman Holmberg, M., see Holmberg, M.S.

Sauneron, S., *Esna*, Vol. 3: *Le temple d'Esna* (Cairo, 1968); Vol. 5: *Les fêtes religieuses d'Esna aux derniers siècles du Paganisme* (Cairo, 1962).

Sauneron, S., *Le Papyrus magique illustré de Brooklyn* (New York, 1970).

Sauneron, S., *Les prêtres de l'ancienne Égypte* (Paris, 1957).

Sauneron, S., and J. Yoyotte, "La naissance du monde selon l'Égypte ancienne," *Sources orientales:* Vol. 1, *La naissance du monde* (Paris, 1959), pp. 19-91.

Schack-Schackenburg, H., *Das Buch von den Zwei Wegen des seligen Toten* (Leipzig, 1903).

Schäfer, H., "Ein Bruchstück altägyptischer Annalen," *Abhandlungen der Königlichen Preussischen Akademie der Wissenschaften, 1902. Phil.-hist. Abh.* (Berlin, 1902), pp. 3-41.

Schäfer, H., *Ägyptische Inschriften aus den Königlichen Museen zu Berlin*, Vol. 1 (Leipzig, 1913).

Schäfer, H., and W. Andrae, *Die Kunst des alten Orients* (Berlin, 1935).

Schlögel, H.A., *Der Gott Tatenen* (Freiburg, Switz., and Göttingen, 1980).

Schweitzer, U., *Das Wesen des Ka im Diesseits und Jenseits der alten Ägypter (Ägyptologische Forschungen*, Heft 19) (Glückstadt/Hamburg/New York, 1956).

Seeber, C., see Hornung, E., and.

Seele, K.C., see Steindorff, G., and.

Sethe, K., "Hitherto Unnoticed Evidence Regarding Copper Works of Art of the Oldest Period of Egyptian History," *JEA*, Vol. 1 (1914), pp. 233-36.

Sethe, K., *Ägyptische Lesestücke* (Leipzig, 1924, 2nd ed., 1928).

Sethe, K., *Die altägyptischen Pyramidentexte*, 4 vols. (Leipzig, 1908-22).

Sethe, K., *Amun und die acht Urgötter von Hermopolis: Eine Untersuchung über Ursprung und Wesen des ägyptischen Götterkönigs* (Berlin, 1929).

Sethe, K., *Dramatische Texte zu altägyptischen Mysterienspielen (Untersuchungen zur Geschichte und Altertumskunde Ägyptens*, Vol. 10) (Leipzig, 1928; repr. Hildesheim, 1964).

Sethe, K., *Erläuterungen zu den ägyptischen Lesestücken* (Leipzig, 1927).

Sethe, K., *Übersetzung und Kommentar zu den altägyptischen Pyramidentexten*, 6 vols. (Hamburg,

1962).

Sethe, K., *Untersuchungen zur Geschichte und Altertumskunde Ägyptens*, Vol. 3: *Beiträge zur ältesten Geschichte Ägyptens* (Leipzig, 1905).

Sethe, K., *Urkunden der 18. Dynastie (Urkunden des ägyptischen Altertums, IV,* [=*Urkunden* IV]), 4 vols. (Leipzig, 1906-09, 2nd ed., 1927-30); continued by W. Helck, Vols. 5-6 (Berlin, 1955-58); paginated continuously.

Sethe, K., *Urkunden des Alten Reichs*, Vol. 1, 2nd ed. (Leipzig, 1933) (=*Urkunden* I).

Simpson, W.K., *The Literature of Ancient Egypt* (New Haven and London, 1971, new ed. 1973).

Simpson, W.K., see W.S. Smith, *The Art and Architecture of Ancient Egypt*.

Smith, W.S., *The Art and Architecture of Ancient Egypt*, revised with additions by W.K. Simpson (Harmondsworth, 1981).

Smith, W.S., *A History of Egyptian Sculpture and Painting in the Old Kingdom*, 2nd ed. (Boston, 1940).

Spencer, A.J., *Death in Ancient Egypt* (Harmondsworth, 1982).

Spencer, P., *The Egyptian Temple: A Lexicographical Study* (London / Boston / Melbourne / Henley, 1984).

Steindorff, G., and K.C. Seele, *When Egypt Ruled the East* (revised ed., Chicago and London, 1957).

Stewart, H.M., "Traditional Egyptian Sun Hymns of the New Kingdom," *Bulletin of the Institute of Archaeology*, University of London, Vol. 6 (1967), pp. 29-74.

Stewart, H.M., *Egyptian Stelae, Reliefs and Paintings from the Petrie Collection. Part Two: Archaic Period to Second Intermediate Period* (Warminster,

1979).

Te Velde, H., *Seth, God of Confusion* (Leiden, 1977).

Terrace, E.L.B., and H.G. Fischer, *Treasures of Egyptian Art from the Cairo Museum* (New York and Los Angeles, 1970).

Thompson, H., see Griffith, F.Ll., and.

Uphill, E.P., *The Temple of Per Ramesses* (Warminster, 1984).

Vandier, J., *Manuel d'archéologie égyptienne*, Vol. 1 (Paris, 1952).

Vandier, J., *La religion égyptienne* (Paris, 1949).

Vareille, A., "L'hymne au soleil des architectes d'Amenophis III Souti et Hor," *BIFAO*, Vol. 41 (1942), pp. 25-30, and plates I-II.

Vernus, P., "Name," *Lexikon der Ägyptologie*, Vol. 4, cc. 320-26.

Volten, A., *Demotische Traumdeutung (Pap. Carlsberg XIII und XIV verso), Analecta aegyptiaca*, Vol. 3 (Copenhagen, 1942).

von Känel, F., *Les prêtres-ouâb de Sekhmet et les conjurateurs de Serket* (Paris, 1984).

Wadell, W.G., *Manetho with an English Translation* (London and Cambridge, Mass., 1940, repr. 1980).

Wainwright, G.A., "The Origin of Amun," *JEA*, Vol. 49 (1963), pp. 21-23.

Wainwright, G.A., Review of K. Sethe's *Amun und die acht Urgötter von Hermopolis* in *JEA*, Vol. 17 (1931), pp. 151-52.

Wainwright, G.A., see Newberry, P.E., and.

Ward, W.A., *Index of Egyptian Administrative and Religious Titles of the Middle Kingdom* (Beirut, 1982).

Weber, M., *Beiträge zur Kenntnis des Schrift- und Buchwesens der alten Ägypter* (Cologne, 1969).

Weigall, A., *A History of the Pharaohs*, Vol. 1 (New York, 1925).

Weill, R., *Le champ des roseaux et le champ des offrandes dans la religion funéraire et la religion générale* (Paris, 1936).

Werbrouck, M., see Capart, J., and.

Westendorf, W., see Grapow, H., and.

Wilkinson, C.K., and M. Hill, *Egyptian Wall Paintings: The Metropolitan Museum of Art's Collection of Facsimiles* (New York, 1983).

Wilson, J.A., "Egypt," in H. and H.A. Frankfort, J.A. Wilson, T. Jacobsen, and W.A. Irwin, *The Intellectual Adventure of Ancient Man* (Chicago and London, 1946; Phoenix ed., 1977).

Wilson, J.A., see Pritchard, J.B.

Wolf, W., "Der Berliner Ptah-Hymnus (P 3048, II-XII.)," *ZÄS*, Vol. 64 (1929), pp. 17-44.

Yoyotte, J., see Sauneron, S., and.

Žabkar, L.V., *A Study of the Ba Concept in Ancient Egyptian Texts* (Chicago, 1968).

Zandee, J., "The Book of Gates," *Studies in the History of Religions (Supplements to Numen)*, XVII: *Liber amicorum, Studies in Honour of Professor Dr. C. J. Bleeker* (Leiden, 1969), pp. 282-324.

Zandee, J., "Das Schöpferwort im alten Ägypten," *Verbum: Essays on Some Aspects of the Religious Function of Words dedicated to Dr. H.W. Obbink* (Rheno-Traiectina, Vol. 6, 1964), pp. 33-66.

Zandee, J., *Hymnen aan Amon van Papyrus Leiden I 350* (Leiden, 1948).

Index of Egyptian Words

In the following index the two kinds of "s" are ordinarily grouped together, the distinction between them being given only when I have made that distinction for "*s̀*" in the text, which is usually in connection with Old-Kingdom words. Note also that when "*z*" has been used by the authors I have quoted, then I have given that transcription under "z". In line with this I have also paid attention to the varying transcription techniques used by other authors whom I have quoted (e.g., see the different forms under *Ij-mr.s* and note that "*j*" is often written for either "*i*" or "*y*"). Further, I have given the few instances when phonetic transcriptions of proper names appear. However, the reader should consult the index of proper names for a much more complete list of such names. When I or other authors have added vowels to Egyptian words, they appear in the other index.

ỉꜥꜥw: 519
ỉwdt: 247
ị̉ht: 217
ỉh, ỉht, ỉhw: 97, 248, 327, 443, 481, 518, 579
ỉh ỉhw: 544
ị̉tt: 6

𓄿, 𓄿𓄿

𝕁

bty: 261
bd.t: 163

□

pỉwtyw: 564
pis: 254
pᶜt (a kind of land?): 248
pᶜt (nobles or mankind): 251, 337
pr: 253
pry: 252
pr-ᶜỉ: 215
Pr-ᶜnḫ: 19
Pr-wr: 121
Pr-wr-sỉḫ: 170
Pr-Bỉ-nb-Ḏdt: 253
Pr-Mśdỉwt: 153, 163
pr mḏỉt: 25, 35
Pr-nw: 121
Pr-nsr: 121
Pr-Ḥśn or *Ḥsn:* 153, 170
pr sbỉ nt sśw: 233
Pr-Śpỉ or *Pr-spỉ:* 153, 163, 166
Pr-śśtt: 153, 164, 166
Pr-ḳd: 166, 170
Pr-Dśw: 153, 163
pḥwy: 348, 611
pḫỉ: 254
psn or *pzn:* 238, 250, 357, 463

fkty: 31

Mṯn: Doc. I.2, pass.
mdꜣt: 472
mdꜣt: 25, 27, 35, 251
mdḫw: 250

~~~~~

n-nbw: 118
Nj-mꜣct-Ḥp (or *Ḥcpj*) or *N-mꜣc.t-ḫꜣp:* 159-61
Nj-swtḫ: 170
nisw: 24, 146
ncr: 4
nw (abyss): 298, 420, 431, 437-38, 445, 459, 461, 467,
      566, 588
nw, nwy(t): 247-48, 261
*nw-*god: 518
nwb: 129
nb: 7
Nb.ś or *Nbsnt* or *Nb-snt:* 168, 170-71
nbi: 316
nbdy: 250
npt(?): 254
Nfrii: 227-28
nfr(t): 252
nmst: 420
Nn-nsw: 253
nhy: 347, 608
nht: 347, 608
Nḥb-kꜣw: 291
nḥḥ: 304, 314, 370, 457, 459, 467, 509, 542, 546,
      553, 557, 563, 566, 575-79, 624,
nḫb (a kind of land): 248, 562
nḫn: 252
Nḫn, also later *Mḫn:* 253

nḥḥ: 417
nḫt-ḥrw, sʲb-nḫt-ḥrw: 146, 152, 163
nḫtw (victory): 610
nḫtw (stiff): 610
nsyt: 248
nsw: 248
Nswt Ḥn: 141
nsw-bit: 9
nswtyw: 152, 154, 161, 168
nt: 254
Ntrj-mw: 129
ntnṭ.t: 20
nṭr: 25, 248, 384, 395
nṭrt: 248
nṭrw: 114-15

⬦

r-pᶜt: 248
rᶜ, Rᶜ: 7, 102, 253, 401
rᶜm: 255
rwt: 179
rbn: 254
rmn: 56
rmwt (tears): 293
rmṭ, rmwṭ (man, men or people): 251, 293, 337
rn(w?): 248
rn-wr: 199
rn-nfr: 199
rn-nḏs: 199
rnpt: 9, 135, 199, 364, 580
Rnpt: 418
rnpt-sp: 193, 200
rnn: 252, 529

ⱃ

✗

—, ⌐

—

š: 5, 247
šꜣꜥ: 254
šꜥy(t): 250
šꜥyt: 254
šw: 247
šmw ꜥꜣ: 608
šmsy-Ḥrw: 104
šmsw Ḥr: 97
šnꜥ: 247
šnwt: 146, 254
šr-Mṯn or šrt-Mṯn: 160, 166
šgr: 248
št-Mṯn: 160
šty: 154
šṯwj: 169
šd: 327
šdt: 248

◿

ḳꜣ: 326
ḳꜣ-sḫm: 516
Ḳꜣw: 238
ḳꜣyt: 248
Ḳis: 253
ḳꜥḥ: 254
ḳbḥ: 255
ḳnbt: 254
ḳnḳn: 255
ḳḥ: 247
Ḳš: 161
ḳd (construct): 320

INDEX OF EGYPTIAN WORDS

ḳd (manner): 337

kꜣ: 177, 245
kꜣw: 398
kꜣ(w)ty: 251
kꜣrj: 245
kꜣkꜣw: 213
km: 245
kmr: 251
kri: 254
ksbt: 415
ksks: 251
kk, kkw: 386, 431, 437-38 445

gꜣš: 255
gnwty (?): 238, 250
Grg.t Mṯn: 161-62, 164, 167

tꜣ: 159, 162
tꜣ-ᶜt sbꜣ: 229
tꜣ mdꜣt imit dwꜣt: 472
tp-rnpt: 580
tp-ḫwt: 254
tp šbn: 252
tf: 427
tm (Atum): 564
tm (complete): 564
Tn(i): 253

tni: 248
tnmw: 299, 431, 437-38, 445
trk: 255

⊃

t'y md't: 251
T'rw: 253
t'ty: 5
Tmḫw, T'-mḫw: 5, 252
Tni: 253
trr: 250
trst: 255
Tḫnw: 252
tkt: 250
ts: 338
tt: 5-6

⊂⊃

Dw'w: 415
d't, dw't: 472, 566
db(w): 248
dm: 580
dmi: 252
dnd: 255
Dndrw: 423
dšr: 337

٦

dt: 304, 314, 370, 457, 459, 467, 503, 509, 542, 557,
    563, 575-77, 579, 599
d't: 607

INDEX OF EGYPTIAN WORDS

Index of Proper Names and Subjects

      I have not indexed the Chronology or the Bibliography since each of those lists is an independent unit so organized as to be easily consulted. I have not indexed in a detailed way subjects in the documents attached to the second chapter (or their introductions) since the subject matter of those documents was described rather thoroughly in the second chapter and hence the reader can readily see references to the proper places in the documents when reading Chapter Two. Nor have I have included all the rather fictional names of caverns and cities in the Netherworld and names or titles of the gods based on their attributes or powers. However, the indexing of the more common proper names in those documents is quite complete. Note that the Index of Egyptian Words can serve as a supplement to the subject indexing presented here in this index. In general, I have not indexed the names of cities or countries when they are incidental to the account. I have not indexed (except in rare instances) the long list of names and things in Doc. I.9, though the Egyptian equivalents when given are indexed in the Index of Egyptian Words. But I have indexed the categories of items found in that document. In this index I ordinarily give only the name of authors cited and not the titles of their works. However, in some cases where there are a great many citations (as with Gardiner, Helck, or Sethe), I have separated the entries by using abbreviated titles, The full titles may be readily identified by consulting the pages cited or the Bibliography.

baboon, 120, 299, 386-87, 492, 524, 541; Baboon of the Netherworld, 524

Babylon, Greek name for Kheraha, 141

Baedeker, K., 113, 243

Baines, J., and J. Málek, 243, 260, 629, 631, 633, 637-38

Bakhtan, 20-23

Baku, 404

Bark of Millions of Years, 371

Bark of Re, see solar boats

Barta, W., 61, 66, 132

Barucq, A., 552

Basahu, 151

Bast, 136

Bastet, a goddess, 55, 86, 136

Baterits (or Bateritn), 74

Baufre, 209

Baumgartel, E.J., 3, 37

Bawerded, noble, 189-90

Bayoumi, A., 403

Beatty Papyrus IV, 221-22, 228

Beckerath, J. von, 66, 68, 104, 119, 132

beetle (scarab), its identification with creation and rebirth, 266, 363-64, 366-67, 421, 473, 475, 479, 483-84, 492, 502, 507, 510, 520

Bell, B., 112

benben-stone, 286, 423; its house, 315, 320, 525, 529, 558

Beni Hasan, 429

Benti, 492

Bentresh, 20-21, 23, 41

Berlin Museum, 115

el-Bersha, 429, 431

Bianchi, R., iv, 345

Brunner, H., 231, 235
Brunner-Traut, E., 534
Bubastis, 43, 86, 136
Buck, A. de, 375, 429, 434-35, 548-49
Budge, E.A.W., 100-01, 398, 456, 618-19
Busiris, 91, 553
Buto, 9, 89, 117, 121, 138, 153, 170
Butzer, K.W., 112

Cairo Museum, 13, 47, 345, 434, 549
Cairo, Old, 109-10
calendars, 49-53, 106-09, 140-41, and see years (regnal), and Counting of the cattle
Caminos, R.A., 37, 223, 231-32
Causeway of Happiness, a celestial location, 357, 410, 420
Cavern of Osiris, 366, 499
Cavern of the Life of Forms, 365
Cavern of the West, 475, 495
cedar, 54, 83-84, 136
celestial motions, 45-46
Cenival, J.L. de, 59, 61, 103, 120
Černý, J., 120, 128, 135-36, 350, 353, 396-97, 402-04
Chabas, F., 615, 618
Champollion, J.F., 373, 398
chaos, see Abyss, existence and nonexistence, and Isfet
Chaos-gods, 264, 298-99, 431, 437-38, 440, 445
Chassinat, E., 44, 126, 391
Cheops (=Khufu), 85, 187, 212, 227, 285, 426; Tales of Wonder at his court, Doc. I.6 passim; and see Khufu
Chephren (=Khafre), King, 65, 205, 285, 345

177, 179-80, 305; Ptah as craftsman, 391; and see trades, and word lists.

creation, various means of, 266; by naming (the doctrine of the creative word), 239-40, 244, 276-79, 290, 293-94, 303-04, 306-12, 315, 318-20, 322, 326-27, 335-36, 361, 398-99, 427, 447, 455, 469, 476, 483, 514, 522, 528, 539-40, 544-45, 558, 564, 566, 572, 579-81, 587-88, 592, 596, 600-01, 617, 623, and see *Memphite Theology*; by Amun and Amon-Re, 313-22; by Atum or by Re, 239, 277, 285-86, 294-97, 303, 459-60, 592; by Khnum, 324-25, 617-18; by Neith, 326-28; by Ptah, 237, 239, 247, 303, 305-12, 389; by the Ogdoad, 297-98, 616; by Thoth, 303-04; of gods by sexual reproduction, 294, 477, 588, 593; of gods from the mouth or from saliva or exhalations, 316, 327-28, 343, 432, 442, 558, 579-80, 592; of man, 292-93, 316, 319, 327, 343, 354, 432, 442, 459-60, 467, 514, 528, 538, 543, 558, 561, 566, 570, 579-80, 584, 588, 592; of reptiles, 293, 593-94; of Shu and Tefenet and other gods by spitting, or by masturbation, 286-87, 311, 420-21, 423-24, 427, 432, 437, 439, 445, 563, 588, 592; snakes and, 291-92, 294, 365-66, 368, 382-83, 405, 432, 442, 475, 505-06, 510; the creation of "being" or "coming into being", 287, 566, 588, 592; views of existence and nonexistence in, 291-92, 370, 441; four creative acts for man, 295-96, 432, 441-42, 448; and see autogenesis, cosmogonies, and entries under the various gods: Amun, Amon-Re, Aten, Atum, Neith, Khnum, Ptah, Re

Crocodile Nome, 149, 154

crocodiles, 45, 203-04, 207, 234, 340, 542, 615, 621

cubits, palms, and fingers, passim in Nile heights given in Doc. I.1, and see numeration and measurement

391, 418, 466, and see Horus falcon; Horian dreams, 604,
613; the Followers of, 12, 97, 357; the Following of, 11,
48, 50-51, 65, 68-71, 75-82, 103-04; Horus as creator
god at Edfu, 264, as primitive creator god, 279-84; on
his throne, 499; the myth of the genealogy of, 282-84,
289-90, 292, 380, 383, 408, 420, 477; the secret forms
of, 478, 500-01; and see mounds of, and Eye of
    Horus at the Head of the Spirits, 423
    Horus Bull, 55, 85
    Horus falcon, 279, 281, 283, 331, 413; king's title
or representation, 3, 5, 7-8, 118-19, 138, 379, 400
    Horus in *Snwt* (or Shenwet), 28, 36, 337
    Horus Mekhantenirti, 293, 588, 593
    Horus-name of the king, 7, 118-19, and Doc. I.1
passim
    Horus of Buto, 138
    Horus of Dunanu, a falcon god, 139
    Horus of the Heron of Djebakherut (or Buto), 89
    Horus-of-Gold name of the king, 93, 118, 129
    Horus of Hasroet, 102
    Horus of Heaven, 55, 77, 130, 281, 283
    Horus of Letopolis, 385
    Horus of Pe, 138
    Horus of Sezmet, 415, 427
    Horus of the East, 415, 427
    Horus of the Gods, 415, 427
    Horus of the Horizon, 281, 427
    Horus of the Netherworld, 421, 499
    Horus the Beḥdetite, 379
    Horus the Elder, 141, 294 (alluded to), 383-84
    Horus the Praiser, 473, 492, 495
    Horus-Seth, 88, 90-91, 119, 131, 138
    hour goddesses, 482
    hour watchers (astronomers), 24, 250

Lake (or lakes) of Hotep, 464

Lake of the Dat, a celestial lake, 359, 409, 420

Lake (or Island) of the Two Knives, 301-02, 388, 454, 459, 575, 617, 625

Lange, H.O., 615, 618-19

lapis lazuli, 51, 81, 83, 85, 177, 180, 342, 537, 561, 617, 624

lates-fish, 326, 329, 578

Lauer, J.-P., 134, 136, 380

Leclant, J., 411

lector-priests, 23-24, 28, 181, 184-85, 188, 203, 205-06, 209-10, 225, 249; their duties and magical concerns, 206

Lefebvre, G., 243-44, 388

Leiden-London magical papyrus, 349-50

Lepsius, R., 143, 146-47, 192, 384-86, 401, 451-52, 456

Lesko, L.H., 362, 404-05, 434

Letopolis, 553

Lexa, F., 618

libraries, at House of Life and temples, 28, 35-36,44-46, 126, 587, and see House of Books

Libya, 84, 280, 422; nome of, 91, 171

Lichtheim, xiii, 41-43, 193, 207, 216-18, 223, 228, 231, 235, 385, 400, 403, 547-48, 551, 597-98, 602

Life of Forms, 495, 499

Life of Souls, name of the solar bark, 497

Life of the Gods, serpent's name, 368, 506

life sign, see $^c n \underline{h}$ in the Index of Egyptian Words

Lisht, 429

Litany of Re, 239, 263, 332, 616, and Doc. II.5

Logos doctrine, 312

Lone Star (=Venus), 413, 425

Lord of the Years or Lord of Time, titles for

hymns to Neith, 547, 551, 578-81; Neith title of the queen, 118

Neithhotep, wife of the Horus Aha, 11, 118

Nekhbet, the vulture goddess, 55, 78, 89-90, 131, 197, 279

Nekheb, 279

Nekhen (Hieraconpolis), 9, 138, 253

Nemathap (=Nimaathapi, q.v.), 160

Nenet, 298, 414, and see Naunet

Neper, 380-81

Nephthys, 30, 42, 141, 205, 289, 293, 336-38, 350, 383-84, 424, 473, 477, 483, 488, 519, 528, 588, 593

nesu-bit (=*insibya*) name of the king, 10-11, 118, 129

Neteren (Netermu), 129, and see Ninetjer

Netherworld, 344, 357, 404, 408, 454, 576, 625; locales and residents in, 359-60, 362-69; voyages of Re through, 319, 332, 359-69, 566, and Doc. II.4 (and its introduction), and Doc. II.5 passim; and see solar boats, and Dat

Netjeri-khet (=Djoser, q.v.), reign of in *Annals*, 81-82

netjeru, the Egyptian word for gods, 329, 395

Netuty, 519, 528

Neugebauer, O., xii, 42, 629, 637

New Moon Feast, 316, 558, 591

New Year's Day, 580

Newberry, P.E., 99, 121, 131

Ni-heb, 67

Niankhsekhmet, see Doc. I.3

Nile, 266, 320, 325, 354, 369, 442, 448, 562, 564-65, 574, 583 (the flood), and see Hapy; its heights and nilometers, 10, 49, 103, 107-13, 351, and Doc. I.1 passim

Nimaathapi, 152, and see Nemathap

Ramesses VI, 531, 551
Ramesses IX, 512, 515, 578
Ramesseum 17, and Onomasticon of, 237, 242
Ramesside kings, 515
Ranebtaui Mentuhotep, 122
Ranke, H., 199
Re, the sun-god, 7, 55-56, 102, 121, 127, 177-78, 182, 184, 204-05, 240-41, 247, 268, 284-88, 294, 305, 309-10, 314-18, 327, 342, 346, 349-50, 355-56, 360-67, 372, 385, 389, 391, 394, 397, 401-02, 405, 408, 413-14, 416-17, 420-21, 423-25, 430, 432, 436-37, 439-41, 459-62, 467, 477, 479-85, 531-34, 560, 563-64, 574-75, 577, 580-81, 591, 600, 612, 615-17, 621-22, and Docs. II.4, II.5, and II.6 passim; children of, 338; creation by Re or the sun-god, 239, 268, 277, 286-88, 294-97, 302, 338, 397-98, 522, Doc. II.5 passim; creation of Re or the sun-god by another god or gods, 268, 303, 313, 327, 391, 397, 475, 538, 543, 564, 577; gifts for Re, 56, 88-89, 91-92, 94; Heart of Re, 389; heir and lord of Eternity, 295, and see Eternity; his daughter Maat, 282, 397, and see Maat; his earthly home in Heliopolis, 355, 564; his integration with Osiris, 511-13, 527; his kas and bas, 333-34, 350, 362-63, 398-99 542, 553, and see ba, and ka; his ram-headed form, 363, 484, 492, 507, 520, 528; his secret name, 331-32, 343, 397-98; Re, Lord of Bas, 525; Re, Lord of Sakhbu, 204; the Emanations of, 31, 33; the Herald of, 127; son of, 109; the staff or Followers of Re, 32-33; the primacy of Re, Amun, and Ptah, 318, 564; the right and left eyes of Re and other creator gods (the sun and moon disks, or the day and night), 241, 281-82, 319, 325, 405, 460, 468, 565-66, 573-74, 583; uncovering his face, 320; vivified by Horus, 281, 415; and see Atum, Eye of Re, Netherworld, solar boats, son of Re designation, sun, Water of Re

Illustrations

Fig. I.1c Statue of Amenhotep, son of Hapu, as a scribe. Dynasty 18. Cairo Museum. After Terrace and Fischer, *Treasures of Egyptian Art from the Cairo Museum*, p. 119.

Fig. I.1b Another seated scribe. Dynasty 5. Louvre Museum. Photograph by Sue Clagett.

Fig. I.1a A seated scribe. Dynasty 6. Pelizaeus Museum, Hildesheim.

Fig. 1.2b   A scribe in the act of recording an inventory. Dynasty 5. Cropped from the photograph in Pirenne, *Histoire de la civilization de l'Égypte ancienne*, Vol. 1, pl. after p. 140, with the permission of Les Éditions de Baconnière, Boudry, Suisse.

Fig. 1.2a   Wooden panels from niches in the west wall of the tomb of Hesyre in Saqqara. Dynasty 3. Cairo Museum.

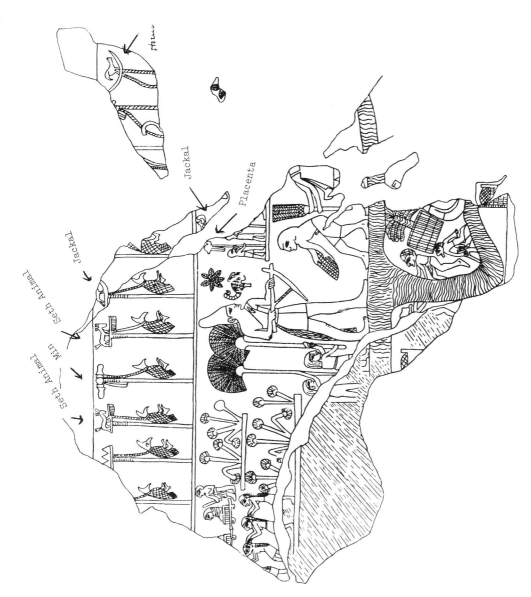

Fig. I.3a Mace-head of King Scorpion. Predynastic. Drawn from the Original in the Ashmolean Museum, Oxford, after Smith, *A History of Egyptian Sculpture and Painting in the Old Kingdom*, Fig. 30. Copied with the permission of the Museum of Fine Arts in Boston.

Fig. I.3b   The Mace-head of King Scorpion reconstructed.

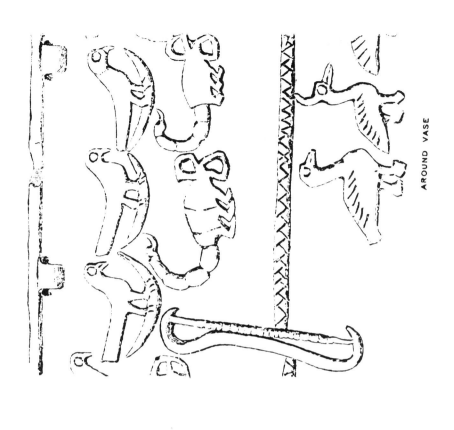

AROUND VASE.

C.R.P.

LIMESTONE VASE.

Fig. I.4a   Vase with the name of the Horus Scorpion. Predynastic. Cf. J. E. Quibell, *Hierakonpolis*, Vol. 1, pl. XIX.

Fig. I.4b Group of early maces and vase. Restored after Quibell, *ibid.*, Vol. 1, pl. XXV.

THE ALPHABET

SIGN	TRANS-LITERATION	OBJECT DEPICTED	APPROXIMATE SOUND-VALUE	REMARKS
	ꜣ	Egyptian vulture	the glottal stop heard at the commencement of German words beginning with a vowel, ex. *der Adler*.	corresponds to Hebrew א *ālēph* and to Arabic ا *'alif hamzatum*.
	ꞽ	flowering reed	usually consonantal *y*; at the beginning of words sometimes identical with ꜣ.	corresponds to Hebrew י *yōdh*, Arabic ي *yā*.
	y	(1) two reed-flowers (2) oblique strokes	*y*	used under specific conditions in the last syllable of words, see § 20.
	ꜥ	forearm	a guttural sound unknown to English	corresponds to Hebrew ע *ᶜayin*, Arabic ع *ᶜain*.
	w	quail chick	*w*	
	b	foot	*b*	
	p	stool	*p*	
	f	horned viper	*f*	
	m	owl	*m*	
	n	water	*n*	corresponds to Hebrew נ *nūn*, but also to Hebrew ל *lāmedh*.
	r	mouth	*r*	corresponds to Hebrew ר *rēsh*, more rarely to Hebrew ל *lāmedh*.
	h	reed shelter in fields	*h* as in English	corresponds to Hebrew ה *hē*, Arabic ه *hā*.
	ḥ	wick of twisted flax	emphatic *h*	corresponds to Arabic ح *ḥā*.
	ḫ	placenta (?)	like *ch* in Scotch *loch*	corresponds to Arabic خ *ḫā*.

(The Alphabet is continued on the next page.)

ḥ	animal's belly with teats	perhaps like *ch* in German *ich*	
s	(1) bolt / (2) folded cloth	s	interchanging early with ⌐, later with ẖ, in certain words. originally two **separate** sounds: (1) z, much like our z, (2) ŝ, unvoiced s, early hardly different from ⌐ ḥ.
š	pool	*sh*	
ḳ	hill-slope	backward *k*; rather like our *q* in *queen*	corresponds to Hebrew ק *qōph*, Arabic ق *ḳāf*.
k	basket with handle	*k*	corresponds to Hebrew כ *kaph*, Arabic ك *kāf*. Written ⌐ in hieratic.
g	stand for jar	hard *g*	
t	loaf	*t*	
ṯ	tethering rope	originally *tsh* (*č* or *tj*)	during Middle Kingdom persists in some words, in others is replaced by ⌐ *t*.
d	hand	*d*	
ḏ	snake	originally *dj* and also a dull emphatic *s* (Hebrew צ)	during Middle Kingdom persists in some words, in others is replaced by ⌐ *d*.

OBS. Later alternative forms are ẹ for *w*, ⌐ for *m*, ⌇ for *n*, and ∫ for *t*. Of these, ẹ arose from an abbreviated form of ⌇ in Middle Kingdom hieratic, so that it appears in our transcriptions of hieratic texts belonging to a time when ẹ was not yet written in hieroglyphic;[2] ⌐ and ∫ originate in the biliteral signs for *im*[3] and *ti* respectively, while ⌇ is taken from the word *nt* 'crown of Lower Egypt'.[4] Note also that ẹ is used for *g* in a few old words.

[1] The form ⌐ usually employed in printed books is not found on the monuments until a quite late period; early detailed forms are ▭ and ▥, early XII Dyn., ex. PETRIE, *Gizeh and Rifeh* 13 g.   [2] *ÄZ*. 29, 47.   [3] As *m* not before Tuthmosis I, *ÄZ*. 35, 170.   [4] Already sporadically as *n* from

Fig. 1.5   The Egyptian alphabet, i.e. glyphs representing single consonants. Copied from Gardiner, *Egyptian Grammar*, p. 27, with the permission of its publisher, the Griffith Institute. See my Index of Egyptian Words for the alphabetical glyphs produced by the computer program Fontrix, and used in the index as headings.

(F 40) *3w*	(G 40) *p3*	(M 16) *h3*	(F 29) *st*
(U 23) *3b, mr*	(O 1) *pr*	(N 41) *hm*	(M 8) *š3*
(. ·) *jw*	(F 22) *ph*	(M 2) *hn*	(H 6) *šw*
(Aa 13, *in, gs*	(U 1) *m3*	(D 2) *hr*	(V 7) *šn*
(Z 11) *jm*	(W 19) *mj*	(W 14) *hs*	(V 6) *šs*
(K 1) *jn*	(Y 5) *mn*	(T 2) *hd*	(Aa 28) *ḳd*
(D 4) *jr*	(N 36) *mr*	(M 12) *h3*	(D 28) *k3*
(M 40) *js*	(U 6) *mr*	(N 28) *hᶜ*	(I 6) *km*
i (O 29) *ᶜ3*	(V 12) *ʿnh*	(M 3) *ht*	(G 28) *gm*
(V 4) *w3*	(F 31) *ms*	(K 4) *h3*	(U 30) *t3*
(T 21) *wᶜ*	(D 52) *mt*	(D 33) *hn*	(U 33) *tj*
(F 13) *wp*	(W 24) *nw*	(F 26) *hn*	(U 15) *tm*
(E 34) *wn*	(U 19) *nw*	(T 28) *hr*	(G 47) *t3*
(M 42) *wn*	(V 30) *nb*	(G 38) *s3*	(U 28) *d3*
(G 36) *wr*	(T 34) *nm*	(Aa 17/18) *s3*	(N 26) *dw*
(V 24) *wd*	(F 20) *ns*	(M 23) *sw*	(M 36) *dr*
(G 29) *b3*	(Aa 27) *nd*	(T 22) *sn*	Also:
(F 18) *bh, hw*	(E 23) *rw*	(V 29) *sk, w3h*	(M 22) *nn*

Fig. I.6   Some biliteral signs. Taken from H. Brunner, *An Outline of Middle Egyptian Grammar*, p. 9, with the permission of Akademische Druck-u. Verlagsanstalt, Graz, Austria. The designations in parentheses are the sign numbers used by Gardiner in his *Egyptian Grammar*. Gardiner's sign list includes not only the classical alphabetical and biliteral signs depicted here and in Fig. I.5 but also the full range of signs used for determinatives and for other biliteral, triliteral, and quadriliteral words current in the Middle Kingdom.

(a)

(b)

Fig. I.7 Three forms of an extract from ''The Story of Sinuhe'' prepared with the versatile and elegant computer program GLYPH written by Jan Buurman and Ed de Moel: (a) in the normal Egyptian writing from right to left (here justified on the right and left), (b) in reverse writing from left to right (here justified only on the left), and (c) in normal columnar writing (with the columns justified at top and bottom and to be read from right to left). It is an ideal program for preparing blocks of texts but cannot yet easily embed glyphs into a sophisticated word-processing program. Notice how glyphs are grouped together on the same and differing levels and face the direction from which the writing begins. These specimens were printed on a Kyocera Laser printer (F-1000A).

(c)

Fig. I.8   The Palette of Narmer, recto and verso. Predynastic or
Dynasty 1. Cairo Museum.

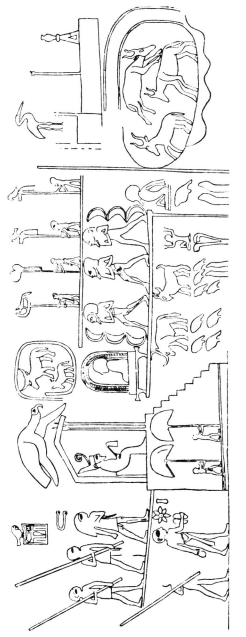

Fig. I.9 The Narmer Mace-Head. According to Quibell, *Hierakonpolis*, Vol. 1, pl. XXVI.B.

ı	1
ıo	10
ıoo	100
1,000	
10,000	
100,000	
1,000,000	

Fig. I.10 Egyptian signs for the powers of 10.

Fig. I.11a   Two statues of Khasekhem from Hieraconpolis. Dynasty 2. The left one in the Ashmolean Museum at Oxford and the right one from the Cairo Museum.

Fig. I.11b   Details from the bases of the above statues giving the number of victims. In each, the Horus name of the King as appearing on the top side of the pedestal is given. See Quibell, *Hierakonpolis*, pl. LX.

Fig. I.12   A predynastic sealing with signs for water *(mw)* according to Emery, *Hor-Aha,* p. 33.

Fig. I.13   Stela of the Horus Djet. Dynasty 1. Louvre Museum. Photograph by Sue Clagett.

-753-

Fig. I.14b   Stela of Queen Meryetneith. Dynasty 1. Cairo Museum.
Photograph by M. Clagett.

Fig. I.14a   Stela of the Horus Nebre. Dynasty 2. The Metropolitan
Museum of Art. Photograph by Sue Clagett.

Fig. I.15a A supposed sealing of Menes (restored). Dynasty 1. Copied from Gardiner, *Egypt of the Pharaohs,* p. 405, with the permission of the Oxford University Press.

Fig. I.15b A Tablet from Naqada. Dynasty 1. From Gardiner, *ibid.,* with the permission of the Oxford University Press.

Fig. I.16   A wooden label from Abydos. Dynasty 1. Drawn after Petrie, *Royal Tombs,* Vol. 2, pl. X,2.

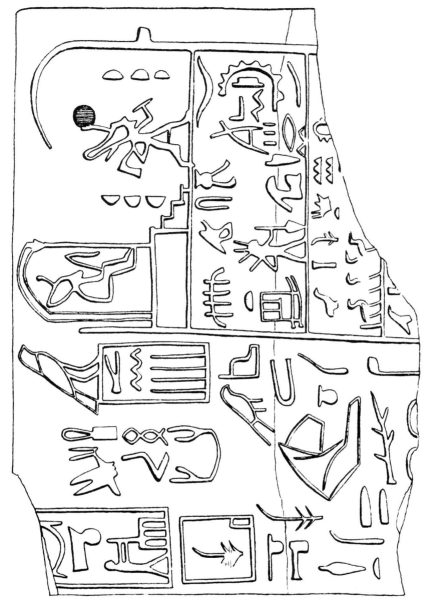

Fig. I.17   An ebony tablet of King Den. Dynasty 1. After Petrie, *Royal Tombs*, Vol. 1, pl. XV,16.

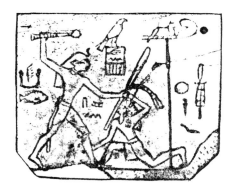

Fig. I.18a An ivory label of King Den in the British Museum. Dynasty 1.

Fig. I.18b An ivory label of King Semerkhet. Dynasty 1. After Petrie, *Royal Tombs,* Vol. 1, pl. XVII,26.

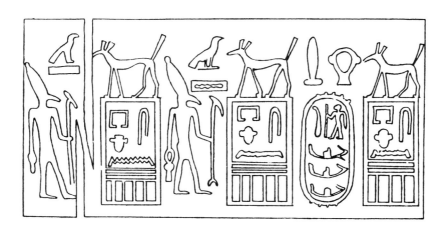

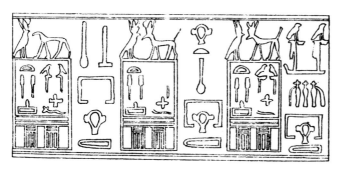

Fig. I.19a   A group of four sealings, the two above found in the Metropolitan Museum, and the two below taken from Petrie, *Royal Tombs,* Vol. 2, pls. XXII,179 and XXIII,197.

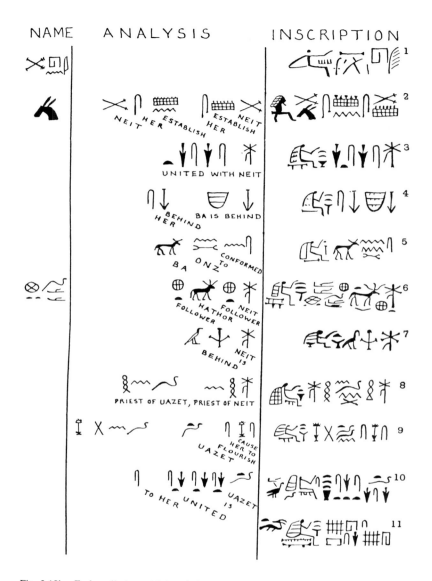

Fig. I.19b Early cylinders with inscriptions. Taken from Petrie, "The Earliest Inscriptions," p. 65.

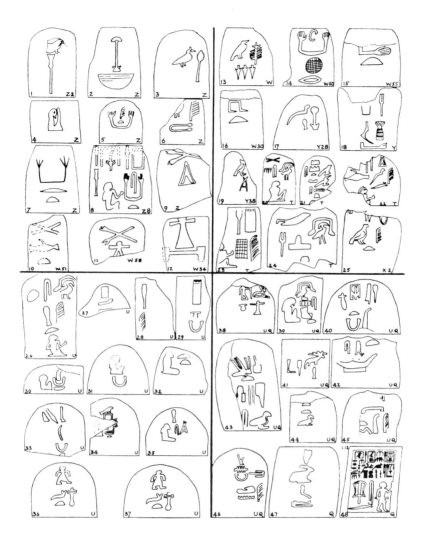

Fig. I.20   Sketches of stelas near the first-dynasty, royal tombs at Abydos. From Petrie, *Royal Tombs,* Vol. 1, pl. XXXI.

Fig. I.22 Archaic stela from Helwan. Reproduced from Vandier, *Manuel d'archéologie égyptienne*, Vol. 1, p. 733, with the permission of A. et J. Picard, Paris.

Fig. I.21 Stela of Sabef from Abydos, End of Dynasty 1. Cairo Museum. Photograph by M. Clagett.

Fig. I.23 Stela of the nobleman Merka. End of
Dynasty 1. Reproduced from Emery, *Archaic Egypt,*
pl. 30a, with the permission of Penguin Books Ltd.

Fig. I.24a   Stone stela of a princess from a tomb niche in Saqqara. Dynasty 2.

Fig. I.24b   Another stone stela from Saqqara (painted). Dynasty 2.

Fig. I.25 Stela of Tetenankh. Dynasty 3. Drawing
from Vandier, *Manuel,* Vol. 1, p. 754. Copied with
the permission of A. et J. Picard, Paris.

Fig. I.26 Slab-stela of Prince Wepemnofret from his Giza tomb. Dynasty 4. Copied with the
permission of the Lowie Museum of Anthropology, University of California at Berkeley, from
Fazzini, *Images for Eternity,* p. 28.

Fig. I.27   Reliefs in an offering-niche of the
tomb of Khabawsekar at Saqqara. Dynasty
3. Cairo Museum.

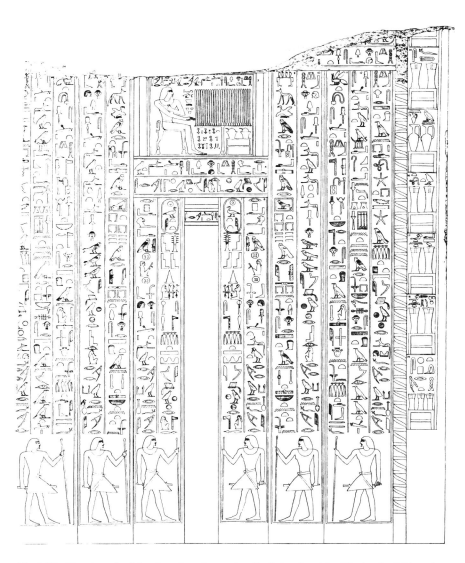

Fig. I.28   Drawing of a false door from the west wall of chamber A8 of the tomb of Mereruka, Saqqara. Dynasty 6. According to Oriental Institute of the University of Chicago, *The Mastaba of Mereruka,* Part I, pl. 62. Courtesy of The Oriental Institute of the University of Chicago.

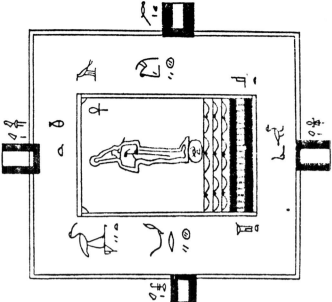

Fig. I.30 A diagram of an ideal House of Life for Abydos. Plate 36, Papyrus Brit. Mus. 10051. According to Gardiner, "House of Life," p. 169.

Fig. I.29 A list of the names of some scribes on a slab from Dynasty 3, now at the Ashmolean Museum. Cf. Garstang, *Tombs of the Third Egyptian Dynasty*, pl. XXVII,2.

Fig. I.31a Seshat on a limestone relief from the mortuary temple of Sesostris I at Lisht (Dynasty XII).
Photograph provided through the courtesy of The Brooklyn Museum, Charles Edwin Wilbour Fund.

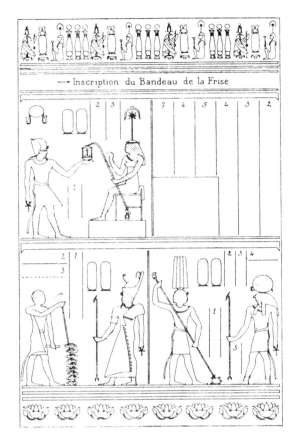

Fig. I.31b Seshat on the east wall of the Library at the Temple of Edfu. According to Chassinat, *Le Temple d'Edfou*, Vol. 3, pl. LXXXII.

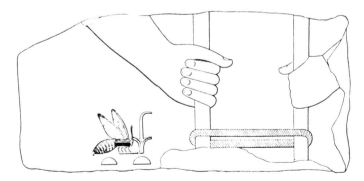

Fig. I.31c The hands of a king and a divinity (Seshat?), stretching the chord, from a block at Khatâna. Most likely Dynasty 12. Taken from Habachi, "Khatâna," pl. VII.

Fig. I.32 The Palermo Stone, recto. H. Schäfer, "Ein Bruchstüch," Tafel I.

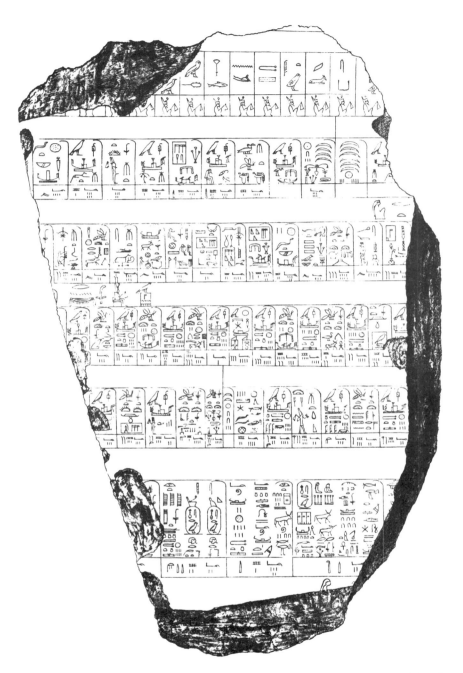

Fig. I.33   The Palermo Stone, recto. Drawing according to Pellegrini, ''Nota sopra un'inscrizione,''
Tavola I.

Fig. I.34   The Palermo Stone, verso. Schäfer, Taf. II.

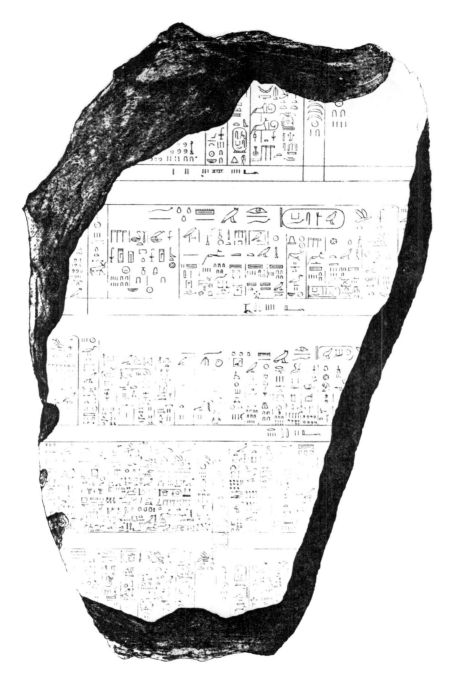

Fig. I.35   Palermo Stone verso. Pellegrini, Tav. II.

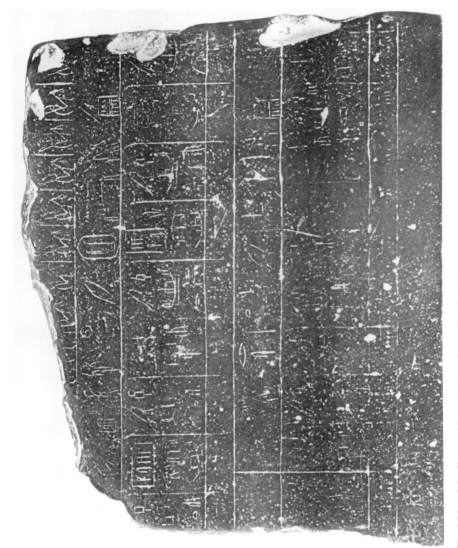

Fig. I.36a Cairo Fragment 1, recto. According to Gauthier, "Quatre nouveaux fragments." pl. XXV.

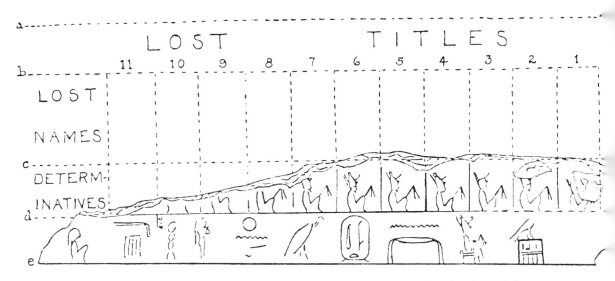

Fig. I.36b   Top lines of Cairo Fragment 1, recto. Drawn according to Breasted, "The predynastic union," pl. I.

Fig. I.37   Cairo Fragment 1, verso (significant detail). Gauthier, pl. XXVII.

Fig. I.38　Cairo Fragments 2 and 3. Gauthier, pl. XXX.

Fig. I.39   Cairo Fragment 4. Gauthier, pl. XXXI.

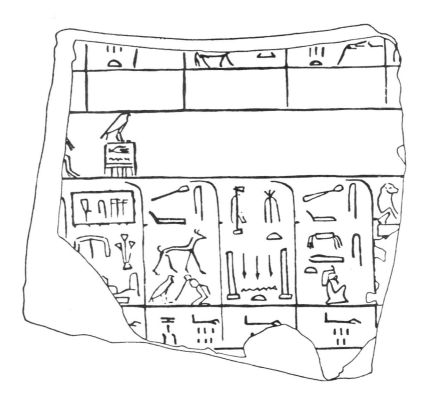

Fig. I.40   Cairo Fragment 5. Taken from de Cenival, ''Un nouveau fragment,'' p. 15.

Fig. I.41a and b   London Fragment, recto and verso. Photographs provided
by the Petrie Collection of University College, London University.

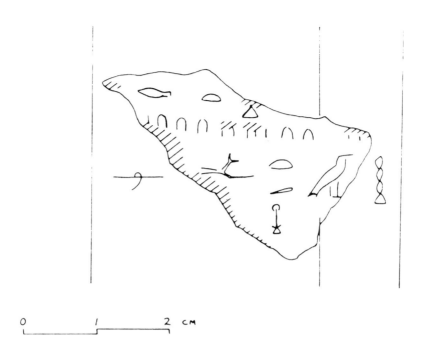

Fig. I.41c   London Fragment, verso. Drawing by Reeves, "A Fragment of Fifth Dynasty Annals," p. 50.

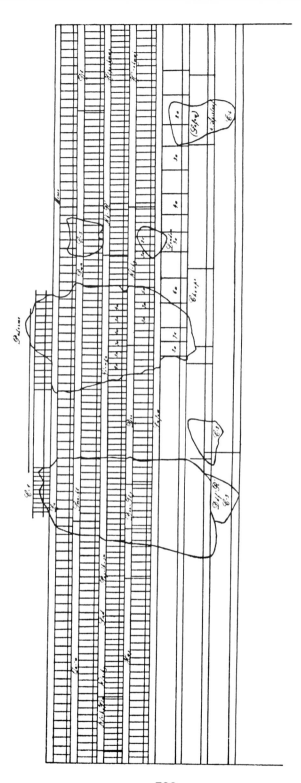

Fig. I.42 Early Annals, recto, showing the location of the fragments. Redrawn according to Helck, "Bemerkungen zum Annalenstein," p. 34. The Fragments are labeled Palermo, C-1 to C-5 (Cairo), and London.

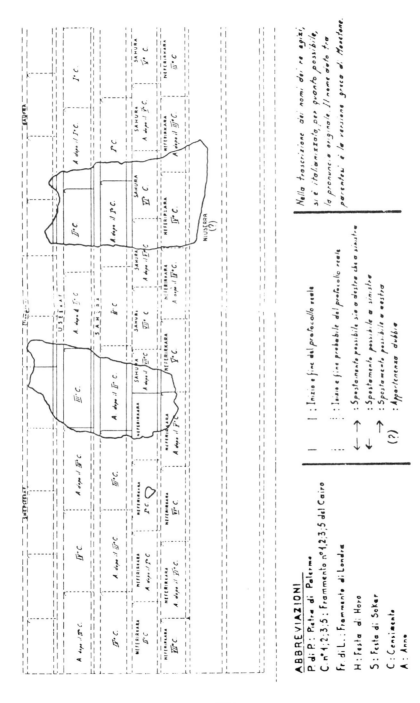

Fig. I.43 Early Annals, verso, showing Fragments Palermo, C-1, and London, according to Giustolisi, ''La 'Pietra di Palermo','' March 1969, pp. 46–47.

A fragment of the Turin Canon of Kings.

Part of the Table of Abydos.

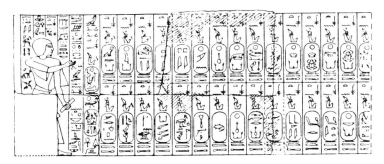

Part of the Table of Saḳḳâra.

Fig. I.44a Samples of three principal king-lists. Copied from Gardiner, *Egypt of the Pharaohs*, p. 49, with the permission of the Oxford University Press.

Fig. I.44b and c   Parts of the king-list from the temple of Seti I at Abydos. Taken from Capart's
Memphis, pp. 115, 146.

Fig. I.45 An inscription of Horus Djer on a
stone vessel. Dynasty I. Taken from Borchardt,
Die Annalen, p. 31.

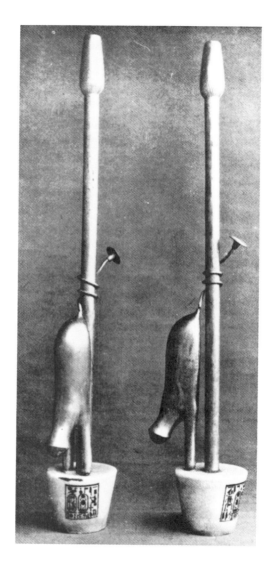

Fig. I.46 Two Anubis Emblems. The pelt on a staff,
read as "He who is in the bandaging room." Taken
from Carter, *Tut-ankh-Amen*. Vol. 2, opposite p. 34.

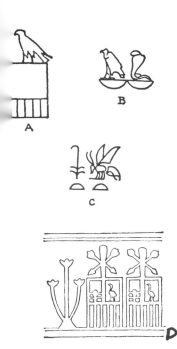

Fig. I.47a The Archaic period. A-C reproduced from Emery, *Archaic Egypt*, Fig. 69, p. 107, and D from *ibid.*, Fig. 8, p. 49, with the permission of Penguin Books Ltd.

Fig. I.47b A relief in the Metropolitan Museum of Art from Dynasty 3, showing the titles: "King of Upper and Lower Egypt" (i.e. He of the Sedge and the Bee) and "He of the Two Ladies" (i.e. the *nebty*: the Vulture-goddess Nekhbet of Upper Egypt and the Cobra-goddess Wadjet of Lower Egypt).

Fig. I.47c The early use of the cartouche around Sneferu's name, but without a "Son of Re" name or title. Dynasty 4. Relief of the Personalized Estates from the Valley temple of the Bent Pyramid in Dashur.

Fig. I.47e   An inscription from Aswan giving the full titulary of Nebhepetre Mentuhotep. Dynasty 11. Taken from Gauthier, *Le Livre des rois,* Vol. 1, p. 229. Notice that all five titles are now used.

Fig. I.47d   A drawing of a diorite stand with the names and titles of Chephren, but without the "Son of Re" title. Dynasty IV. In the Metropolitan Museum of Art.

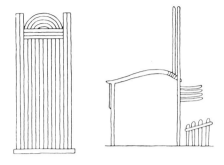

Fig. I.48 The Dual Shrines. The figure on the left represents the Per-nw of Lower Egypt at Buto and that on the right the Per-wer of Upper Egypt at Hieraconpolis. The drawings are those of Frankfort, *Kingship and the Gods,* Fig. 30 after p. 212. Copied with the permission of the University of Chicago Press.

PROBLEM 48

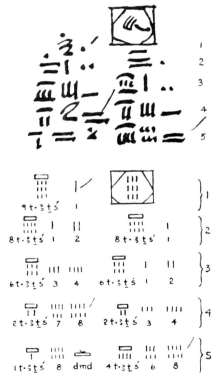

Fig. I.49 Problem 48 of the Rhind Mathematical Papyrus. From Chace, *The Rhind Mathematical Papyrus,* Vol. 2, pl. 70.

	Cubit-Areas (▭)	Avouras (– ℓ –)
ḫ₃ (𝍩)	1000	10
ꜣt₃ꜥ (▭ or – ℓ – or ▭)	100	1*
rmn (◁)	50	1/2
ḥsb (✗)	25	1/4
ꜣ₃ (✗)	12-1/2	1/8
mḥ (◦▭)	1	1/100

*1 aroura (or ꜣt₃ꜥ) = 2728.7 sq. meters

Fig. I.50   Area measurements on the Palermo Stone.

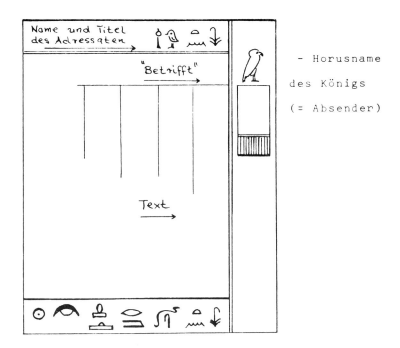

Fig. I.51   The form of a royal, written command in the Old Kingdom. Taken from Gödecken, *Eine Betrachtung*, p. 3, with the kind permission of Verlag Otto Harrassowitz.

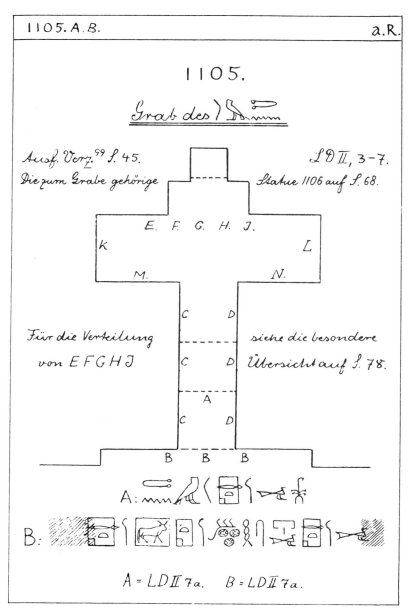

Fig. I.52 Location of inscriptions in the tomb of Metjen. Taken from H. Schäfer, *Ägyptische Inschriften*, as are the hieroglyphic transcriptions here and in the remaining figures. The photographs of Inscriptions A and B on the next page and those of the remaining inscriptions in the succeeding figures are reproduced from the plates appended to Goedicke, "Die Laufbahn des *Mtn*," with the author's kind permission. The references to LD are to Lepsius' *Denkmäler*.

Inscription A.

Inscription B.

C.

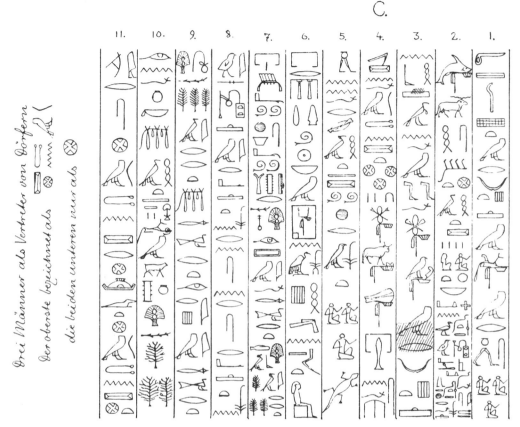

Drei Männer als Vertreter von Dörfern

Der oberste bezeichnet als

die beiden unteren nur als

Vielleicht ist die Inschrift C an D anzuschliessen. C = LD II, 6 und 7ª.

Fig. I.53   Inscription C.

Drei Frauen als Vertreterinnen von Dörfern

Die oberste bezeichnet als [Hieroglyphen] so!

die beiden unteren ohne Bezeichnung.

a. so? (¿en) hat wohl nie da-gestanden.

D. (Forts.)

11. 10. 9. 8. 7. 6.

D.

5. 4. 3. 2. 1.

D = LD II 5 und 75.

Fig. I.54 Inscription D.

E–I.

Übersicht:

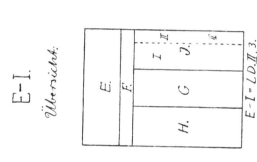

E–I = LD.II. 3.

E.

1.

2.

3.

4.

5.

6.

7.

8.

9.

E. (Forts.)

10.

11.

12.

13.

14.

E. (Forts.)

15.

16.

17.

18.

Fig. 1.55   Arrangement of Inscriptions E-I. Inscription E.

F.

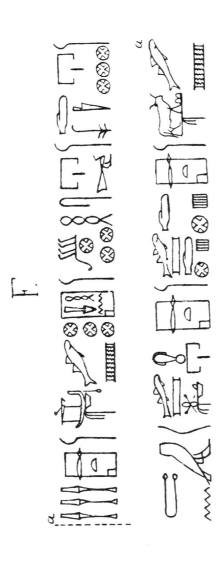

Fig. I.56 Inscription F.

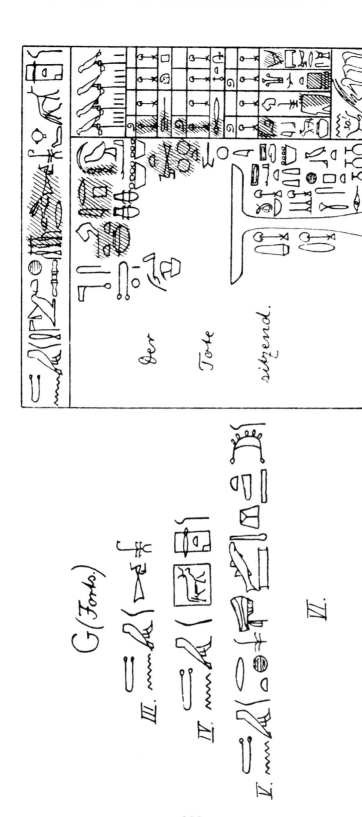

(See page 805 for photographs.)

-803-

G.

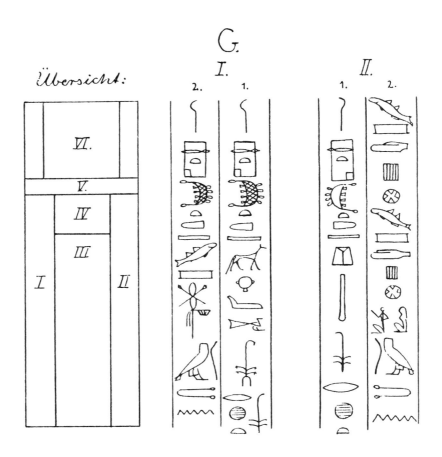

Fig. I.57   Inscriptions G I-II, IV-VI.

G (Forts.)

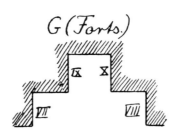

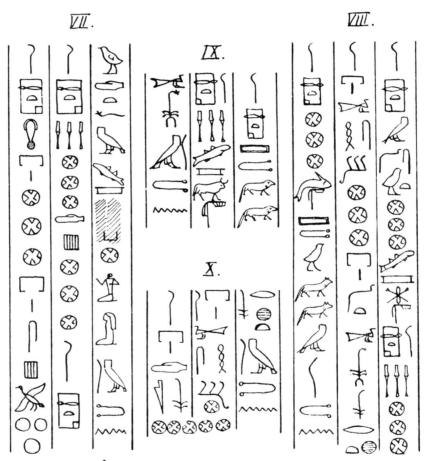

$\mathcal{G}\ \overline{VII} - \overline{X} = LD\ \overset{..}{II}\ 5\ links\ u.\ 6\ rechts.$

Fig. I.58   Inscriptions G III, VII-VIII, X.

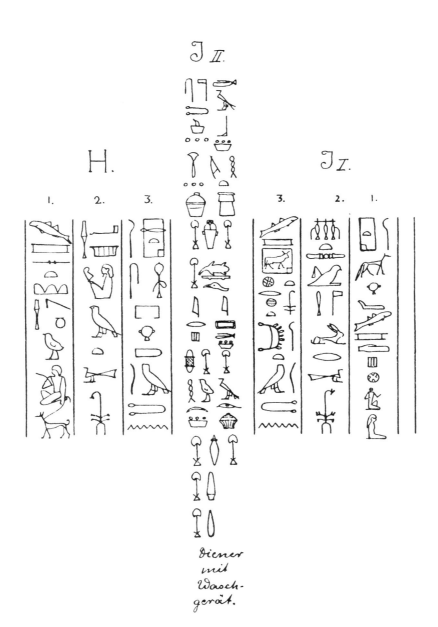

J II.

H.

J I.

1. 2. 3. 3. 2. 1.

Diener
mit
Wasch-
gerät.

Inschrift J

Inschrift H

Fig. I.59 Inscriptions H and J.

M. [LD II, 4 rechts]

Unten: Der Verstorbene stehend nach l. Vor ihm zwei Männer mit gefangenen Antilopen (die Darstellung schliesst an
K an. Über ihm:

Mitte: 4 Männer mit Gaben, nach r. Über ihnen:

Mann mit Gefäss und Sanda-len.	Mann mit Waschge-rät.	Mann mit Wasser krügen.	Mann mit ei-nem Kasten und

Oben: Leerer Raum.

N. [LD II, 4 links]

Unten: Der Tote stehend, nach r.
Über ihm:

Vor ihm 2 Männer über einander.
Der obere ein als ⌂ be-
zeichneter Trie- ster.
Über ihm:

Der untere ein Diener
mit Waschgerät. Über
und vor ihm:

Mitte: 4 Männer mit Gaben, nach links. Über ihnen:

Mann mit einem Futteral.	Mann mit Stab, Sandalen und Halsband(?)	Mann mit Se-stell der Form	Mann mit Ge-fäss und mit Kopfstütze.

Oben: 4 Männer beim Schlachten eines Rindes 5, 4. 3. 2. 1
No 1. Aufseher . No 3 schneidet. No 2 und 4 ziehen an den
Beinen des Rindes. No 5 trägt eine Schale mit Herz und
Beischriften: Rippen fort.

No 3. ohne
Beischrift. zu No 2.   zu No 1.

zu No 5 und 4 # zum Rind.

-810-

Kultkammer des *Mṯn*: Inschrift M

Fig. I.60   Inscriptions M and N.

Fig. I.61   The backwall of Metjen's Cult-Chamber. Taken from Gödecken, *Eine Betrachtung,* Abbildung I after p. 168, with the kind permission of Verlag Otto Harrassowitz.

ABBILDUNG I

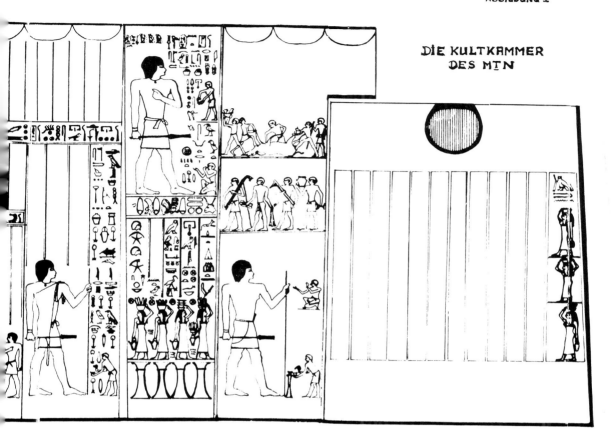

DIE KULTKAMMER
DES MTN

Fig. I.62 Relief from a tomb wall at Saqqara. Dynasty 19. It pictures famous men, among whom in the top row are the sages Imhotep and Kaires, while in the bottom row there are more sages, including Khety and Khakheperreseneb. Reproduced from Simpson, *The Literature of Ancient Egypt*, Fig. 6, with permission of the Yale University Press.

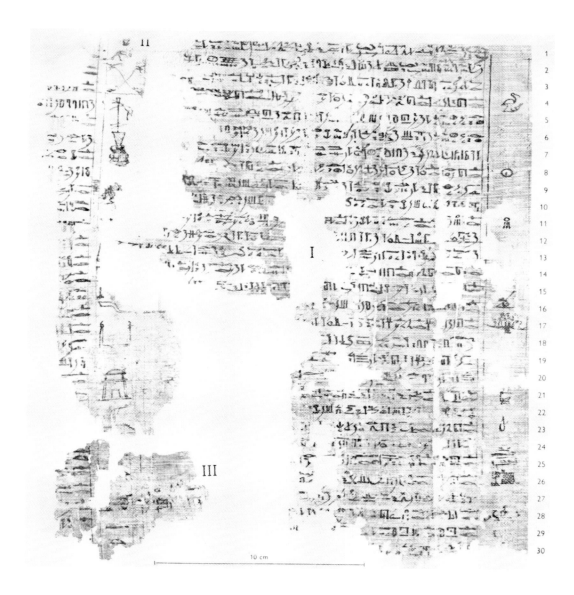

Fig. I.63 Fragments from a hieroglyphic dictionary. Taken from Iversen, *Papyrus Carlsberg Nr. VII*, following p. 31.

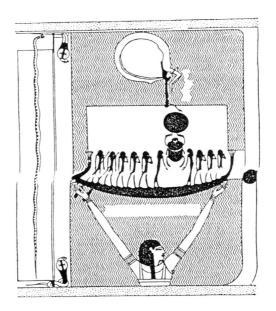

Fig. II.1   The rebirth of the sun as the Khepri beetle.
Schematized drawing from a papyrus in the British Museum. Cf. E.A.W. Budge, *The Gods of the Egyptians*,
Vol. 1 (Chicago and London, 1904), p. 204.

Fig. II.2a   Nut as the Divine or Celestial Cow. A watercolor by Robert Hay in the British Library (Add. Mss. 29820, f. 97), reproduced in E. Hornung, *Der ägyptische Mythos von der Himmelskuh (Orbis biblicus et orientalis,* Vol. 42), Fribourg, Switzerland, Abb. 2. Copied with the permission of the Éditions Universitaires, Fribourg, Suisse.

Fig. II.2b   Nut as the sky being supported by Shu (or Maat?), with Geb as the earth below. Taken from Erman, *A Handbook of Egyptian Religion*, p. 29.

Fig. II.2c   Nut from the Tomb of Ramesses II. After Piankoff, *The Tomb of Ramesses VI*, Vol. 1, Facsimile facing page 33. Reproduced with the permission of Princeton University Press.

Fig. II.3 An Ivory comb with the name of the Horus
Wadji. Dynasty 1. In the Cairo Museum. Copied from
Frankfort, *Kingship and the Gods,* Fig. 17, with the per-
mission of the University of Chicago Press.

Fig. II.4   Nun in human form and called "Father of the Gods."
According to Lanzone, *Dizionario di mitologia,* Tav. CLXVI.

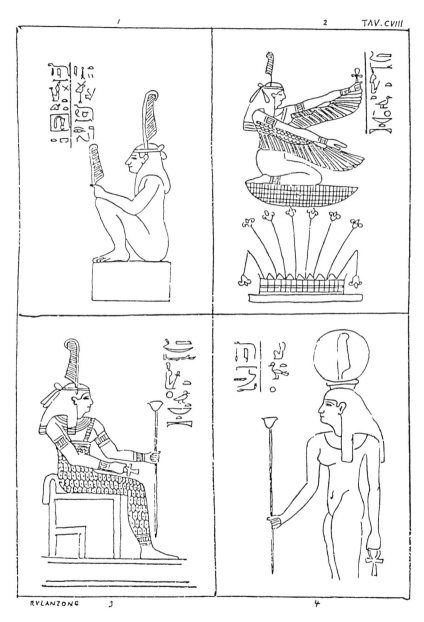

Fig. II.5 Several depictions of Maat, Daughter of Re, Mistress in the Land of Silence (i.e. the Otherworld), reproduced by Lanzone, *Dizionario di mitologia,* Tav. CVIII.

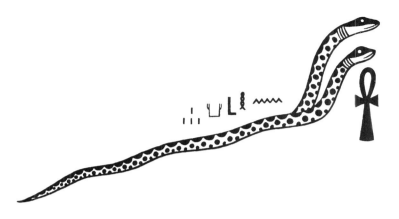

Fig. II.6a "The Provider of Attributes." Double-headed cosmic snake. Copied from Clark, *Myth and Symbol in Ancient Egypt,* Fig. 7, p. 52, with the permission of Thames and Hudson Ltd.

Fig. II.6b The sun-god as a child enclosed in the snake with his tail in his mouth, the Ouroboros. Mortuary Papyrus of Hirweben A. Cairo Museum. Dynasty 21. Reproduced from Hornung, *Conceptions of God in Ancient Egypt,* p. 164.

Fig. II.7   Khnum modeling man on his potter's
wheel. According to Lanzone, *Dizionario di mitologia,* Tav. CCCXXXVI,3.

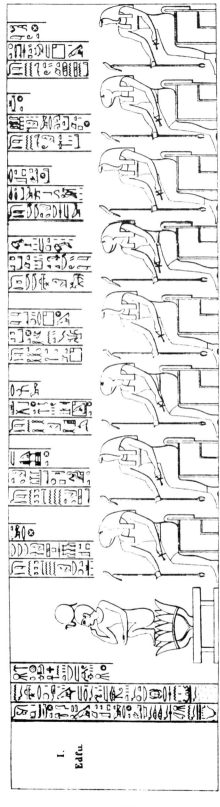

Fig. II.8 The eight chaos-gods depicted as watching the young sun-god rising from the Lotus flower. The male gods bear the heads of frogs. The female gods have rearing cobras for heads. Taken from Lepsius, *Über die Götter der vier Elemente*, Taf. I.

Fig. II.9 Thoth as an Ibis facing a small figure of Maat. Dynasty 26. Kestner Museum, Hannover.

Fig. II.10 Amon-Re-Atum, the Ibis-headed Thoth, and Seshat writing King Ramesses II's titulary on the leaves of the Sacred Tree. Reproduced from Erman, *Life in Ancient Egypt*, p. 347.

Fig. II.11 The Emergence of the young sun-god from a lotus blossom on the primeval hill of Hermopolis. Saitic period. Pelizaeus Museum at Hildesheim (nr. 60-Bronze).

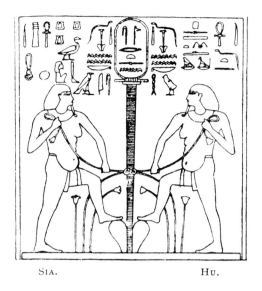

SIA.   HU.

Fig. II.12a Hu and Sia as Nile-gods tying the South
and the North together. Reproduced from Gautier and
Jéquier, *Fouilles de Licht,* p. 34.

Fig. II.12b The bark of Re with Sia (the second anthropomorphic figure) and Hu (the
eighth), their names appearing above them. Reproduced by Jéquier from the *Book of Amduat*
in his *Considérations sur les religions égyptiennes,* p. 39, and copied with the permission
of Les Éditions de la Baconnière, Boudry, Suisse. In my figures from the *Amduat* given
below, the names are not included.

Fig. II.13   Two early representations of Ptah: the first on an alabaster bowl from Tarkhan dating from about Dynasty 1 and the second from the Palermo Stone. See Sandman Holmberg, *The God Ptah*, p. 65*.

Fig. II.14a Ptah on the small shrine of Sesostris I at Karnak. Here Ptah seems to be standing in an open chapel as in early representations.

4.

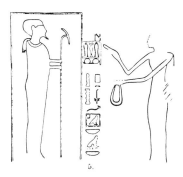

5.

Fig. II.14b Two other Middle-Kingdom depictions of Ptah, where he is in a closed chapel. See Sandman Holmberg, p. 67*.

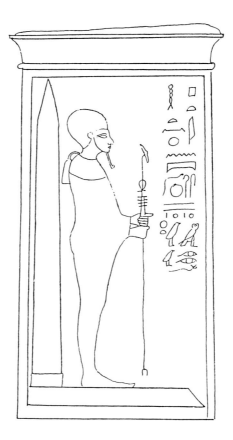

Fig. II.15   Another depiction of Ptah in a closed
chapel or naos, as reproduced by Lanzone, *Di-*
zionario di mitologia, Tav. XCII.

Fig. II.16   Relief from the shrine of Sesostris I at Karnak showing a very early representation of Amun in his ithyphallic form.

Fig. II.17   A predynastic palette in the form of a fish, with perhaps some religious significance. In the Lowie Museum of Anthropology. Copied from Fazzini, *Images for Eternity*, p. 12, with the permission of The Lowie Museum of Anthropology, The University of California at Berkeley.

Fig. II.18a   A predynastic vase with a ship bearing a standard of
possible religious significance, in the Lowie Museum of Anthropology. Copied from Fazzini, p. 11, with the permission of The Lowie
Museum of Anthropology, The University of California at Berkeley.

Fig. II.18b   A predynastic figure of a dog or jackal in the Lowie Museum. Copied from Fazzini,
p. 14, with the permission of The Lowie Museum of Anthropology, The University of California
at Berkeley.

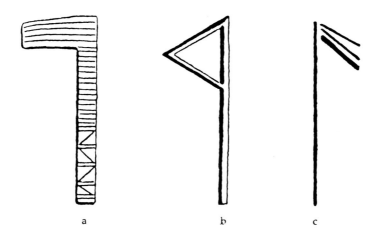

Fig. II.19a   Forms of the hieroglyph for "god" *(nṯr)*, in reverse chronological order. From Hornung, *Conceptions of God*, p. 34.

Fig. II.19b   Other hieroglyphs for "god." From Hornung, *Conceptions of God*, p. 39.

Fig. II.20   Some figures of gods on early dynastic objects. From Hornung, *Conceptions of God*, p. 109.

Fig. II.21   Hathor from the tomb of Tuthmosis IV. Taken from Hornung, *Conceptions of God*, p. 111.

Fig. II.22   Hathor head in the Metropolitan Museum. Dynasty 30. Photograph by M. Clagett.

Fig. II.23   Royal family of Amarna: Akhenaten, Nefertiti, and daughters. With hands at the end of rays holding ankh-signs to the noses of the royal couple. It is in the style of the early Amarna period. Ägyptisches Museum, Berlin (West).

Fig. II.24 The "trio" of Amun, Mut, and Tutankhamen, with the king as the divine child between the gods. The facial similarity between father and son is evident. Cairo Museum.

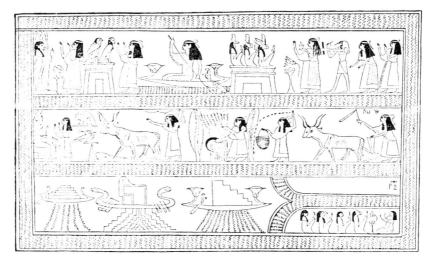

Fig. II.25 A vignette for Spell 110 of the *Book of the Dead* (Papyr. Berlin No. 3008), depicting the Field of Rushes. Taken from Erman, *A Handbook of Egyptian Religion*, p. 92.

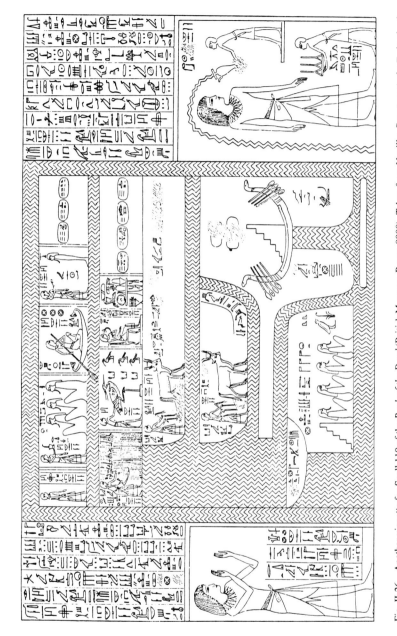

Fig. II.26 Another vignette for Spell 110 of the *Book of the Dead* (British Museum, Papyr. 9900). Taken from Naville, *Das ägyptische Todtenbuch*, Vol. 1, p. CXXIII.

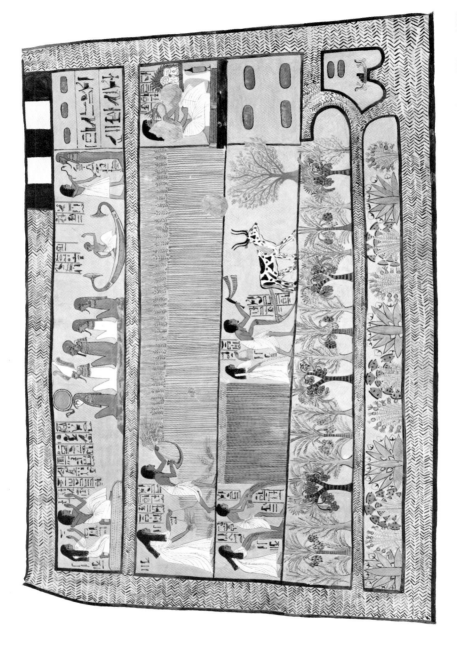

Fig. II.27 Facsimile of 'The Field of Rushes from the tomb of Sennedjem in Western Thebes (T 1). Dynasty 19. Copy by C. K. Wilkinson, 1922. Photograph supplied by The Metropolitan Museum of Art.

-838-

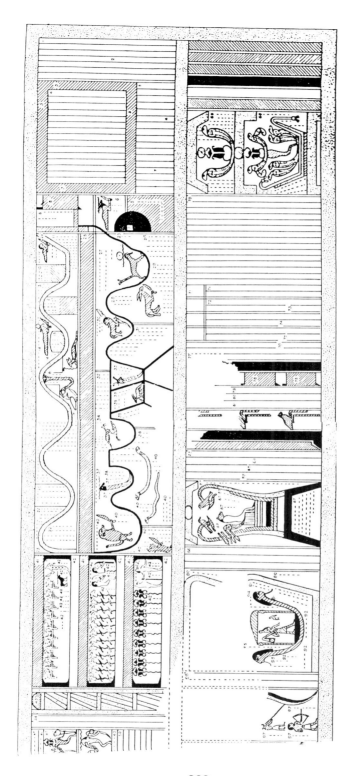

Fig. II.28 A plan showing the routes about which texts are displayed for the *Book of Two Ways*. It appears on a coffin from el-Barsha, now in the Cairo Museum (28083). Reproduced from de Buck, *The Egyptian Coffin Texts*, Vol. 7, Plan 1 (right side). Courtesy of the Oriental Institute of the University of Chicago.

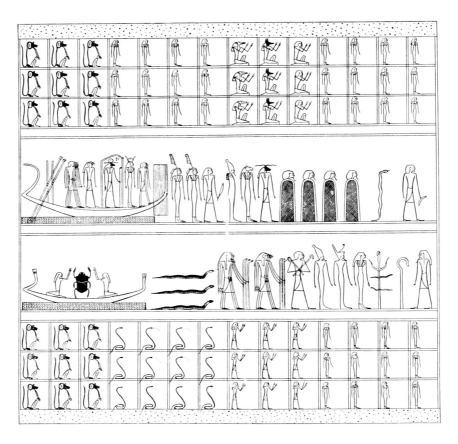

Fig. II.29 Diagram of the First Division (or Hour) of the *Amduat*. This and the next eleven diagrams (Figs. II.30-II.40) are taken from Piankoff, *The Tomb of Ramesses VI*, pp. 227–318, and reproduced with the permission of Princeton University Press. The texts that appear with these diagrams have been omitted from the diagrams but are translated partially in Doc. II.4.

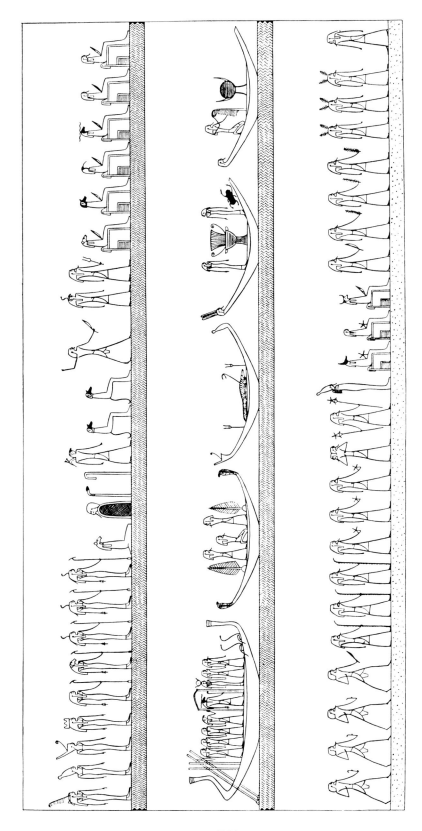

Fig. II.30   Diagram of the Second Division of the *Amduat*.

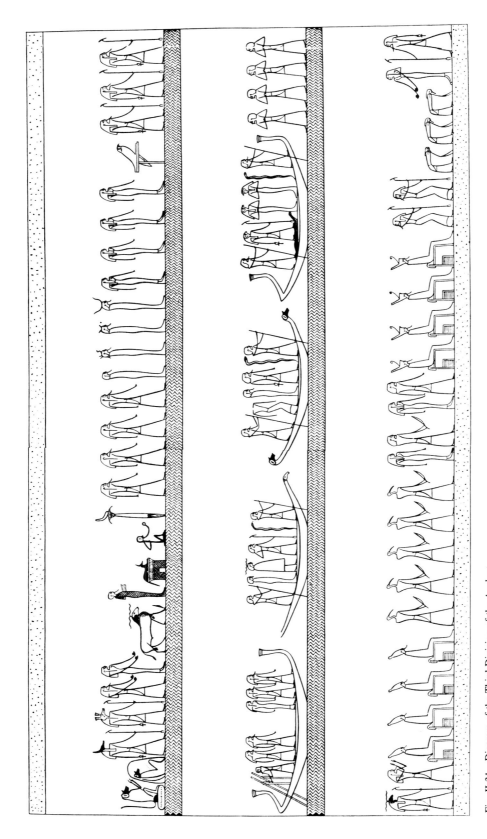

Fig. II.31   Diagram of the Third Division of the *Amduat*.

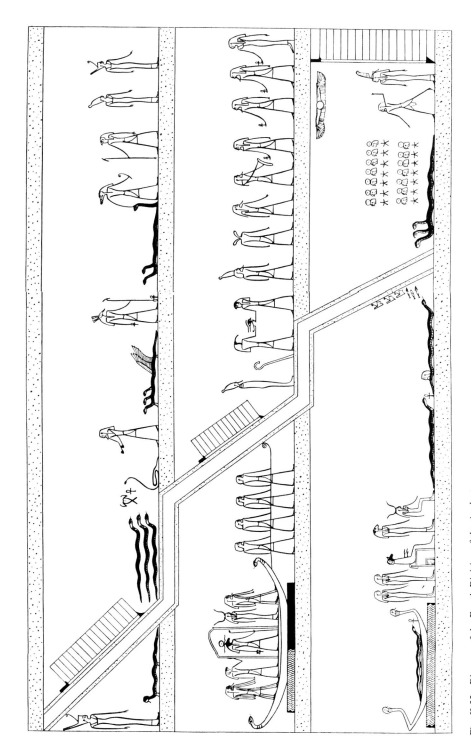

Fig. II.32 Diagram of the Fourth Division of the *Amduat*.

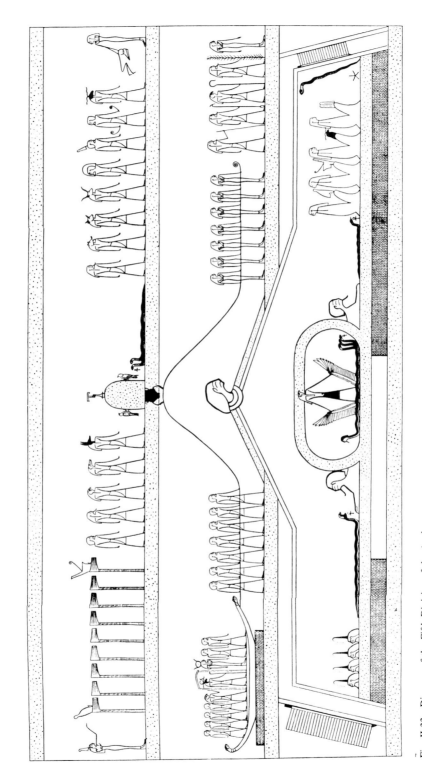

Fig. II.33 Diagram of the Fifth Division of the *Amduat*.

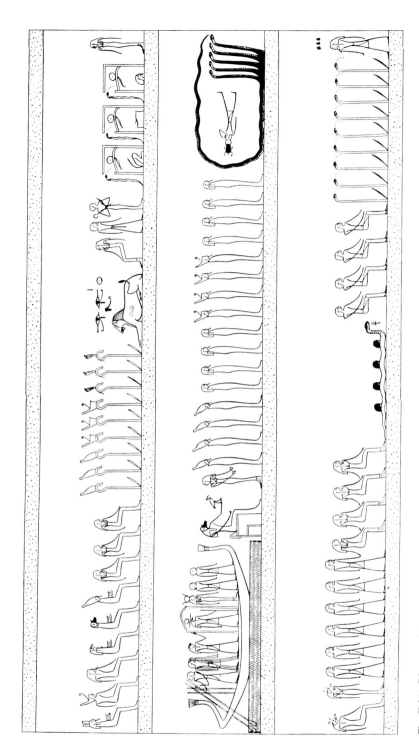

Fig. II.34 Diagram of the Sixth Division of the *Amduat*.

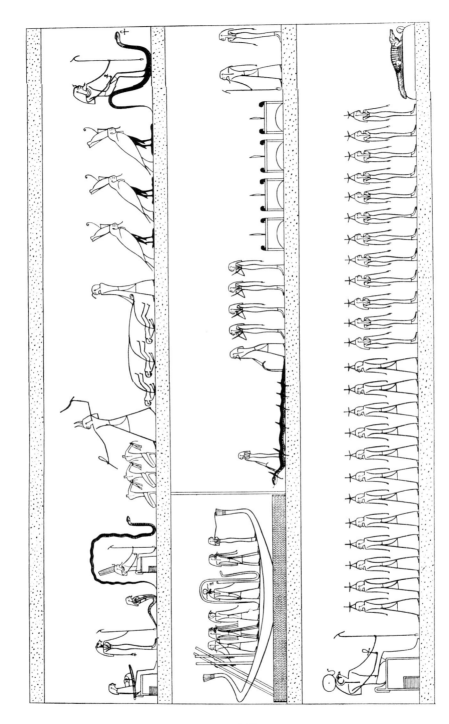

Fig. II.35 Diagram of the Seventh Division of the *Amduat*.

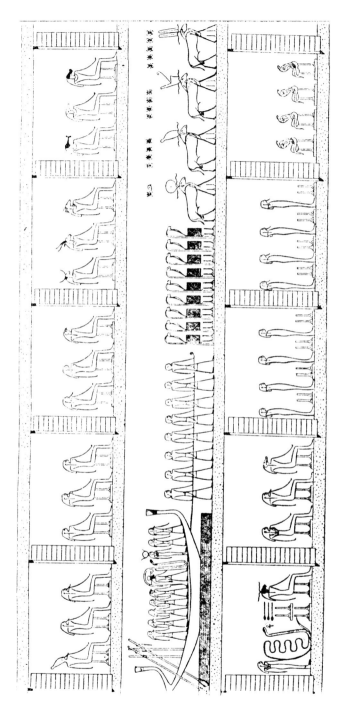

Fig. II.36 Diagram of the Eighth Division of the *Amduat*.

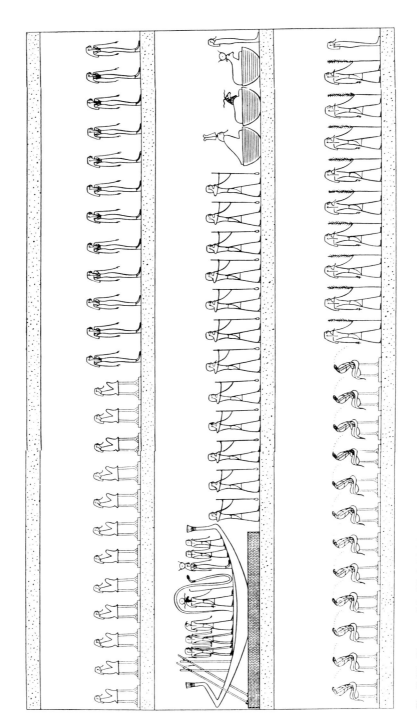

Fig. II.37   Diagram of the Ninth Division of the *Amduat*.

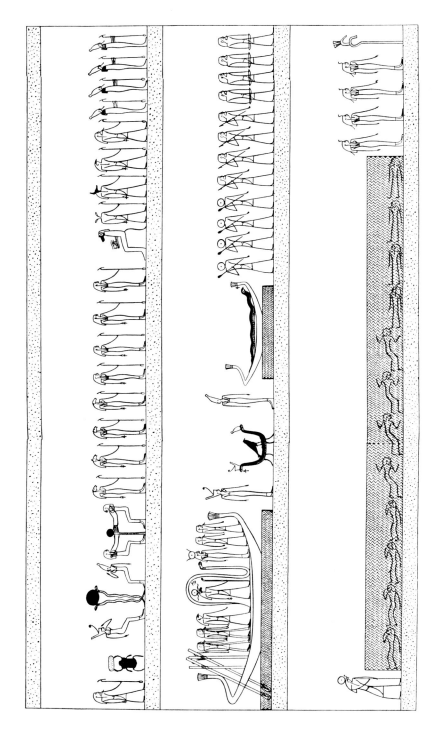

Fig. II.38 Diagram of the Tenth Division of the *Amduat*.

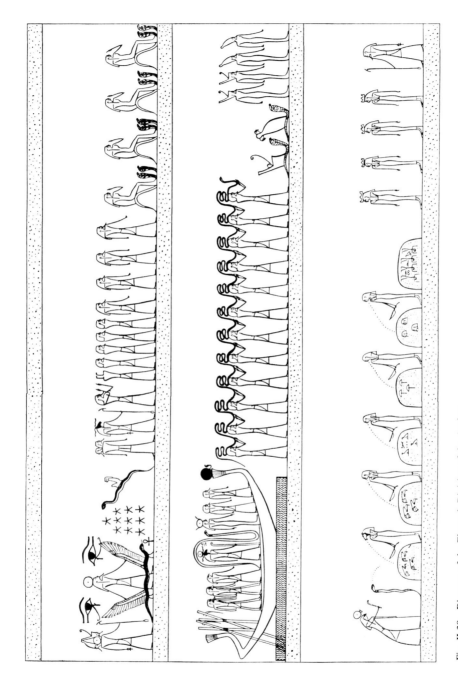

Fig. II.39 Diagram of the Eleventh Division of the *Amduat*.

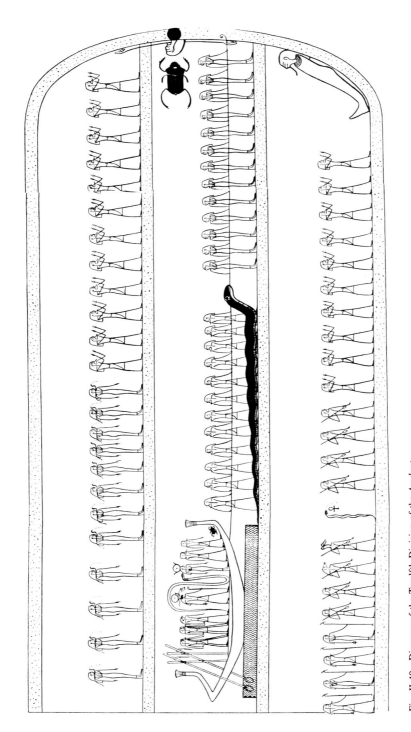

Fig. II.40   Diagram of the Twelfth Division of the *Amduat*.

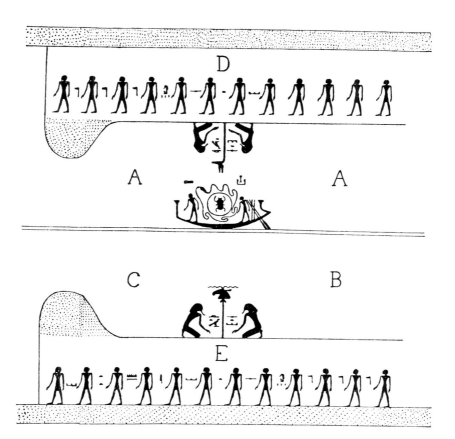

Fig. II.41 First Hour of the *Book of Gates*. Taken from Maystre and Piankoff, *Le Livre des portes,* Vol. 1, p. 2.

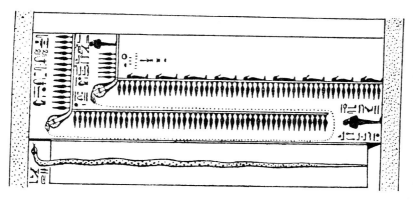

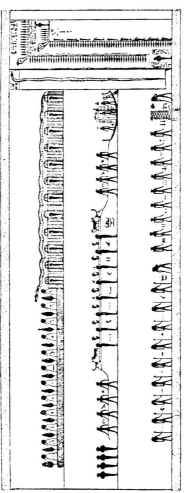

Fig. II.42   The gate at the end of the Third Hour of the *Book of the Gates* (called the ''IIe Division'') by Maystre and Piankoff, Vol. 1, pp. 70, 75).

-853-

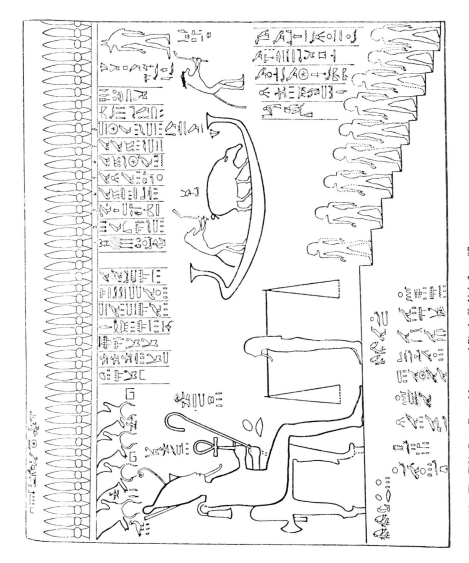

Fig. II.43   The Fifth Gate. From Maystre and Piankoff, Vol. 2, p. 87.

Fig. II.44 Eleventh Hour of the *Book of Gates*. Designated by Hornung as "Abbildung 62 75. und 76. Szene" in *Ägyptische Unterweltsbücher*, p. 288. Copied with the permission of Artemis Publishers in Zurich.

Fig. II.45 Another drawing from the Eleventh Hour (called by Piankoff: "Xᵉ Division"). Designated by Hornung as "Abbildung 64 80. Szene (Ausschnitt)." He uses the last term because he gives only 8 star-crowned goddesses. The figure here is taken from Maystre and Piankoff, Vol. 3, p. 88.

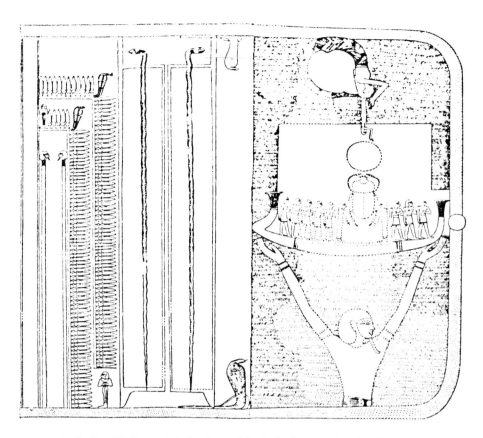

Fig. II.46   The Twelfth Gate and closing figure of the *Book of Gates*. Taken from Hornung, *Das Buch von den Pforten,* Teil II, p. 290, with the permission of Aegyptiaca Helvetica, Éditions de Belles Lettres, Geneva, Switzerland.

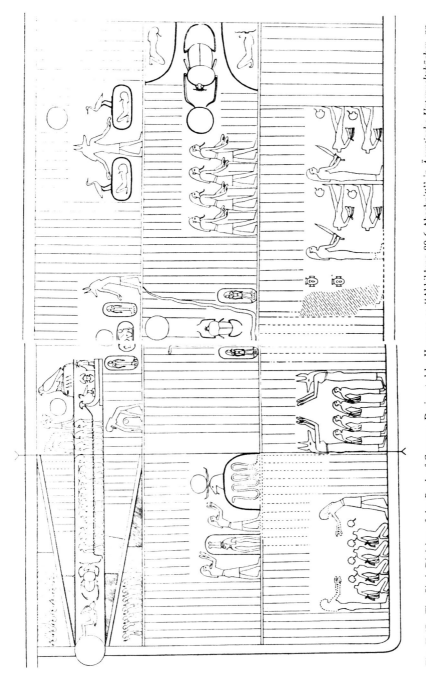

Fig. II.47 The Sixth Division of the *Book of Caverns*. Designated by Hornung as "Abbildung 80 6.Abschnitt" in *Ägyptische Unterweltsbücher*, pp. 404–05. Copied with the permission of Artemis Publishers Zürich.

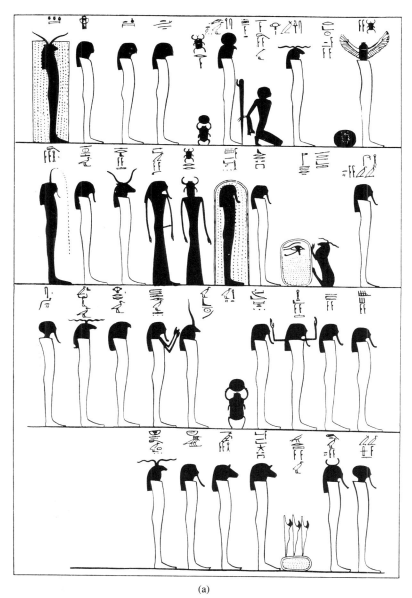

(a)

Fig. II.48 Figures (a) and (b) comprise the Forms of Re in the Tomb of Tuthmosis III. According to Piankoff, *The Litany of Re,* pp. 14–15, and reproduced with the permission of Princeton University Press.

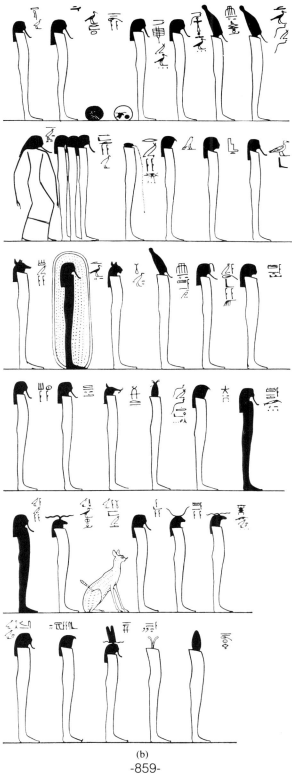

(b)

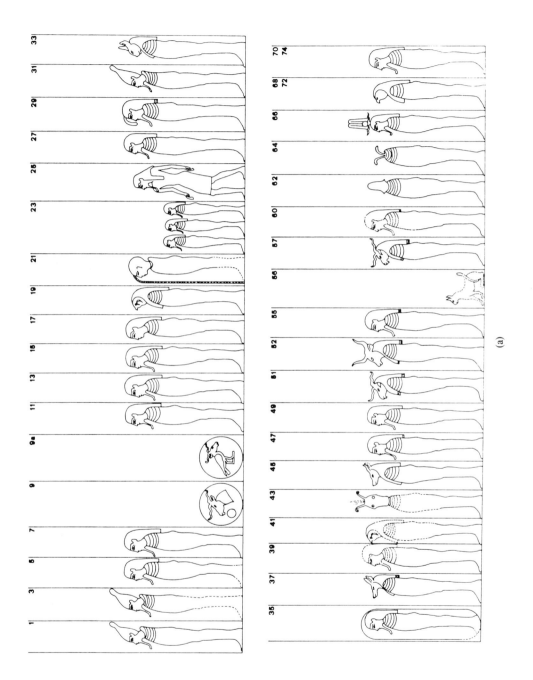

(a)

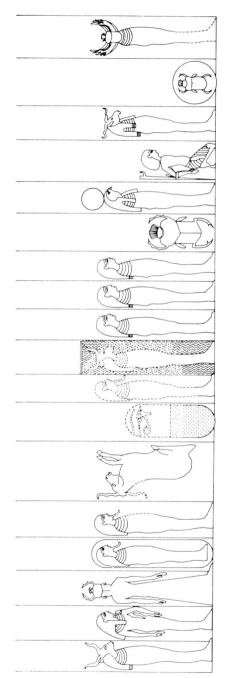

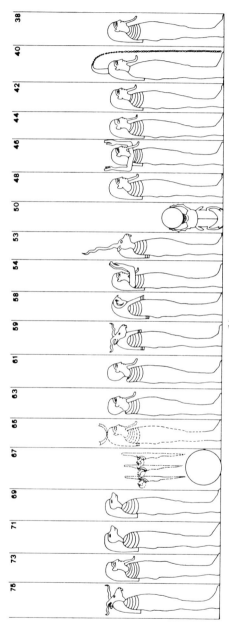

(b)

Fig. II.49 Figures (a) and (b) show the Forms of Re as given on the left and right walls of the second corridor of the tomb of Seti I. Drawings by A. Brodbeck according to the photographs and sketches of C. Castioni and reproduced in Hornung, *Das Buch der Anbetung des Re im Westen*, Vol. 2, pp. 58–59. Copied with the permission of Aegyptiaca, Éditions de Belles-Lettres.

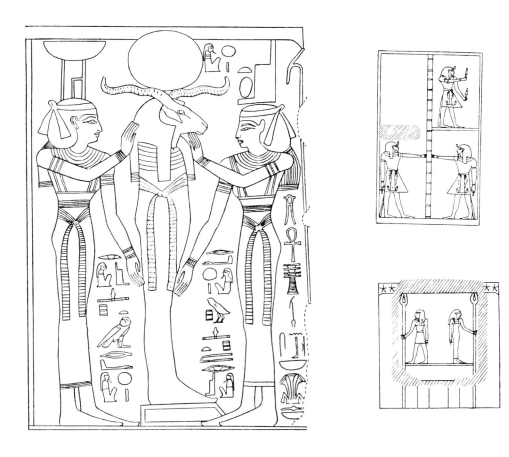

Fig. II.50   Representation of Re-Osiris, i.e. the United One, from the tomb of Queen Nefertari in the Valley of the Queens. Dynasty 19. Taken from Piankoff and Rambova, *The Tomb of Ramesses VI*, p. 34, with the permission of Princeton University Press.

Fig. II.51   The king helping to support the sky. From Hornung, *Der Ägyptische Mythos von der Himmelskuh*, p. 86. Courtesy of Éditions Universitaires, Fribourg, Switzerland.

Fig. II.52   Eternity *(nḥḥ)* and Everlastingness *(ḏt)* as personified gods supporting the sky. From Hornung, *ibid.* Courtesy of Éditions Universitaires, Fribourg, Switzerland.

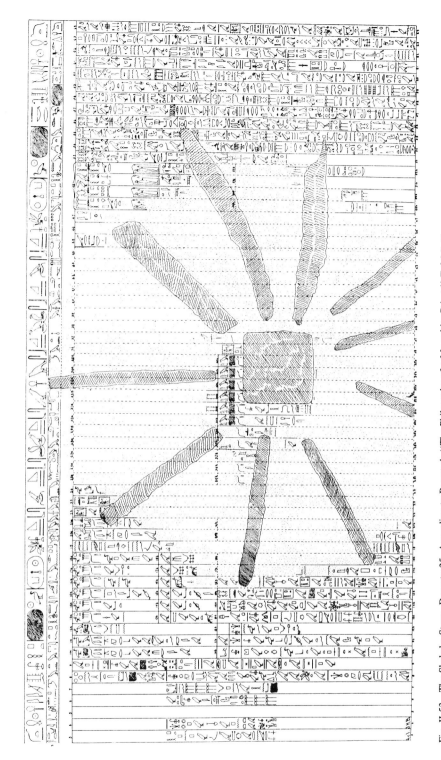

Fig. II.53   The Shabaka Stone. Dynasty 25. According to Breasted, "The Philosophy of a Memphite Priest," Tafel I. II.